# CALCULATIONS FOR A-LEVEL CHEMISTRY

**Second Edition**

## E N Ramsden BSc, PhD, DPhil

| NAME | FORM | YEAR |
|------|------|------|
| K. WOOLF | VII Pd | 1989-91 |
|  |  |  |
|  |  |  |
|  |  |  |
|  |  |  |

**Stanley Thornes (Publishers) Ltd**

First published in 1982 by Stanley Thornes (Publishers) Ltd, Old Station Drive, Leckhampton, Cheltenham GL53 0DN

Reprinted 1983
Reprinted 1984
Reprinted 1985

Second Edition 1987
Reprinted 1988

British Library Cataloguing in Publication Data

Ramsden, E.N.
    Calculations for A-level chemistry.—
    2nd ed.
    1. Chemistry—Mathematics—
    Examinations, questions, etc.
    I. Title
    540'.1'51        QD 39.3.M3

    ISBN 0-85950-755-6

# ACKNOWLEDGEMENTS

I thank the following examination boards for permission to print questions from recent A-level papers:

The Associated Examining Board; The Joint Matriculation Board; The Northern Ireland Schools Examinations Council; The Oxford and Cambridge Schools Examination Board; The Oxford Delegacy of Local Examinations; The Southern Universities Joint Board; The University of Cambridge Schools Local Examinations Syndicate; The University of London School Examinations Council; The Welsh Joint Education Committee.

Many numerical values have been taken from the *Chemistry Data Book* by J G Stark and H G Wallace (published by John Murray). For help with definitions and physical constants, reference has been made to *Physico-chemical Quantities and Units* by M L McGlashan (published by The Royal Institute of Chemistry).

I have been fortunate in receiving excellent advice from Professor R R Baldwin, Professor R P Bell, FRS, Dr G H Davies, Professor W C E Higginson, Dr K A Holbrook, Dr R B Moyes and Dr J R Shorter. I thank these chemists for the help they have given me. I am indebted to Mr J P D Taylor for checking the answers to the problems in the First Edition and for many valuable comments and corrections.

I thank Stanley Thornes (Publishers) for their collaboration and my family for their encouragement.

*E N Ramsden*
Hull 1987

Typeset by Tech-Set, Gateshead, Tyne & Wear.
Printed and bound in Great Britain at The Bath Press, Avon.

# Contents

# List of Exercises

Questions from A-level papers are on the immediately preceding topic(s). Each question is appended with the name of the Examination Board and the year (80 = 1980 etc). p indicates a part question, S an S-level question and N a Nuffield syllabus. The most difficult (often S-level) questions are also denoted by an asterisk.

## ABBREVIATIONS OF EXAMINATION BOARDS

| | |
|---|---|
| AEB | Associated Examining Board |
| C | University of Cambridge Schools Local Examinations Syndicate |
| JMB | Joint Matriculation Board |
| L | University of London Schools Examinations Council |
| NI | Northern Ireland Schools Examinations Council |
| O | Oxford Delegacy of Local Examinations |
| O & C | Oxford and Cambridge Schools Examinations Board |
| SUJB | Southern Universities' Joint Board |
| WJEC | Welsh Joint Education Committee |

# Foreword

It is a common complaint of university and college teachers of physical sciences that many of their incoming students are unable to carry out even simple calculations, although they may appear to have a satisfactory grasp of the underlying subject matter. Moreover, this is by no means a trivial complaint, since inability to solve numerical problems nearly always stems from a failure to understand fundamental principles, rather than from mathematical or computational difficulties. This situation is more likely to arise in chemistry than in physics, since in the latter subject it is much more difficult to avoid quantitative problems and at the same time produce some semblance of understanding.

In attempting to remedy this state of affairs teachers in schools often feel the lack of a single source of well-chosen calculations covering all branches of chemistry. This gap is admirably filled by Dr Ramsden's collection of problems. The brief mathematical introduction serves to remind the student of some general principles, and the remaining sections cover the whole range of chemistry. Each section contains a theoretical introduction, followed by worked examples and a large number of problems, some of them from past examination papers. Since answers are also given, the book will be equally useful in schools and in home study. It should make a real contribution towards improving the facility and understanding of students of chemistry in their last years at school and in the early part of their university or college courses.

*R P Bell FRS*
Honorary Research Professor, University of Leeds, and
formerly Professor of Chemistry, University of Stirling

# Preface

Many topics in Chemistry involve numerical problems. Textbooks are not long enough to include sufficient problems to give students the practice which they need in order to acquire a thorough mastery of calculations. This book aims to fill that need.

Chapter 1 is a quick revision of mathematical techniques, with special reference to the use of the calculator, and some hints on how to tackle chemical calculations. With each topic, a theoretical background is given, leading to worked examples and followed by a large number of problems and a selection of questions from past examination papers. The theoretical section is not intended as a full treatment, to replace a textbook, but is included to make it easier for the student to use the book for individual study as well as for class work. The inclusion of answers is also an aid to private study.

The material will take students up to GCE A- and S-level examinations. It will also serve the needs of students preparing for the Ordinary National Diploma. A few of the topics covered are not in the A-level syllabuses of all the Examination Boards, and it is expected that students will be sufficiently familiar with the syllabus they are following to omit material outside their course if they wish. S-level topics and the more difficult calculations are marked with an asterisk.

In the Second Edition, the selection of questions from past papers has been updated and a few topics which no longer feature in A-level examinations have been omitted.

*E N Ramsden*
Hull 1987

# 1 Basic Mathematics

## INTRODUCTION

Calculations are a part of your chemistry course. To some students, they are a source of distress and dismay. To other students, they give the tremendous satisfaction of knowing that a problem has been solved, a correct answer obtained, and full marks gained for that question – a feat which it is not easy to achieve on other types of question!

Calculations are not just an extra activity: the time you spend on calculations will be richly rewarded. Your perception of chemistry will become at the same time deeper and more precise. No one can come to an understanding of science without acquiring the sharp, logical approach that is needed for solving numerical problems.

To succeed in solving numerical problems you need two things. The first is an understanding of the chemistry involved. The second is some facility in simple mathematics. Calculations are a perfectly straightforward matter. A numerical problem gives you some data and asks you to obtain some other numerical values. The connection between the data you are given and the information you are asked for is a chemical relationship. You will need to know your chemistry to recognise what that relationship is.

This introduction is a reminder of some of the mathematics which you studied earlier in your school career. It is included for the sake of students who are not studying mathematics concurrently with their chemistry course. A few problems are included to help you to brush up your mathematical skills before you go on to tackle the chemical problems.

## WORKING WITH NUMBERS IN STANDARD FORM

You are accustomed to writing numbers in decimal notation, for example 123 677.54 and 0.001 678. In working with large numbers and small numbers, you will find it convenient to write them in a different way, known as *scientific notation* or *standard form*. This means writing a number as a product of two factors. In the first factor, the decimal point comes after the first digit. The second factor is a multiple of ten. For example, $2123 = 2.123 \times 10^3$ and $0.000\,167 = 1.67 \times 10^{-4}$. $10^3$ means $10 \times 10 \times 10$, and $10^{-4}$ means $1/(10 \times 10 \times 10 \times 10)$. The number 3 or $-4$ is called the exponent, and the number 10 is the base. $10^3$ is referred to as '10 to the power 3' or '10 to the third power'. You will have noticed that, if the exponent is increased by 1, the decimal point must be moved one place to the left.

$$2.5 \times 10^3 = 0.25 \times 10^4 = 25 \times 10^2 = 250 \times 10^1 = 2500 \times 10^0$$

Since $10^0 = 1$, this last factor is normally omitted.

When you multiply numbers in standard form, the exponents are added. The product of $2 \times 10^4$ and $6 \times 10^{-2}$ is given by

$$(2 \times 10^4) \times (6 \times 10^{-2}) = (2 \times 6) \times (10^4 \times 10^{-2})$$
$$= 12 \times 10^2 = 1.2 \times 10^3$$

In division, the exponents are subtracted:

$$\frac{1.44 \times 10^6}{4.50 \times 10^{-2}} = \frac{1.44}{4.50} \times \frac{10^6}{10^{-2}} = 0.320 \times 10^8 = 3.20 \times 10^7$$

In addition and subtraction, it is convenient to express numbers using the same exponents. An example of addition is

$$(6.300 \times 10^2) + (4.00 \times 10^{-1}) = (6.300 \times 10^2) + (0.00400 \times 10^2)$$
$$= 6.304 \times 10^2$$

An example of subtraction is

$$(3.60 \times 10^{-3}) - (4.20 \times 10^{-4}) = (3.60 \times 10^{-3}) - (0.420 \times 10^{-3})$$
$$= 3.18 \times 10^{-3}$$

## ESTIMATING YOUR ANSWER

One advantage of standard form is that very large and very small numbers can be entered on a calculator. Another advantage is that you can easily estimate the answer to a calculation to the correct order of magnitude (i.e. the correct power of 10).

For example,

$$\frac{2456 \times 0.0123 \times 0.00414}{5223 \times 60.7 \times 8.51}$$

Putting the numbers into standard form gives

$$\frac{2.456 \times 10^3 \times 1.23 \times 10^{-2} \times 4.14 \times 10^{-3}}{5.223 \times 10^3 \times 6.07 \times 10 \times 8.51}$$

This is approximately

$$\frac{2 \times 1 \times 4}{5 \times 6 \times 8} \times \frac{10^3 \times 10^{-2} \times 10^{-3}}{10^3 \times 10} = \frac{1}{30} \times 10^{-6} = 3 \times 10^{-8}$$

By putting the numbers into standard form, you can estimate the answer very quickly. A complete calculation gives the answer $4.64 \times 10^{-8}$. The rough estimate is sufficiently close to this to reassure you that you have not made any slips with exponents of ten.

## LOGARITHMS

The logarithm (or 'log') of a number $N$ is the power to which 10 must be raised to give the number.

If $N = 1$,      then since    $10^0 = 1$,      $\lg N = 0$.

If $N = 100$,    then since    $10^2 = 100$,    $\lg N = 2$.

If $N = 0.001$,  then since    $10^{-3} = 0.001$, $\lg N = -3$.

We say that the logarithm of 100 to the base 10 is 2 or $\lg 100 = 2$.

There is another widely used set of logarithms to the base e. They are called natural logarithms as e is a significant quantity in mathematics. It has the value 2.71828 . . . Natural logarithms are written as $\ln N$. The relationship between the two systems is

$$\ln N = \ln 10 \times \lg N$$

Since $\ln 10 = 2.3026$, for most purposes it is sufficiently accurate to write

$$\ln N = 2.303 \lg N$$

Whenever scientific work gives an equation in which $\ln N$ appears, you can substitute 2.303 times the value of $\lg N$.

To obtain the log of a number, enter the number on your calculator and press the log key. The value of the log will appear in the display. This will happen whether you enter the number in standard form or another form. For example, $\lg 12\,345 = 4.0915$, whether you enter the number as 12 345 or as $1.2345 \times 10^4$. However, there is a limit to the number of digits your calculator will accept, and you need to enter very large and very small numbers in standard notation.

Operations on logarithms are:

*Multiplication.* The logs of the numbers are added:

$$\lg (A \times B) = \lg A + \lg B$$

*Division.* The logs are subtracted:

$$\lg (P/Q) = \lg P - \lg Q$$

*Powers.* This is a special case of multiplication.

$$\lg A^2 = \lg A + \lg A = 2 \lg A$$
$$\lg A^{-3} = -3 \lg A$$

*Roots.* It is easy to show that $\lg\sqrt{B} = \frac{1}{2}\lg B$.

Since
$$B = B^{1/2} \times B^{1/2}$$
$$\lg B = \lg B^{1/2} + \lg B^{1/2}$$
$$\lg B^{1/2} = \tfrac{1}{2}\lg B$$

Similarly,
$$\lg\sqrt[3]{B} = \tfrac{1}{3}\lg B$$

## ANTILOGARITHMS

Your calculator will give you the antilog of a number. You should consult the manual to find out the procedure for your own model of calculator.

Most calculators will give you reciprocals, squares and other powers, square roots and other roots directly. If you have a simpler form of calculator, you can obtain powers and roots by using logarithms.

## ROUNDING OFF NUMBERS

Often your calculator will display an answer containing more digits than the numbers you fed into it. Suppose you are given the information that $18.6\,cm^3$ of sodium hydroxide solution exactly neutralise $25.0\,cm^3$ of a solution of hydrochloric acid of concentration $0.100\,mol\,dm^{-3}$. You want to find the concentration of sodium hydroxide solution, and you put the numbers $(25.0 \times 0.100)/18.6$ into your calculator and obtain a value of $0.134\,408\,6\,mol\,dm^{-3}$. The concentration of the solution is not known as accurately as this, however, because you cannot read the burette as accurately as this. Since you read the burette to three figures, you quote your answer to three figures. In the number $0.134\,408\,6$, the figures you are sure of are termed the *significant figures*. The significant figures are retained, and the insignificant figures are dropped. This operation is called *rounding off*. If the first number had been $0.134\,708\,6$, it would have been rounded off to $0.135$. If the first of the insignificant figures being dropped is 5 or greater, the last of the significant figures is rounded up to the next digit. If the first of the dropped figures is less than 5, the last significant figure is left unaltered.

Some calculations involve several stages. It is sound practice to give one more significant figure in your answer at each stage than the number of significant figures in the data. Then, in the final stage, the answer is rounded off.

If the calculation were $(25.0 \times 0.100)/26.2 = 0.095\,419\,84\,mol\,dm^{-3}$, would you still round off to 3 significant figures? This would make the answer $0.0954\,mol\,dm^{-3}$. Stated in this way, the answer is claiming an accuracy of 1 part in 954 — about 1 part in 1000. Since the hydrochloric acid concentration is known to about 1 part in 100, the answer cannot be stated to a higher degree of accuracy. You have to use the 3-significant-figure rule sensibly, and say that an error of $\pm 1$ in 95 is about as significant as an error of $\pm 1$ in 134. The answer should therefore be quoted as $0.095 \, \text{mol} \, \text{dm}^{-3}$.

The number of significant figures is the number of figures which is accurately known. The number 123 has 3 significant figures. The number $1.23 \times 10^4$ has 3 significant figures, but 12 300 has 5 significant figures because the final zeros mean that each of these digits is known to be zero and not some other digit. The number 0.001 23 has 3 significant figures. The number 25.1 has 3 significant figures, and the number 25.10 has 4 significant figures as the final 0 states that the value of this number is known to an accuracy of 1 part in 2500.

In addition, the sum is known with the accuracy of the least reliable numbers in the sum. For example, the sum of

$$
\begin{array}{r}
1.4167 \, \text{g} \\
+\,100.5 \quad\;\; \text{g} \\
+\quad 7.12 \quad\; \text{g} \\
\hline
\end{array}
$$

is         $109.0367 \, \text{g}$

Since 1 figure is known to only 1 place after the decimal point, the sum also is known to 1 place after decimal point and should be written as 109.0 g. The same guideline is used for subtraction.

In multiplication and division the product or quotient is rounded off to the same number of significant figures as the number with the fewest significant figures. For example, $12\,340 \times 2.7 \times 0.003\,65 = 121.6107$. The product is rounded off to 2 significant figures, $1.2 \times 10^2$.

## CHOICE OF A CALCULATOR

The functions which you need in a calculator for the problems in this book are:

- Addition, Subtraction, Multiplication and Division
- Squares and other powers ($x^2$ and $x^y$ keys)
- Square roots and other roots ($\sqrt{x}$ and $x^{1/y}$ keys)
- Reciprocals
- $\text{Log}_{10}$ and antilog$_{10}$ ($10^x$)
- Natural logarithms, $\ln_e$ and antiln$_e$ ($e^x$)
- Exponent key and $+/-$ key
- Brackets
- Memory

A variety of scientific calculators have these functions and others (such as sin, cos, tan and $\Sigma x$) which will be useful to you in physics and mathematics problems.

## UNITS

There are two sets of units currently employed in scientific work. One is the CGS system, based on the centimetre, gram and second. The other is the Système Internationale (SI) which is based on the metre, kilogram, second and ampere. SI units were introduced in 1960, and in 1979 the Association for Science Education published a booklet called *Chemical Nomenclature, Symbols and Terminology for Use in School Science* that recommended that schools and colleges adopt this system.

Listed below are the SI units for the seven fundamental physical quantities on which the system is based and also a number of derived quantities and their units.

Chemists are still using some of the CGS units. You will find mass in g; volume in $cm^3$ and $dm^3$; concentrations in $mol\,dm^{-3}$ or $mol\,litre^{-1}$; conductivity in $\Omega^{-1}cm$ as well as $\Omega^{-1}m$. Pressure is sometimes given in mm mercury and temperatures in $°C$.

### Basic SI Units

| *Physical Quantity* | *Name of Unit* | *Symbol* |
|---|---|---|
| Length | metre | m |
| Mass | kilogram | kg |
| Time | second | s |
| Electric current | ampere | A |
| Temperature | kelvin | K |
| Amount of substance | mole | mol |
| Light intensity | candela | cd |

### Derived SI Units

| *Physical Quantity* | *Name of Unit* | *Symbol* | *Definition* |
|---|---|---|---|
| Energy | joule | J | $kg\,m^2\,s^{-2}$ |
| Force | newton | N | $J\,m^{-1}$ |
| Electric charge | coulomb | C | $A\,s$ |
| Electric potential difference | volt | V | $J\,A^{-1}s^{-1}$ |
| Electric resistance | ohm | $\Omega$ | $V\,A^{-1}$ |
| Area | square metre | | $m^2$ |
| Volume | cubic metre | | $m^3$ |
| Density | kilogram per cubic metre | | $kg\,m^{-3}$ |
| Pressure | newton per square metre or pascal | | $N\,m^{-2}$ or Pa |
| Molar mass | kilogram per mole | | $kg\,mol^{-1}$ |

With all these units, the following prefixes (and others) may be used:

| Prefix | Symbol | Meaning |
|--------|--------|---------|
| deci | d | $10^{-1}$ |
| centi | c | $10^{-2}$ |
| milli | m | $10^{-3}$ |
| micro | $\mu$ | $10^{-6}$ |
| nano | n | $10^{-9}$ |
| kilo | k | $10^{3}$ |
| mega | M | $10^{6}$ |
| giga | G | $10^{9}$ |
| tera | T | $10^{12}$ |

It is very important when putting values for physical quantities into an equation to be consistent in the use of units. If you are, then the units can be treated as factors in the same way as numbers. Suppose you are asked to calculate the volume occupied by 0.0110 kg of carbon dioxide at 27 °C and a pressure of $9.80 \times 10^4 \, \mathrm{N\,m^{-2}}$. You know that the gas constant is $8.31 \, \mathrm{J\,mol^{-1}K^{-1}}$ and that the molar mass of carbon dioxide is $44.0 \, \mathrm{g\,mol^{-1}}$. Use the ideal gas equation:

$$PV = nRT$$

The pressure $\quad P = 9.80 \times 10^4 \, \mathrm{N\,m^{-2}}$

The constant $\quad R = 8.31 \, \mathrm{J\,K^{-1}mol^{-1}}$

The temperature $T = 27 + 273 = 300 \, \mathrm{K}$

The number of moles

$$n = \text{Mass/Molar mass}$$
$$= 0.0110 \, \mathrm{kg}/44.0 \times 10^{-3} \, \mathrm{kg\,mol^{-1}}$$
$$= 0.250 \, \mathrm{mol}$$

Then $\quad V = \dfrac{0.250 \, \mathrm{mol} \times 8.31 \, \mathrm{J\,K^{-1}mol^{-1}} \times 300 \, \mathrm{K}}{9.80 \times 10^4 \, \mathrm{N\,m^{-2}}}$

$$= 6.34 \times 10^{-3} \, \mathrm{J\,N^{-1}m^2}$$

Since $\quad \mathrm{J} = \mathrm{N\,m} \quad$ (1 joule = 1 newton metre)

$$V = 6.34 \times 10^{-3} \, \mathrm{N\,m\,N^{-1}m^2}$$
$$= 6.34 \times 10^{-3} \, \mathrm{m^3}$$

Volume has the unit of cubic metre. This calculation illustrates what people mean when they say that SI units form a *coherent system of units.* You can convert from one unit to another by multiplication and division, without introducing any numerical factors.

## SOLUTION OF QUADRATIC EQUATIONS

A quadratic equation is the name for an equation where the largest exponent of the unknown quantity is 2. In the equation

$$ax^2 + bx + c = 0$$

$x$ is the unknown quantity, $a$ and $b$ are the coefficients of $x$, and $c$ is a constant. The solution of this equation is given by

$$x = \frac{-b \pm \sqrt{b^2 - 4ac}}{2a}$$

There are two solutions to the equation. Often you will be able to decide that one solution cannot be allowed. You may be calculating some physical quantity that cannot possibly be negative, so that a negative solution can be disregarded and a positive solution adopted.

## DRAWING GRAPHS

Here are some hints for drawing graphs.

a) Whenever possible, data should be plotted in a form that gives a straight line graph. It is easier to draw the best straight line through a set of points than to draw a curve.

If the dimensions $x$ and $y$ are related by the expression $y = ax + b$, then a straight line will result when experimental values of $y$ are plotted against the corresponding values of $x$. The values of $x$ are plotted against the horizontal axis (the $x$-axis or abscissa), and the corresponding values of $y$ are plotted along the vertical axis (the $y$-axis or ordinate). The gradient of the straight line obtained $= a$, and the intercept on the $y$-axis $= b$.

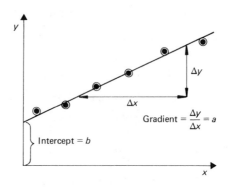

Fig. 1.1    Plotting a graph

b) Choose a scale which will allow the graph to cover as much of the piece of graph paper as possible. There is no need to start at zero. If the points lie between 90 and 100, to start at zero would cramp your graph into a small section at the top of the page.

Choose a scale which will make plotting the data and reading the graph as simple as possible.

c) Label the axes with the dimensions and the units. Make the scale units as simple as possible. Instead of plotting as scale units $1 \times 10^{-3}\,\text{mol dm}^{-3}$, $2 \times 10^{-3}\,\text{mol dm}^{-3}$, $3 \times 10^{-3}\,\text{mol dm}^{-3}$, etc., plot 1, 2 and 3, etc., and label the axis as Concentration/$\text{mol dm}^{-3} \times 10^{3}$.

d) When you come to draw a straight line through the points, draw the best straight line you can, to pass through, or close to, as many points as possible. Owing to experimental error, not all the points will fall on to the line. A graph of experimental results gives you a better accuracy than calculating a value from just one point. If you are drawing a curve, draw a smooth curve. Do not join up the points with straight lines. The curve may not pass through every point, but it is more reliable than any one of the points.

## A FINAL POINT

Always look critically at your answer. Ask yourself whether it is a reasonable answer. Is it of the right order of magnitude for the data? Is it in the right units? Many errors can be detected by an assessment of this kind.

## EXERCISE 1    Practice with Calculations

1. Convert the following numbers into standard form:

   a) 23 678      b) 437.6      c) 0.0169
   d) 0.000 345   e) 672 891

2. Convert each of the following numbers into standard form, enter into your calculator, and multiply by 237. Give the answers in standard form.

   a) 246.8    b) 11 230    c) 267 831    d) 0.051    e) 0.567

3. Find the following quotients:

   a) 2360/0.000 71    b) 28 780/0.106    c) 85.42/460 000
   d) 58/900 670       e) 0.000 88/0.144

4. Find the following sums and differences:
   a) $(2.000 \times 10^4) + (0.10 \times 10^2)$
   b) $48.0 + (5.600 \times 10^3)$
   c) $(1.23 \times 10^5) + (6.00 \times 10^3)$
   d) $(4.80 \times 10^{-4}) - (1.6 \times 10^{-3})$
   e) $(6.300 \times 10^4) - (4.8 \times 10^2)$

5. Make an approximate estimate of the answers to the following:
   a) $\dfrac{4.0 \times 10^3 \times 5.6 \times 10^{-2} \times 7.1 \times 10^6}{8.2 \times 10^{-6} \times 4.9 \times 10^3}$
   b) $567 \times 4183 \times 0.001\,27 \times 0.107$
   c) $\dfrac{496 \times 7124 \times 83\,000 \times 4.7}{7260 \times 41 \times 0.0075}$
   d) $\dfrac{1480 \times 6730 \times 0.173 \times 0.0097}{0.15 \times 0.0088 \times 100\,860 \times 0.10}$
   e) $\dfrac{208 \times 100\,490}{560 \times 0.005\,5 \times 0.000\,49}$

6. Find the logarithms of the following numbers:
   a) 4735
   b) $5.072 \times 10^3$
   c) 0.001 327
   d) 10.076
   e) $2.314 \times 10^{-6}$

7. Find the antilogarithms of the following:
   a) 3.4567
   b) 0.0549
   c) 7.8432
   d) $-6.4712$
   e) $-2.0571$

8. Find the reciprocals of the following:
   a) 234.5
   b) 3 488 123
   c) 0.002 477
   d) $4.865 \times 10^9$
   e) $2.645 \times 10^{-5}$

9. Solve the quadratic equations:
   a) $(x + 3)^2 = 100$
   b) $5x^2 - 8 = 18x$
   c) $(y + 4)^2 = 18y - 9$
   d) $(z - 6)^2 = 4z + 8$
   e) $(2x + 4)^2 = 7 - 4x$

# 2  The Mole

## DEFINITIONS

The masses of elements are very small, from $10^{-24}$ to $10^{-22}$ grams. Instead of using the actual masses of atoms, we use the *relative atomic masses*. Since hydrogen is the lightest of atoms, it was decided to compare the masses of other atoms with that of hydrogen. Then:

$$\text{Original relative atomic mass} = \frac{\text{Mass of one atom of an element}}{\text{Mass of one atom of hydrogen}}$$

On this scale, the hydrogen atom has a mass of 1 atomic mass unit $(m_u)$ and hydrogen has a relative atomic mass of $1.000\,00$ ($1m_u = 1.6605 \times 10^{-27}$ kg).

Since relative atomic masses are now determined by mass spectrometry, and since volatile carbon compounds are much used in mass spectrometry, the mass of an atom of $^{12}_{6}$C is now taken as the standard of reference ($^{12}_{6}$C is an atom of the isotope of carbon containing 6 protons and 6 neutrons). Thus:

$$\text{Modern relative atomic mass} = \frac{\text{Mass of one atom of an element}}{\frac{1}{12}\,\text{Mass of one atom of carbon-12}}$$

The difference between the two scales is small. On the carbon-12 scale, the relative atomic mass of carbon is $12.000\,00$, and the relative atomic mass of hydrogen is $1.008\,97$.

You can see from this ratio that, if an atom of carbon is 12 times as heavy as an atom of hydrogen, then 12 g of carbon contain the same number of atoms as 1 g of hydrogen. The same is true of the relative atomic mass expressed in grams for any element: 63.5 g of copper and 27 g of aluminium contain the same number of atoms. This number is $6.022 \times 10^{23}$, and is called the Avogadro constant. We call $6.022 \times 10^{23}$ units of any species a *mole* of that species. Thus, $6.022 \times 10^{23}$ atoms of copper are a mole of copper; $6.022 \times 10^{23}$ hydrogen molecules are a mole of hydrogen molecules; and $6.022 \times 10^{23}$ electrons are a mole of electrons. The symbol for mole is mol.

*The mole is defined as the amount of a substance which contains as many elementary entities as there are atoms in 12 grams of carbon-12.*

The *relative molecular mass* of a compound is defined by the expression:

$$\text{Relative molecular mass} = \frac{\text{Mass of one molecule of the compound}}{\frac{1}{12} \text{ Mass of one atom of carbon-12}}$$

The relative molecular mass of a compound expressed in grams is a mole of the compound; thus 44 g of carbon dioxide is a mole of carbon dioxide, and contains $6.022 \times 10^{23}$ molecules. In the case of ionic compounds, which do not consist of molecules, you refer to a 'formula unit' of the compound. A formula unit of sodium sulphate is $2Na^+SO_4^{2-}$. Then, the relative formula mass in grams, 142 g, is a mole of sodium sulphate.

The term *relative molar mass* embraces relative molecular mass, relative formula mass and relative atomic mass:

$$\text{Relative molar mass} = \frac{\text{Mass of one mole of substance}}{\frac{1}{12} \text{ Mass of one mole of carbon-12}}$$

Since this equation divides mass by mass, relative molar mass is a ratio and does not have units.

The mass of one mole of an element or compound is referred to as its *molar mass*. The molar mass of sodium hydroxide is $40 \, \text{g mol}^{-1}$. The molar mass of copper is $63.5 \, \text{g mol}^{-1}$. If you have $m$ grams of a substance which has a molar mass of $M \, \text{g mol}^{-1}$, then the number of moles of substance, $n$, is given by:

$$n \, (\text{mol}) = \frac{m \, (\text{g})}{M \, (\text{g mol}^{-1})}$$

The *number of moles of substance* is referred to simply as the *amount of substance*:

$$\text{Amount of substance (number of moles)} = \frac{\text{Mass}}{\text{Molar mass}}$$

## CALCULATION OF MOLAR MASS

The relative molar mass of a compound is the sum of the relative atomic masses of the atoms in a molecule or formula unit of the compound.

**EXAMPLE**  What is the molar mass of glucose?

Formula $= C_6H_{12}O_6$

Relative molar mass $= (6 \times 12) + (12 \times 1) + (6 \times 16) = 180$

**ANSWER**  The relative molar mass is 180, and the molar mass is $180\,g\,mol^{-1}$.

## EMPIRICAL FORMULAE

The formula of a compound is composed of the symbols which show which elements are present together with numbers, as subscripts, which show how many atoms of each element are present in a molecule of the compound or in a formula unit of the compound. The empirical formula is the simplest formula which represents the composition of a compound; it shows the ratio of numbers of atoms present. The molecular formula is a simple multiple of the empirical formula. If the empirical formula is $CH_2O$, the molecular formula may be $CH_2O$ or $C_2H_4O_2$ or $C_3H_6O_3$ and so on. You can tell which molecular formula is correct by finding out which gives the correct molar mass.

To find the empirical formula, you find the ratio of moles of each element present. Remember:

$$\text{Number of moles} = \frac{\text{Mass}}{\text{Molar mass}}$$

**EXAMPLE 1**  A chloride of iron contains 65.5% chlorine. What is its empirical formula?

**METHOD**

| Elements present | Iron | Chlorine |
|---|---|---|
| Percentage by mass | 34.5% | 65.5% |
| Relative atomic mass | 56 | 35.5 |
| Number of moles in 100 g of compound | 34.5/56 = 0.616 | 65.5/35.5 = 1.85 |
| Ratio of moles | $\dfrac{0.616}{0.616}$ : | $\dfrac{1.85}{0.616}$ |
| | = 1 : | 3 |
| Ratio of atoms | = 1 : | 3 |

**ANSWER**  The empirical formula is $FeCl_3$.

**EXAMPLE 2**  Find $n$ in the formula $MgSO_4 \cdot nH_2O$. A sample of 7.38 g of magnesium sulphate crystals lost 3.78 g of water on heating.

| METHOD | Compounds present | Magnesium sulphate | Water |
|---|---|---|---|
| | Mass | 3.60 g | 3.78 g |
| | Molar mass | 120 g mol$^{-1}$ | 18 g mol$^{-1}$ |
| | Number of moles | 3.60/120 | 3.78/18 |
| | | = 0.030 | = 0.21 |
| | Ratio of moles | $\dfrac{0.030}{0.030}$ : | $\dfrac{0.21}{0.030}$ |
| | | = 1 : | 7 |

ANSWER     The empirical formula is $MgSO_4 \cdot 7H_2O$.

## MOLECULAR FORMULAE

The molecular formula can be found from the empirical formula of the compound if the molar mass is known. For methods of finding molar masses, see Chapter 4. The molar mass is a multiple of the empirical formula mass.

EXAMPLE     A compound has the empirical formula $CH_2O$ and molar mass 180 g mol$^{-1}$. What is its molecular formula?

METHOD     Empirical formula mass = 30 g mol$^{-1}$

Molar mass = 180 g mol$^{-1}$

The molar mass is 6 times the empirical formula mass. Therefore the molecular formula is 6 times the empirical formula. Therefore:

ANSWER     The empirical formula is $C_6H_{12}O_6$.

## PERCENTAGE COMPOSITION

From the formula of a compound and the relative atomic masses of the elements in it, the percentage of each element in the compound can be calculated. This is called the *percentage composition by mass.*

EXAMPLE 1  Calculate the percentage composition by mass of magnesium oxide.

METHOD     Relative atomic masses are Mg = 24; O = 16

Relative molar mass of MgO = 24 + 16 = 40

Percentage of Mg = $\dfrac{24}{40} \times 100 = 60\%$

Percentage of O = $\dfrac{16}{40} \times 100 = 40\%$

ANSWER     The percentage composition of MgO by mass is 60% Mg, 40% O.

EXAMPLE 2  Calculate the percentage of water of crystallisation in copper(II) sulphate-5-water.

**METHOD**   Formula is $CuSO_4 \cdot 5H_2\dot{O}$

Relative atomic masses are Cu, 63.5; S, 32; O, 16; H, 1

Relative molar mass $= 63.5 + 32 + (4 \times 16) + (5 \times 18) = 249.5$

Percentage of water $= \dfrac{90}{249.5} \times 100 = 36.1\%$

**ANSWER**   The percentage by mass of water of crystallisation is 36.1%.

**EXERCISE 2**   Problems on Empirical and Molecular Formulae and Percentage Composition

1. Calculate the percentage by mass of the named element in the compound listed:

   a) Na in $Na_2SO_4 \cdot 10H_2O$           b) Ca in $Ca(CN)_2$
   c) N in $(NH_4)_2CO_3$              d) U in $UO_2(NO_3)_2 \cdot 6H_2O$
   e) Al in $KAl(SO_4)_2 \cdot 12H_2O$

2. Find the empirical formulae of the compounds formed in the reactions described below:

   a) 10.800 g magnesium form 18.000 g of an oxide
   b) 3.400 g calcium form 9.435 g of a chloride
   c) 3.528 g iron form 10.237 g of a chloride
   d) 2.667 g copper form 4.011 g of a sulphide
   e) 4.662 g lithium form 5.328 g of a hydride.

3. Calculate the empirical formulae of the compounds with the following percentage composition:                          BY mass

   a) 77.7% Fe   22.3% O        b) 70.0% Fe   30.0% O
   c) 72.4% Fe   27.6% O        d) 40.2% K   26.9% Cr   32.9% O
   e) 26.6% K   35.4% Cr   38.0% O   f) 92.3% C   7.6% H
   g) 81.8% C   18.2% H

4. Samples of the following hydrates are weighed, heated to drive off the water of crystallisation, cooled and reweighed. From the results obtained, calculate the values of $a$–$f$ in the formulae of the hydrates:

   a) 0.869 g of $CuSO_4 \cdot aH_2O$ gave a residue of 0.556 g
   b) 1.173 g of $CoCl_2 \cdot bH_2O$ gave a residue of 0.641 g
   c) 1.886 g of $CaSO_4 \cdot cH_2O$ gave a residue of 1.492 g
   d) 0.904 g of $Pb(C_2H_3O_2)_2 \cdot dH_2O$ gave a residue of 0.774g
   e) 1.144 g of $NiSO_4 \cdot eH_2O$ gave a residue of 0.673 g
   f) 1.175 g of $KAl(SO_4)_2 \cdot fH_2O$ gave a residue of 0.639 g.

   207 +
   2(24 + 3 + 4

5. An organic compound, X, which contains only carbon, hydrogen and oxygen, has a molar mass of about $85 \text{ g mol}^{-1}$. When 0.43 g of X is burnt in excess oxygen, 1.10 g of carbon dioxide and 0.45 g of water are formed.

   a) What is the empirical formula of X?
   b) What is the molecular formula of X?

6. A liquid, Y, of molar mass $44\,\text{g mol}^{-1}$ contains 54.5% carbon, 36.4% oxygen and 9.1% hydrogen.
   a) Calculate the empirical formula of Y, and
   b) deduce its molecular formula.

7. An organic compound contains 58.8% carbon, 9.8% hydrogen and 31.4% oxygen. The molar mass is $102\,\text{g mol}^{-1}$.
   a) Calculate the empirical formula, and
   b) deduce the molecular formula of the compound.

8. An organic compound has molar mass $150\,\text{g mol}^{-1}$ and contains 72.0% carbon, 6.67% hydrogen and 21.33% oxygen. What is its molecular formula?

## CALCULATIONS BASED ON CHEMICAL EQUATIONS

Equations tell us not only what substances react together but also what amounts of substances react together. The equation for the action of heat on sodium hydrogencarbonate

$$2NaHCO_3(s) \longrightarrow Na_2CO_3(s) + CO_2(g) + H_2O(g)$$

tells us that 2 moles of $NaHCO_3$ give 1 mole of $Na_2CO_3$. Since the molar masses are $NaHCO_3 = 84\,\text{g mol}^{-1}$ and $Na_2CO_3 = 106\,\text{g mol}^{-1}$, it follows that 168 g of $NaHCO_3$ give 106 g of $Na_2CO_3$.

The amounts of substances undergoing reaction, as given by the balanced chemical equation, are called the *stoichiometric* amounts. *Stoichiometry* is the relationship between the amounts of reactants and products in a chemical reaction. If one reactant is present in excess of the stoichiometric amount required for reaction with another of the reactants, then the excess of one reactant will be left unused at the end of the reaction.

**EXAMPLE 1**    How many moles of iodine can be obtained from $\frac{1}{6}$ mole of potassium iodate(V)?

**METHOD**    The equation

$$KIO_3(aq) + 5KI(aq) + 6H^+(aq) \longrightarrow 3I_2(aq) + 6K^+(aq) + 3H_2O(l)$$

tells us that 1 mole of $KIO_3$ gives 3 moles of $I_2$. Therefore:

**ANSWER**    $\frac{1}{6}$ mole of $KIO_3$ gives $\frac{1}{6} \times 3$ moles of $I_2 = \frac{1}{2}$ mole of $I_2$.

**EXAMPLE 2**   What is the maximum mass of ethyl ethanoate that can be obtained from 0.1 mole of ethanol?

**METHOD**   Write the equation:

$$C_2H_5OH(l) + CH_3CO_2H(l) \longrightarrow CH_3CO_2C_2H_5(l) + H_2O(l)$$

1 mole of $C_2H_5OH$ gives 1 mole $CH_3CO_2C_2H_5$
0.1 mole of $C_2H_5OH$ gives 0.1 mole $CH_3CO_2C_2H_5$
The molar mass of $CH_3CO_2C_2H_5$ is $88\,g\,mol^{-1}$. Therefore:

**ANSWER**   0.1 mole of ethanol gives 8.8 g ethyl ethanoate.

**EXAMPLE 3**   A mixture of 5.00 g of sodium carbonate and sodium hydrogen-carbonate is heated. The loss in mass is 0.31 g. Calculate the percentage by mass of sodium carbonate in the mixture.

**METHOD**   On heating the mixture, the reaction

$$2NaHCO_3(s) \longrightarrow Na_2CO_3(s) + CO_2(g) + H_2O(g)$$

takes place. The loss in mass is due to the decomposition of $NaHCO_3$.
Since 2 mol $NaHCO_3$ form 1 mol $CO_2$ + 1 mol $H_2O$
2 × 84 g $NaHCO_3$ form 44 g $CO_2$ and 18 g $H_2O$
168 g $NaHCO_3$ lose 62 g in mass.
The observed loss in mass of 0.31 g is due to the decomposition of

$$\frac{0.31}{62} \times 168\,g\ NaHCO_3 = 0.84\,g$$

The mixture contains 0.84 g $NaHCO_3$
The difference, $5.00 - 0.84 = 4.16\,g\ Na_2CO_3$.

**ANSWER**   Percentage of $Na_2CO_3 = \dfrac{4.16}{5.00} \times 100 = 83.2\%$.

**EXAMPLE 4**   A mixture of $MgSO_4 \cdot 7H_2O$ and $CuSO_4 \cdot 5H_2O$ is heated at $120\,°C$ until a mixture of the anhydrous salts is obtained. If 5.000 g of the mixture give 3.000 g of the anhydrous salts, calculate the percentage by mass of $MgSO_4 \cdot 7H_2O$ in the mixture.

**METHOD**   Both salts lose water of crystallisation on heating:

$$MgSO_4 \cdot 7H_2O(s) \longrightarrow MgSO_4(s) + 7H_2O(g)$$

246 g of crystals $\longrightarrow$ 120 g of anhydrous salt + 126 g of steam

$$CuSO_4 \cdot 5H_2O(s) \longrightarrow CuSO_4(s) + 5H_2O(g)$$

249.5 g of crystals $\longrightarrow$ 159.5 g of anhydrous salt + 90.0 g of steam

Let $a$ g $=$ Mass of $MgSO_4 \cdot 7H_2O$ in the mixture.

Then $(5.000 - a)$ g $=$ Mass of $CuSO_4 \cdot 5H_2O$

$a$ g of $MgSO_4 \cdot 7H_2O$ loses $\dfrac{a}{246} \times 126$ g steam

$(5.000 - a)$ g of $CuSO_4 \cdot 5H_2O$ loses $\dfrac{(5.000 - a)}{249.5} \times 90.0$ g steam

Total loss in mass $=$ 2.000 g

Therefore $\left(\dfrac{a}{246} \times 126\right) + \dfrac{(5.000 - a) \times 90}{249.5} = 2.000$

Mass of $MgSO_4 \cdot 7H_2O = 1.298$ g

Mass of $CuSO_4 \cdot 5H_2O = 3.702$ g

Percentage of $MgSO_4 \cdot 7H_2O = \dfrac{1.298}{5.000} \times 100 = 25.96\%$

**ANSWER**     The mixture contains 26.0% by mass of $MgSO_4 \cdot 7H_2O$.

## EXERCISE 3     Problems on Reacting Masses of Solids

1. What mass of glucose must be fermented to give 5.00 kg of ethanol?

    $$C_6H_{12}O_6(aq) \longrightarrow 2C_2H_5OH(aq) + 2CO_2(g)$$

2. What mass of sulphuric acid can be obtained from 1000 tonnes of an ore which contains 32.0% of $FeS_2$?

3. What mass of silver chloride can be precipitated from a solution which contains $1.000 \times 10^{-3}$ moles of silver ions?

4. The pollutant, sulphur dioxide, can be removed from the air by the reaction

    $$2CaCO_3(s) + 2SO_2(g) + O_2(g) \longrightarrow 2CaSO_4(s) + 2CO_2(g)$$

    What mass of calcium carbonate is needed to remove 10.0 kg of $SO_2$?

5. When potassium iodate(V) is allowed to react with acidified potassium iodide, iodine is formed.

    $$KIO_3(aq) + KI(aq) + 6H^+(aq) \longrightarrow 3I_2(aq) + 6K^+(aq) + 3H_2O(l)$$

    What mass of $KIO_3$ is required to give 10.00 g of iodine?

6. How many tonnes of iron can be obtained from 10.00 tonnes of a) $Fe_2O_3$ and b) $Fe_3O_4$?

7. What mass of sodium carbonate can be obtained by heating 100 g of sodium hydrogencarbonate?

    $$2NaHCO_3(s) \longrightarrow Na_2CO_3(s) + CO_2(g) + H_2O(g)$$

8. What mass of quicklime can be obtained by heating 75.0 g of limestone, which is 86.8% calcium carbonate?

9. What mass of barium sulphate can be precipitated fro~~m~~
which contains 4.000 g of barium chloride?

10. Calculate the mass of each of the reactants needed to prod~~uce~~
of phosphorus by the reaction

$$2Ca_3(PO_4)_2(s) + 6SiO_2(s) + 5C(s) \longrightarrow P_4(g) + 6CaSiO_3(s) + 5CO_2(g)$$

11. A mixture of calcium and magnesium carbonates weighing 10.0000 g
was heated until it reached a constant mass of 5.0960 g. Calculate the
percentage composition of the mixture, by mass.

12. A mixture of anhydrous sodium carbonate and sodium hydrogencarbonate weighing 10.0000 g was heated until it reached a constant
mass of 8.7080 g. Calculate the composition of the mixture in grams
of each component.

# REACTING VOLUMES OF GASES

In stating the volume of a gas, one needs to state the temperature
and pressure at which the volume was measured. It is usual to give
the volume at $0\,^\circ$C and 1 atmosphere pressure (273 K and $1.01 \times 10^5\,N\,m^{-2}$). These conditions are called standard temperature and
pressure (s.t.p.). Chapter 5 deals with calculating the volume of a gas
at s.t.p. from the volume measured under experimental conditions.

A mole of gas occupies $22.4\,dm^3$ at s.t.p. One can say that *the gas
molar volume is 22.4 dm³ at s.t.p.* This makes calculations on reacting
volumes of gases very simple. An equation which shows how many
moles of different gases react together also shows the ratio of the
volumes of the different gases that react together. For example, the
equation

$$2NO(g) + O_2(g) \longrightarrow 2NO_2(g)$$

tells us that 2 moles of NO + 1 mole of $O_2$ form 2 moles of $NO_2$

∴  $44.8\,dm^3$ of NO + $22.4\,dm^3$ of $O_2$ form $44.8\,dm^3$ of $NO_2$

In general, 2 volumes of NO + 1 volume of $O_2$ form 2 volumes of $NO_2$.

EXAMPLE 1  What is the volume of oxygen needed for the complete combustion
of $2\,dm^3$ of propane?

METHOD  Write the equation:

$$C_3H_8(g) + 5O_2(g) \longrightarrow 3CO_2(g) + 4H_2O(g)$$

1 mole of $C_3H_8$ needs 5 moles of $O_2$

1 volume of $C_3H_8$ needs 5 volumes of $O_2$. Therefore:

ANSWER  $2\,dm^3$ of propane need $10\,dm^3$ of oxygen.

**EXAMPLE 2**   What volume of hydrogen is obtained when 3.00 g of zinc react with an excess of dilute sulphuric acid at s.t.p.?

**METHOD**   Write the equation:

$$Zn(s) \; + \; H_2SO_4(aq) \longrightarrow H_2(g) \; + \; ZnSO_4(aq)$$

1 mole of Zn forms 1 mole of $H_2$

65 g of Zn form 22.4 dm$^3$ of $H_2$ (at s.t.p.)

3.00 g of Zn form $\dfrac{3.00}{65} \times 22.4$ dm$^3$ = 1.03 dm$^3$ $H_2$

**ANSWER**   3.00 g of zinc give 1.03 dm$^3$ of hydrogen at s.t.p.

**EXAMPLE 3**   A 250 cm$^3$ sample of ozonised oxygen is treated with an unsaturated hydrocarbon. A contraction in volume occurs. When a 250 cm$^3$ sample of ozonised oxygen is heated to 200 °C and then cooled, an expansion occurs, which is half the contraction observed in the previous treatment. Assuming the formula of oxygen is $O_2$, what is the formula for ozone?

**METHOD**   Let the formula of ozone be $O_x$.

Let the contraction on treatment with alkene be $a$ cm$^3$

Then the volume of ozone in the mixture is $a$ cm$^3$

When ozone is converted to oxygen an expansion of $a/2$ cm$^3$ occurs

Thus, $a$ cm$^3$ of ozone must form $1.5a$ cm$^3$ of oxygen

and 2 volumes of ozone form 3 volumes of oxygen

and 2 molecules of ozone form 3 molecules of oxygen.

Thus, $2O_x \longrightarrow 3O_2$.

Therefore $x = 3$.

**ANSWER**   Ozone has the formula $O_3$.

**EXAMPLE 4**   10 cm$^3$ of a hydrocarbon, $C_4H_x$ were allowed to react with an excess of oxygen at 150 °C and 1 atmosphere. There was an expansion of 10 cm$^3$. Deduce the value of $x$.

**METHOD**   Write the equation:

$$C_4H_x(g) \; + \; (4 + x/4)O_2(g) \longrightarrow 4CO_2(g) \; + \; x/2 \, H_2O(g)$$

Volume of hydrocarbon = 10 cm$^3$

Volume of oxygen = $10(4 + x/4)$ = $(40 + 5x/2)$ cm$^3$

Volume of carbon dioxide = 40 cm$^3$

Volume of steam = $x/2 \times 10$ cm$^3$ = $5x$ cm$^3$

Let $a$ cm$^3$ = Volume of unused oxygen.

Since there is an expansion of $10 \text{ cm}^3$,

$$\text{Final volume} = 10 + \text{Initial volume}$$
$$40 + 5x + a = 10 + 10 + 40 + 5x/2 + a$$

and
$$5x/2 = 20$$
$$x = 8$$

**ANSWER**   The formula is $C_4H_8$.

**EXAMPLE 5**   $10 \text{ cm}^3$ of a hydrocarbon, $C_aH_b$, are exploded with an excess of oxygen. A contraction of $35 \text{ cm}^3$ occurs, all volumes being measured at room temperature and pressure. On treatment of the products with sodium hydroxide solution, a contraction of $40 \text{ cm}^3$ occurs. Deduce the formula of the hydrocarbon.

**METHOD**   Write the equation:

$$C_aH_b(g) + (a + b/4)O_2(g) \longrightarrow aCO_2(g) + b/2\ H_2O(l)$$

Volume of hydrocarbon $= 10 \text{ cm}^3$

Volume of $CO_2 = a \times$ Volume of $C_aH_b$

From reaction with NaOH, volume of $CO_2 = 40 \text{ cm}^3$

Therefore $a = 4$

Let the volume of unused oxygen be $c \text{ cm}^3$.

Final volume $=$ Initial volume $- 35 \text{ cm}^3$

*Note* that $H_2O(l)$ is a liquid at room temperature and pressure, and does not contribute to the final volume of gas.

$$40 + c = 10 + 40 + 5b/2 + c - 35$$
$$25 = 5b/2$$
$$b = 10$$

**ANSWER**   The formula is $C_4H_{10}$.

**EXAMPLE 6**   $20 \text{ cm}^3$ of ammonia are burned in an excess of oxygen at $110 \,^\circ C$. $10 \text{ cm}^3$ of nitrogen and $30 \text{ cm}^3$ of steam are formed. Deduce the formula for ammonia, given that the formula of nitrogen is $N_2$, and the formula of steam is $H_2O$.

**METHOD**   Let the formula of ammonia be $N_aH_b$.

The equation for combustion is

$$N_aH_b + b/4\ O_2 \longrightarrow a/2\ N_2 + b/2\ H_2O$$

Volume of $N_aH_b = 20 \text{ cm}^3$

Volume of $N_2 = a/2 \times 20 = 10a \text{ cm}^3 = 10 \text{ cm}^3 \quad \therefore a = 1$

Volume of $H_2O(g) = b/2 \times 20 = 10b \text{ cm}^3 = 30 \text{ cm}^3 \quad \therefore b = 3$

**ANSWER**   The formula is $NH_3$.

**EXAMPLE 7**  25 cm³ of a mixture of methane and ethane were completely oxidised by 72.5 cm³ of oxygen, measured at the same temperature and pressure. What was the composition of the mixture?

**METHOD**  Let $a$ cm³ = Volume of methane

Then $(25 - a)$ cm³ = Volume of ethane

The equation

$$CH_4(g) + 2O_2(g) \longrightarrow CO_2(g) + 2H_2O(g)$$

tells us that $a$ cm³ of methane need $2a$ cm³ of oxygen.

The equation

$$2C_2H_6(g) + 7O_2(g) \longrightarrow 4CO_2(g) + 6H_2O(g)$$

tells us that $(25 - a)$ cm³ of ethane need $\frac{7}{2}(25 - a)$ cm³ of oxygen.

Thus      Volume of oxygen $= 2a + \frac{7}{2}(25 - a) = 72.5$

$$4a + 175 - 7a = 145$$

$$3a = 30$$

$$a = 10$$

**ANSWER**  The mixture consists of 10 cm³ methane and 15 cm³ of ethane.

## EXERCISE 4      Problems on Reacting Volumes of Gases

1. What volume of oxygen (at s.t.p.) is required to burn exactly:
   a) 1 dm³ of methane, according to the reaction
   $$CH_4(g) + 2O_2(g) \longrightarrow CO_2(g) + 2H_2O(g)$$
   b) 500 cm³ of hydrogen sulphide, according to the reaction
   $$2H_2S(g) + 3O_2(g) \longrightarrow 2SO_2(g) + 2H_2O(g)$$
   c) 250 cm³ of ethyne, according to the equation
   $$2C_2H_2(g) + 5O_2(g) \longrightarrow 4CO_2(g) + 2H_2O(g)$$
   d) 750 cm³ of ammonia, according to the reaction
   $$4NH_3(g) + 5O_2(g) \longrightarrow 4NO(g) + 6H_2O(g)$$
   e) 1 dm³ of phosphine, according to the reaction
   $$PH_3(g) + 2O_2(g) \longrightarrow H_3PO_4(s)?$$

2. 1 dm³ of $H_2S$ and 1 dm³ of $SO_2$ were allowed to react, according to the equation
   $$2H_2S(g) + SO_2(g) \longrightarrow 2H_2O(l) + 3S(s)$$
   What volume of gas will remain after the reaction?

3. 100 cm³ of a mixture of ethane and ethene at s.t.p. were treated with bromine. 0.357 g of bromine was used up. Calculate the percentage by volume of ethene in the mixture.

4. 10 cm$^3$ of a hydrocarbon $C_xH_y$ were exploded with an excess of oxygen. There was a contraction of 30 cm$^3$. When the product was treated with a solution of sodium hydroxide, there was a further contraction of 30 cm$^3$. Deduce the formula of the hydrocarbon. All gas volumes are at s.t.p.

5. 10 cm$^3$ of a hydrocarbon $C_aH_b$ were exploded with excess oxygen. A contraction of 25 cm$^3$ occurred. On treating the product with sodium hydroxide, a further contraction of 40 cm$^3$ occurred. Deduce the values of $a$ and $b$ in the formula of the hydrocarbon. All measurements of gas volumes are at s.t.p.

6. 10 cm$^3$ of a hydrocarbon $C_4H_8$ were exploded with an excess of oxygen. A contraction of $a$ cm$^3$ occurred. On adding sodium hydroxide solution, a further contraction of $b$ cm$^3$ occurred. What are the volumes, $a$ and $b$? All gas volumes are at s.t.p.

7. Hydrogen sulphide burns in oxygen in accordance with the following equation:

$$2H_2S(g) + 3O_2(g) \longrightarrow 2H_2O(g) + 2SO_2(g)$$

If 4 dm$^3$ of $H_2S$ are burned in 10 dm$^3$ of oxygen at 1 atmosphere pressure and 120 °C, what is the final volume of the mixture?

a 6 dm$^3$   b 8 dm$^3$   c 10 dm$^3$   (d) 12 dm$^3$   e 14 dm$^3$

**EXERCISE 5**   Problems on Reactions Involving Solids and Gases

1. In the Solvay process

$$NaCl(aq) + NH_3(g) + H_2O(l) + CO_2(g) \longrightarrow NaHCO_3(s) + NH_4Cl(aq)$$

what volume of carbon dioxide (at s.t.p.) is required to produce 1.00 kg of sodium hydrogencarbonate?

2. What volume of ethyne (at s.t.p.) can be prepared from 10.0 g of calcium carbide by the reaction

$$CaC_2(s) + 2H_2O(l) \longrightarrow Ca(OH)_2(aq) + C_2H_2(g)?$$

3. What mass of phosphorus is required for the preparation of 200 cm$^3$ of phosphine (at s.t.p.) by the reaction

$$P_4(s) + 3NaOH(aq) + 3H_2O(l) \longrightarrow 3NaH_2PO_4(aq) + PH_3(g)?$$

4. Calculate the mass of ammonium chloride required to produce 1.00 dm$^3$ of ammonia (at s.t.p.) in the reaction

$$2NH_4Cl(s) + Ca(OH)_2(s) \longrightarrow 2NH_3(g) + CaCl_2(s) + 2H_2O(g)$$

5. What mass of potassium chlorate(V) must be heated to give 1.00 dm$^3$ of oxygen at s.t.p.? The reaction is

$$2KClO_3(s) \longrightarrow 2KCl(s) + 3O_2(g)$$

½ mole of $Cl_2$ produced.

6. What volume of chlorine (at s.t.p.) can be obtained from the electrolysis of a solution containing 60.0 g of sodium chloride?

7. What volume of oxygen (at s.t.p.) is needed for the complete combustion of 1.00 kg of octane? The reaction is

$$2C_8H_{18}(l) + 25O_2(g) \longrightarrow 16CO_2(g) + 18H_2O(g)$$

## PERCENTAGE YIELD

There are many reactions which do not go to completion. Reactions between organic compounds do not often give a 100% yield of product. The actual yield is compared with the yield calculated from the molar masses of the reactants. The equation

$$\text{Percentage yield} = \frac{\text{Actual mass of product}}{\text{Calculated mass of product}} \times 100$$

is used to give the percentage yield.

**EXAMPLE**  From 23 g of ethanol are obtained 44 g of ethyl ethanoate by esterification with ethanoic acid in the presence of concentrated sulphuric acid. What is the percentage yield of the reaction?

**METHOD**  Write the equation:

$$CH_3CO_2H(l) + C_2H_5OH(l) \longrightarrow CH_3CO_2C_2H_5(l) + H_2O(l)$$

46 g of $C_2H_5OH$ forms 108 g of $CH_3CO_2C_2H_5$

23 g of $C_2H_5OH$ should give $\dfrac{23}{46} \times 108\,g = 54\,g$ of $CH_3CO_2C_2H_5$

Actual mass obtained $= 44\,g$

$$\text{Percentage yield} = \frac{\text{Actual mass of product}}{\text{Calculated mass of product}} \times 100$$

**ANSWER**  Percentage yield $= \dfrac{44}{54} \times 100 = 81.5\%$.

## EXERCISE 6      Problems on Percentage Yield

1. Phenol, $C_6H_5OH$, is converted to trichlorophenol, $C_6H_2Cl_3OH$. If 488 g of product are obtained from 250 g of phenol, calculate the percentage yield.

2. 29.5 g of ethanoic acid, $CH_3CO_2H$, are obtained from the oxidation of 25.0 g of ethanol, $C_2H_5OH$. What percentage yield does this represent?

3. 0.8500 g of hexanone, $C_6H_{12}O$, is converted to its 2,4-dinitrophenylhydrazone. After isolation and purification, 2.1180 g of product, $C_{12}H_{18}N_4O_4$, are obtained. What percentage yield does this represent?

4. Benzaldehyde, $C_7H_6O$, forms a hydrogensulphite compound of formula $C_7H_7SO_4Na$. From 1.210 g of benzaldehyde, a yield of 2.181 g of the product was obtained. Calculate the percentage yield.

5. 100 cm³ of barium chloride solution of concentration 0.0500 mol dm⁻³ were treated with an excess of sulphate ions in solution. The precipitate of barium sulphate formed was dried and weighed. A mass of 1.1558 g was recorded. What percentage yield does this represent?

## LIMITING REACTANT

In a chemical reaction, the reactants are often added in amounts which are not stoichiometric. One or more of the reactants is in excess and is not completely used up in the reaction. The amount of product is determined by the amount of the reactant that is not in excess and is used up completely in the reaction. This is called the *limiting reactant*. You first have to decide which is the limiting reactant before you can calculate the amount of product formed.

**EXAMPLE** 5.00 g of iron and 5.00 g of sulphur are heated together to form iron(II) sulphide. Which reactant is present in excess? What mass of product is formed?

**METHOD** Write the equation:

$$Fe(s) + S(s) \longrightarrow FeS(s)$$

1 mole of Fe + 1 mole of S form 1 mole of FeS

∴ 56 g Fe and 32 g S form 88 g FeS

5.00 g Fe is 5/56 mol = 0.0893 mol Fe

5.00 g S is 5/32 mol = 0.156 mol S

There is insufficient Fe to react with 0.156 mol S; iron is the limiting reactant.

0.0893 mol Fe forms 0.0893 mol FeS = 0.0893 × 88 g = 7.86 g

**ANSWER** Mass formed = 7.86 g.

## EXERCISE 7 Problems on Limiting Reactant

1. In the blast furnace, the overall reaction is

$$2Fe_2O_3(s) + 3C(s) \longrightarrow 3CO_2(g) + 4Fe(s)$$

What is the maximum mass of iron that can be obtained from 700 tonnes of iron(III) oxide and 70 tonnes of coke? (1 tonne = 1 000 kg.)

2. In the manufacture of calcium carbide

$$CaO(s) + 3C(s) \longrightarrow CaC_2(s) + CO(g)$$

What is the maximum mass of calcium carbide that can be obtained from 40 kg of quicklime and 40 kg of coke?

3. In the manufacture of the fertiliser ammonium sulphate

$$H_2SO_4(aq) \ + \ 2NH_3(g) \ \longrightarrow \ (NH_4)_2SO_4(aq)$$

What is the maximum mass of ammonium sulphate that can be obtained from 2.0 kg of sulphuric acid and 1.0 kg of ammonia?

4. In the Solvay process, ammonia is recovered by the reaction

$$2NH_4Cl(s) \ + \ CaO(s) \ \longrightarrow \ CaCl_2(s) \ + \ H_2O(g) \ + \ 2NH_3(g)$$

What is the maximum mass of ammonia that can be recovered from $2.00 \times 10^3$ kg of ammonium chloride and 500 kg of quicklime?

5. In the Thermit reaction

$$2Al(s) \ + \ Cr_2O_3(s) \ \longrightarrow \ 2Cr(s) \ + \ Al_2O_3(s)$$

Calculate the percentage yield when 180 g of chromium are obtained from a reaction between 100 g of aluminium and 400 g of chromium(III) oxide.

# DERIVING THE EQUATION FOR A REACTION

If you know the mass of each solid or the volume of each gas taking part in a reaction, you can calculate the number of moles of each substance taking part, and this will tell you the equation for the reaction.

**EXAMPLE**     When 100 cm³ of a hydrocarbon X burn in 500 cm³ of oxygen, 50 cm³ of oxygen are unused, 300 cm³ of carbon dioxide are formed, and 300 cm³ of steam are formed. Deduce the equation for the reaction and the formula of the hydrocarbon.

**METHOD**
$$X \ + \ O_2(g) \ \longrightarrow \ CO_2(g) \ + \ H_2O(g)$$
$$\text{100 cm}^3 \quad \text{450 cm}^3 \qquad\qquad \text{300 cm}^3 \qquad \text{300 cm}^3$$

The volumes of gases reacting tell us that

$$X \ + \ 4\tfrac{1}{2}O_2(g) \ \longrightarrow \ 3CO_2(g) \ + \ 3H_2O(g)$$

To balance the equation, X must be $C_3H_6$. Then,

$$C_3H_6(g) \ + \ 4\tfrac{1}{2}O_2(g) \ \longrightarrow \ 3CO_2(g) \ + \ 3H_2O(g)$$

**ANSWER**
$$2C_3H_6(g) \ + \ 9O_2(g) \ \longrightarrow \ 6CO_2(g) \ + \ 6H_2O(g)$$

# EXERCISE 8     Problems on Deriving Equations

1. To a solution containing 2.975 g of sodium persulphate, $Na_2S_2O_8$, is added an excess of potassium iodide solution. A reaction occurs, in which sulphate ions are formed and 3.175 g of iodine are formed. Deduce the equation for the reaction.

2. Amidosulphuric acid, $H_2NSO_3H$, reacts with warm sodium hydroxide solution to give ammonia and a solution which contains sulphate ions. 0.540 g of the acid, when treated with an excess of alkali, gave 153 cm$^3$ of ammonia at 60 °C and 1 atm. Deduce the equation for the reaction.

3. A solution containing $5.00 \times 10^{-3}$ mol of sodium thiosulphate was shaken with 1 g of silver chloride. 0.7175 g of silver chloride dissolved, and analysis showed that $5.00 \times 10^{-3}$ mol of chloride ions were present in the resulting solution. Derive an equation for the reaction.

4. An unsaturated hydrocarbon of molar mass 80 g mol$^{-1}$ reacts with bromine. If 0.250 g of hydrocarbon reacts with 1.00 g of bromine, what is the equation for the reaction?

5. Given that 1.00 g of phenylamine, $C_6H_5NH_2$, reacts with 5.16 g of bromine, derive an equation for the reaction.


## EXERCISE 9    Questions from A-level Papers

1. a) When 0.203 g of hydrated magnesium chloride, $MgCl_m \cdot nH_2O$, was dissolved in water and titrated with 0.1 M silver nitrate ($AgNO_3$) solution, 20.0 cm$^3$ of the latter were required. A sample of the hydrated chloride lost 53.2% of its mass when heated in a stream of hydrogen chloride, leaving a residue of anhydrous magnesium chloride. From these figures calculate the values of $m$ and $n$.

   b) When the hydrated chloride was heated in air instead of hydrogen chloride, the loss in mass was greater than 53.2% and both HCl and $H_2O$ were evolved. What reaction do you think might have occurred and what was the solid product?                    (L80)

2. On repeated sparking 10 cm$^3$ of a mixture of carbon monoxide and nitrogen required 3 cm$^3$ oxygen for combustion. What was the volume of nitrogen in the mixture? (All volumes were measured at room temperature and pressure.)

   a $3\frac{1}{3}$ cm$^3$          b 4 cm$^3$          c 5 cm$^3$
   d 7 cm$^3$          e $8\frac{1}{2}$ cm$^3$                    (C80)

3. A sample of gallium chloride weighing 0.1 g was vaporised and it was found to occupy 16 cm$^3$ at a temperature of 415 °C and at a pressure of 1 atmosphere.

   a) Calculate a value for the relative formula mass of gallium chloride under the conditions of the experiment. (Molar volume of a gas at 0 °C and 1 atmosphere = 22.4 dm$^3$ mol$^{-1}$ *or* $R = 0.082$ atm dm$^3$ K$^{-1}$ mol$^{-1}$ = 8.314 J K$^{-1}$ mol$^{-1}$.)

   b) Suggest a molecular formula for gallium chloride in the vapour state, indicating how you arrive at your answer. (Relative atomic mass: Cl = 35.5; Ga = 69.7.)                    (L(N)82, p)

4. What is the theoretical yield of trichloroethanoic acid when 10.65 g of chlorine reacts with 8.20 g of ethanoic acid under suitable conditions, one reactant being in excess and the reaction going to completion? ($H = 1, C = 12, O = 16, Cl = 35.5$.)

$$3Cl_2 + CH_3COOH \longrightarrow CCl_3COOH + 3HCl$$

a  7.99 g           b  8.18 g           c  16.0 g
d  18.9 g           e  22.3 g           (NI82)

5. a) Explain briefly the difference between *empirical formula* and *molecular formula*.

   b) The percentage composition by mass of a carboxylic acid $X$ was found to be C, 54.5; H, 9.1 and O, 36.4. Calculate the molecular formula of $X$ given that $M_r = 88$.

   c) State briefly what is meant by *structural isomerism*.

   d) Draw the structures of two isomers having the molecular formula $C_3H_6O$ and indicate how you would distinguish between them by means of a simple chemical test.          (JMB82)

6. A piece of tin, of mass 1.527 g, was placed in a solution of iodine in trichloromethane and left for several days in a stoppered flask. At the end of that time the tin, which had been in excess, was removed, washed with trichloromethane and dried. Its mass was now 1.170 g. The tin washings were added to the original solution and the trichloromethane evaporated, leaving an orange residue of mass 1.881 g.

   a) Calculate the empirical formula of the compound of tin and iodine that was formed.

   b)  i) Explain the nature of the bonding in the iodine molecule.
       ii) What type of bonding is present in the tin compound formed? Give a reason for your answer.

   c) State reasons why, in this experiment, trichloromethane rather than water or aqueous potassium iodide is used as the solvent.
                                                                    (AEB82)

7. When excess aqueous barium chloride was added to a solution containing 9.60 g of sulphate ions, 23.34 g of barium sulphate were precipitated. The percentage by mass of barium in barium sulphate is given by

a  $\dfrac{9.60 \times 100}{23.34}$     b  $\dfrac{23.34 \times 100}{9.60}$     c  $\dfrac{13.74 \times 100}{23.34}$

d  $\dfrac{23.34 \times 100}{13.74}$     e  $\dfrac{9.60 \times 100}{13.74}$                (AEB81)

8. If 0.5 mol of a hydrated salt contains 63 g of water, how many moles of water of crystallisation are contained in 1 mol of salt? ($M_r(H_2O) = 18$.)

a  1               b  2               c  3
d  7               e  10                             (AEB81)

9. In a certain compound 7 g of nitrogen are combined with 12 g of oxygen. $(A_r(N) = 14; A_r(O) = 16.)$

The empirical formula of the compound is

a  NO          b  $NO_2$          c  $N_2O$

d  $N_2O_3$      e  $N_2O_5$                    (AEB81)

10. 10.00 cm³ of a gaseous hydrocarbon was exploded with 100.0 cm³ of oxygen. The total volume after the explosion was 75.0 cm³ which decreased to 25.0 cm³ on shaking the gaseous mixture with aqueous sodium hydroxide. (All volumes measured at room temperature and pressure.)

a) Say why there was a decrease in volume on shaking with alkali. Write an equation.

b) Deduce the volume of carbon dioxide formed in the explosion.

c) Deduce the volume of oxygen that *actually reacted* with 10.0 cm³ of the hydrocarbon.

d) Why was the hydrocarbon mixed with *excess* of oxygen before the explosion?

e) Calculate the molecular formula of the hydrocarbon setting out your work clearly.

f) Write the structural formulae of four possible compounds the hydrocarbon could be and give them their I.U.P.A.C. names.

g) One of the isomers from part f) is dissolved in tetrachloromethane and ozonised oxygen is bubbled through the solution. The product of this reaction is reduced with hydrogen and platinum catalyst. The final reaction yields a mixture of ethanal and propanone. Identify the isomer and explain both reactions using appropriate equations to illustrate your answer.                    (SUJB84)

11. When 25 cm³ of the gaseous hydrocarbon, $A$, were exploded with 200 cm³ of oxygen, the residual gases occupied 150 cm³. After shaking the residual gases with excess aqueous sodium hydroxide, the final volume was 50 cm³.

(All volumes were measured at room temperature and pressure.)

a) Why was there a decrease in volume when the residual gases were shaken with aqueous sodium hydroxide? Give an equation.

b) Calculate the molecular formula of $A$. Explain your working.

c) Write the structural formulae of six possible compounds (cyclic and non-cyclic) which $A$ could be and give the systematic names for each of the six formulae.

d) Which of the six structural formulae show compounds which are

i) structural isomers;

ii) stereoisomers;

iii) optical isomers?                    (SUJB81)

12. 0.61 g of metal hydroxide, $M(OH)_2$, exactly neutralises 20.0 cm$^3$ of a solution of hydrochloric acid containing 18.25 g of hydrogen chloride per dm$^3$. Write an equation for the reaction and calculate the relative molecular mass of the metal hydroxide; hence determine the relative atomic mass of M.                                (AEB80,p)

13. An experiment was carried out in order to determine the number of moles of water of crystallisation in one mole of hydrated barium chloride crystals, $BaCl_2 \cdot xH_2O$. The following results were obtained:

a)  mass of crucible + lid                                              =    15.35 g
b)  mass of crucible + lid + hydrated barium chloride      =    27.55 g
c)  mass of crucible + lid + residue after heating            =    25.75 g
d)  mass of crucible + lid + residue after further heating  =    25.75 g

Why was step d) necessary?

Calculate the value of $x$ in the formula $BaCl_2 \cdot xH_2O$.             (AEB80,p)

# 3 Volumetric Analysis

## CONCENTRATION

Chemical reactions are often carried out between substances in solution. The concentration of a solution is measured in terms of the number of moles of solute contained in a cubic decimetre of solution.

$$\text{Concentration (mol dm}^{-3}) = \frac{\text{Amount in moles of solute}}{\text{Volume (dm}^3) \text{ of solution}}$$

Another way of writing this is:

$$\left(\begin{array}{l}\text{Amount in moles}\\\text{of solute}\end{array}\right) = \text{Concentration (mol dm}^{-3}) \times \text{Volume (dm}^3)$$

The concentration in $\text{mol dm}^{-3}$ used to be referred to as the *molarity* of a solution. (In strict SI units, concentration is expressed in $\text{mol m}^{-3}$.)

A solution of known concentration is called a *standard solution*. Such a solution can be used to find the concentrations of solutions of other reagents.

In *volumetric analysis*, the concentration of a solution is found by measuring the volume of solution that will react with a known volume of a standard solution. The procedure of adding one solution to another in a measured way until the reaction is complete is called *titration*. Volumetric analysis is often referred to as *titrimetric analysis* or *titrimetry*.

## ACID-BASE TITRATIONS

A standard solution of acid can be used to find the concentration of a solution of alkali. A known volume of alkali is taken by pipette, a suitable indicator is added, and the alkali is titrated against the standard acid until the equivalence point is reached. The number of moles of acid used can be calculated and the equation used to give the number of moles of alkali neutralised.

**EXAMPLE 1** *Standardising sodium hydroxide solution*
What is the concentration of a solution of sodium hydroxide, $25.0 \text{ cm}^3$ of which requires $20.0 \text{ cm}^3$ of hydrochloric acid of concentration $0.100 \text{ mol dm}^{-3}$ for neutralisation?

**METHOD**   a) Write the equation:

$$NaOH(aq) + HCl(aq) \longrightarrow NaCl(aq) + H_2O(l)$$

1 mole of NaOH needs 1 mole of HCl for neutralisation.

b) Find the number of moles of the reagent of known concentration, in this case HCl.

$$\begin{aligned} \text{Amount (mol) of HCl} &= \text{Volume (dm}^3) \times \text{Concn (mol dm}^{-3}) \\ &= 20.0 \times 10^{-3} \times 0.100 = 2.00 \times 10^{-3}\,\text{mol} \end{aligned}$$

From equation: No. of moles of NaOH = No. of moles of HCl

$$= 2.00 \times 10^{-3}\,\text{mol}$$

But:  Amount (mol) of NaOH = Volume (dm$^3$) $\times$ Concn (mol dm$^{-3}$)

$$= 25.0 \times 10^{-3} \times c$$

(where $c$ = concn)

Equate these two values: $2.00 \times 10^{-3} = 25.0 \times 10^{-3} \times c$

$$c = (2.00 \times 10^{-3})/(25.0 \times 10^{-3})$$

$$= 0.080\,\text{mol dm}^{-3}$$

**ANSWER**   The concentration of sodium hydroxide is 0.080 mol dm$^{-3}$.

**EXAMPLE 2**   *Standardising hydrochloric acid*
Sodium carbonate (anhydrous) is used as a primary standard in volumetric analysis. A solution of sodium carbonate of concentration 0.100 mol dm$^{-3}$ is used to standardise a solution of hydrochloric acid. 25.0 cm$^3$ of the standard solution of sodium carbonate require 35.0 cm$^3$ of the acid for neutralisation. Calculate the concentration of the acid.

**METHOD**   a) Write the equation:

$$Na_2CO_3(aq) + 2HCl(aq) \longrightarrow 2NaCl(aq) + CO_2(g) + H_2O(l)$$

1 mole of Na$_2$CO$_3$ neutralises 2 moles of HCl.

b) Find the number of moles of the standard reagent used.

$$\begin{aligned} \text{Amount (mol) of Na}_2\text{CO}_3\text{(aq)} &= \text{Volume (dm}^3) \times \text{Concn (mol dm}^{-3}) \\ &= 25.0 \times 10^{-3} \times 0.100 \\ &= 2.50 \times 10^{-3}\,\text{mol} \end{aligned}$$

From equation: No. of moles of HCl = 2 $\times$ No. of moles of Na$_2$CO$_3$

$$= 5.00 \times 10^{-3}\,\text{mol}$$

But:  Amount (mol) of HCl(aq) $=$ Volume $(dm^3) \times$ Concn $(mol\, dm^{-3})$

$$= 35.0 \times 10^{-3} \times c$$

(where $c =$ concn)

Equate these two values:  $5.00 \times 10^{-3} = 35.0 \times 10^{-3} \times c$

$$c = (5.00 \times 10^{-3})/(35.0 \times 10^{-3})$$

$$= 0.143\, mol\, dm^{-3}$$

**ANSWER**    The concentration of hydrochloric acid is $0.143\, mol\, dm^{-3}$.

**EXAMPLE 3**  *Calculating the percentage of sodium carbonate in washing soda crystals*

5.125 g of washing soda crystals are dissolved and made up to $250\, cm^3$ of solution. A $25.0\, cm^3$ portion of the solution requires $35.8\, cm^3$ of $0.0500\, mol\, dm^{-3}$ sulphuric acid for neutralisation. Calculate the percentage of sodium carbonate in the crystals.

**METHOD**    a)  Write the equation:

$$Na_2CO_3(aq) + H_2SO_4(aq) \longrightarrow Na_2SO_4(aq) + CO_2(g) + H_2O(l)$$

1 mole of $Na_2CO_3$ neutralises 1 mole of $H_2SO_4$.

b)  Calculate the amount, in moles, of the standard reagent.

Amount (mol) of $H_2SO_4 = 35.8 \times 10^{-3} \times 0.0500 = 1.79 \times 10^{-3}\, mol$

Amount (mol) of $Na_2CO_3 = 1.79 \times 10^{-3}\, mol$

But:   Amount of $Na_2CO_3 = 25.0 \times 10^{-3} \times c\, mol$

(where $c =$ concn)

Equate these two values:   $1.79 \times 10^{-3} = 25.0 \times 10^{-3} \times c$

$$c = (1.79 \times 10^{-3})/(25.0 \times 10^{-3})$$

$$= 0.0716\, mol\, dm^{-3}$$

Amount (mol) of $Na_2CO_3$ in whole solution $=$ Volume $\times$ Concn

$$= 250 \times 10^{-3} \times 0.0716$$

$$= 0.0179\, mol$$

Mass of $Na_2CO_3 =$ Amount (mol) $\times$ Molar mass $= 0.0179 \times 106\, g$

$$= 1.90\, g$$

% of $Na_2CO_3 = \dfrac{\text{Mass of sodium carbonate}}{\text{Mass of crystals}} \times 100$

$$= \frac{1.90}{5.125} \times 100 = 37.1\%$$

**ANSWER**    Washing soda crystals are 37.1% sodium carbonate.

**EXAMPLE 4**  *Estimating ammonium salts*

A sample containing ammonium sulphate was warmed with $250 \, cm^3$ of $0.800 \, mol \, dm^{-3}$ sodium hydroxide solution. After the evolution of ammonia had ceased, the excess of sodium hydroxide solution was neutralised by $85.0 \, cm^3$ of hydrochloric acid of concentration $0.500 \, mol \, dm^{-3}$. What mass of ammonium sulphate did the sample contain?

**METHOD**  a) There are two reactions taking place:

i) the reaction between the ammonium salt and the alkali:

$$(NH_4)_2SO_4(s) + 2NaOH(aq) \longrightarrow 2NH_3(g) + Na_2SO_4(aq) + 2H_2O(l)$$

ii) the reaction between the excess alkali and the hydrochloric acid:

$$NaOH(aq) + HCl(aq) \longrightarrow NaCl(aq) + H_2O(l)$$

b) Pick out the substance for which you have the information you need to calculate the number of moles. As you know its volume and concentration, you can calculate the number of moles of HCl. This will tell you the number of moles of NaOH left over after reaction i). Subtract this from the number of moles of NaOH added to the ammonium salt to obtain the number of moles of NaOH used in reaction i). This will give you the number of moles of $(NH_4)_2SO_4$ with which it reacted.

Amount (mol) of HCl $= 85.0 \times 10^{-3} \times 0.500 = 0.0425 \, mol$

Amount (mol) of NaOH left over from reaction i) $= 0.0425 \, mol$

Amount (mol) of NaOH added $= 250 \times 10^{-3} \times 0.800 = 0.200 \, mol$

Amount (mol) of NaOH used in reaction i) $= 0.200 - 0.0425$

$$= 0.1575 \, mol$$

No. of moles of $(NH_4)_2SO_4 = 0.5 \times$ No. of moles of NaOH

$$= 0.0788 \, mol$$

Molar mass of $(NH_4)_2SO_4 = 132 \, g \, mol^{-1}$

Mass of ammonium sulphate $= 0.0788 \times 132 = 10.4 \, g$

**ANSWER**  The sample contained $10.4 \, g$ of ammonium sulphate.


**EXAMPLE 5**  *Determination of the amounts of sodium hydroxide and sodium carbonate in a mixture*

Different indicators change colour at different values of pH. When a solution of sodium carbonate is titrated against hydrochloric acid, two reactions occur:

$$Na_2CO_3(aq) + HCl(aq) \longrightarrow NaHCO_3 aq) + NaCl(aq)$$

$$NaHCO_3(aq) + HCl(aq) \longrightarrow NaCl(aq) + CO_2(g) + H_2O(l)$$

The indicator phenolphthalein changes from pink to colourless at the end of the first reaction, when the pH of the solution is 9. The indicator methyl orange changes from yellow to orange at the end of the second reaction when the pH is 7. This two-stage titration can be used to estimate sodium hydroxide and sodium carbonate in a mixture of both.

25.0 cm$^3$ of a solution containing sodium hydroxide and sodium carbonate were titrated against 0.100 mol dm$^{-3}$ hydrochloric acid, using phenolphthalein as indicator. After 18.50 cm$^3$ of acid had been added, the indicator was decolourised. Methyl orange was added, and a further 7.50 cm$^3$ of hydrochloric acid were needed to turn the indicator orange. Calculate the concentrations of sodium hydroxide and sodium carbonate in the solution.

**METHOD**    The second stage, using methyl orange as indicator, gives

$$HCO_3^-(aq) \; + \; H^+(aq) \longrightarrow CO_2(g) \; + \; H_2O(l)$$

Amount (mol) of HCl(aq) used $= 7.50 \times 10^{-3} \times 0.100$

$\qquad\qquad\qquad\qquad\qquad = 0.750 \times 10^{-3} \, mol$

$\qquad\qquad\qquad\qquad\qquad =$ Amount (mol) of $HCO_3^-(aq)$ formed in first stage,

$\qquad\qquad\qquad\qquad\qquad =$ Amount (mol) of $CO_3^{2-}(aq)$ in the 25.0 cm$^3$ of solution.

$\therefore$  25.0 cm$^3$ solution contains $0.750 \times 10^{-3}$ moles of $CO_3^{2-}(aq)$.

$\therefore$  Concentration of $CO_3^{2-}(aq) = \dfrac{0.750 \times 10^{-3}}{25.0 \times 10^{-3}} = 0.0300 \, mol \, dm^{-3}$

Volume of acid needed by NaOH(aq) $= 11.00 \times 10^{-3} \times 0.100$

$\qquad\qquad\qquad\qquad\qquad\qquad = 1.10 \times 10^{-3} \, mol$

Amount (mol) of NaOH(aq) in 25 cm$^3$ of solution $= 1.10 \times 10^{-3} \, mol$

Concn of NaOH(aq) $= \dfrac{1.10 \times 10^{-3}}{25.0 \times 10^{-3}} = 0.0440 \, mol \, dm^{-3}$

**ANSWER**    The concentrations are 0.0300 mol dm$^{-3}$ sodium carbonate and 0.0440 mol dm$^{-3}$ sodium hydroxide.

**EXAMPLE 6**  *Estimation of sodium carbonate and sodium hydrogencarbonate in a solution*
The same technique, employing two indicators, can be used to estimate a mixture of sodium carbonate and sodium hydrogencarbonate. Using phenolphthalein as indicator, the reaction measured is

$$CO_3^{2-}(aq) \; + \; H^+(aq) \longrightarrow HCO_3^-(aq)$$

When methyl orange is added, and titration is continued, the reaction measured is

$$HCO_3^-(aq) \; + \; H^+(aq) \longrightarrow CO_2(g) \; + \; H_2O(l)$$

A 25 cm$^3$ aliquot of a solution containing sodium carbonate and sodium hydrogencarbonate needed 27.5 cm$^3$ of a solution of hydrochloric acid of concentration 0.0800 mol dm$^{-3}$ to decolourise phenolphthalein, which had been added as indicator. On addition of methyl orange, a further 32.5 cm$^3$ of the acid were needed to turn this indicator to its neutral colour.

**METHOD**     In the first stage, the $CO_3^{2-}$(aq) in the mixture is converted to $HCO_3^-$(aq).

Amount (mol) of HCl(aq) needed to convert $CO_3^{2-}$(aq) to $HCO_3^-$(aq)
$$= 27.5 \times 10^{-3} \times 0.0800 = 2.20 \times 10^{-3} \, mol$$
Amount (mol) of $CO_3^{2-}$(aq) $= 2.20 \times 10^{-3} \, mol$
Concentration of $CO_3^{2-}$(aq) $= (2.20 \times 10^{-3})/(25.0 \times 10^{-3})$
$$= 8.80 \times 10^{-2} \, mol \, dm^{-3}$$

The volume of acid needed to neutralise the total $HCO_3^-$(aq) is 32.5 cm$^3$.

Of this, 27.5 cm$^3$ are needed to neutralise the $HCO_3^-$(aq) formed from $CO_3^{2-}$(aq) in the first stage. The remaining 5.00 cm$^3$ neutralise the $HCO_3^-$(aq) present in the original solution.

Amount (mol) of HCl(aq) needed for $HCO_3^-$(aq)
$$= 5.00 \times 10^{-3} \times 0.0800$$
$$= 0.0400 \times 10^{-3} \, mol$$
Amount (mol) of $HCO_3^-$(aq) $= 0.0400 \times 10^{-3} \, mol$
Concentration of $HCO_3^-$(aq) $= (0.0400 \times 10^{-3})/(25.0 \times 10^{-3})$
$$= 1.60 \times 10^{-3} \, mol \, dm^{-3}$$

**ANSWER**     The concentrations are $8.80 \times 10^{-2} \, mol \, dm^{-3}$ sodium carbonate and $1.60 \times 10^{-3} \, mol \, dm^{-3}$ sodium hydrogencarbonate.

## EXERCISE 10     Problems on Neutralisation

1. What mass of the solute must be used in order to prepare the required solutions listed below?
   a)  500 cm$^3$ of 0.100 mol dm$^{-3}$ $H_6C_4O_4$(aq) from $H_6C_4O_4$(s)
   b)  250 cm$^3$ of 0.200 mol dm$^{-3}$ $Na_2CO_3$(aq) from $Na_2CO_3$(s)
   c)  750 cm$^3$ of 0.100 mol dm$^{-3}$ $H_2C_2O_4$(aq) from $H_2C_2O_4 \cdot 2H_2O$(s)
   d)  2.50 dm$^3$ of 0.200 mol dm$^{-3}$ NaHCO$_3$(aq) from NaHCO$_3$(s)
   e)  500 cm$^3$ of 0.100 mol dm$^{-3}$ $Na_2B_4O_7$(aq) from $Na_2B_4O_7 \cdot 10H_2O$(s)

2. What volumes of the following concentrated solutions are required to give the stated volumes of the more dilute solutions?
   a)  2.00 dm$^3$ of 0.500 mol dm$^{-3}$ $H_2SO_4$(aq) from 2.00 mol dm$^{-3}$ $H_2SO_4$(aq)
   b)  1.00 dm$^3$ of 0.750 mol dm$^{-3}$ HCl(aq) from 10.0 mol dm$^{-3}$ HCl(aq)

c) 250 cm$^3$ of 0.250 mol dm$^{-3}$ NaOH(aq) from 5.50 mol dm$^{-3}$ NaOH(aq)

d) 500 cm$^3$ of 1.25 mol dm$^{-3}$ HNO$_3$(aq) from 3.25 mol dm$^{-3}$ HNO$_3$(aq)

e) 250 cm$^3$ of 2.00 mol dm$^{-3}$ KOH(aq) from 2.60 mol dm$^{-3}$ KOH(aq)

3. a) What mass of the following will react completely with 50.0 cm$^3$ of 0.100 mol dm$^{-3}$ sulphuric acid?
    i) zinc
    ii) solid potassium hydroxide
    iii) solid copper(II) carbonate
    iv) sodium carbonate-10-water crystals

b) What volumes of the following solutions will react with 50.0 cm$^3$ of 0.100 mol dm$^{-3}$ sulphuric acid?
    i) 0.200 mol dm$^{-3}$ potassium hydroxide solution
    ii) 0.0500 mol dm$^{-3}$ sodium carbonate solution
    iii) 0.500 mol dm$^{-3}$ lead nitrate solution
    iv) 0.250 mol dm$^{-3}$ sodium hydrogencarbonate solution

4. What volume of 0.125 mol dm$^{-3}$ sodium hydroxide solution is needed to titrate 25.0 cm$^3$ of 0.085 mol dm$^{-3}$ sulphuric acid?

5. Arsenic(V) acid, H$_3$AsO$_4$, is a tribasic acid. 25.0 cm$^3$ of a solution of the acid require 35.7 cm$^3$ of a solution of sodium hydroxide of concentration 0.100 mol dm$^{-3}$ for neutralisation. What is the concentration of the acid?

6. Soda lime is 85.0% NaOH and 15.0% CaO. What volume of 0.500 mol dm$^{-3}$ nitric acid is needed to neutralise 2.50 g of soda lime?

7. Carbon dioxide is prepared by the action of 2.00 mol dm$^{-3}$ hydrochloric acid on marble. What volume of acid is required to give 10.0 dm$^3$ of carbon dioxide, measured at $9.9 \times 10^4$ N m$^{-2}$ and 26 °C?

8. 0.500 g of impure ammonium chloride is warmed with an excess of sodium hydroxide solution. The ammonia liberated is absorbed in 25.0 cm$^3$ of 0.200 mol dm$^{-3}$ sulphuric acid. The excess of sulphuric acid requires 5.64 cm$^3$ of 0.200 mol dm$^{-3}$ sodium hydroxide solution for titration. Calculate the percentage of ammonium chloride in the original sample.

9. A 1.00 g sample of limestone is allowed to react with 100 cm$^3$ of 0.200 mol dm$^{-3}$ hydrochloric acid. The excess acid required 24.8 cm$^3$ of 0.100 mol dm$^{-3}$ sodium hydroxide solution. Calculate the percentage of calcium carbonate in the limestone.

10. An impure sample of barium hydroxide of mass 1.6524 g was allowed to react with 100 cm³ of hydrochloric acid of concentration 0.200 mol dm⁻³. When the excess of acid was titrated against sodium hydroxide, 10.9 cm³ of sodium hydroxide solution were required. 25.0 cm³ of the sodium hydroxide required 28.5 cm³ of the hydrochloric acid in a separate titration. Calculate the percentage purity of the sample of barium hydroxide.

11. A solution contains sodium carbonate and sodium hydrogencarbonate. After the addition of a few drops of phenolphthalein, 25.0 cm³ of the solution were titrated against a 0.200 mol dm⁻³ solution of hydrochloric acid, until the indicator turned colourless. Methyl orange was added, and titration was continued until the indicator changed colour. Fig. 3.1 shows how the pH changed during the titration and the pH ranges of the two indicators. Calculate the concentrations of sodium carbonate and sodium hydrogencarbonate in the solution.

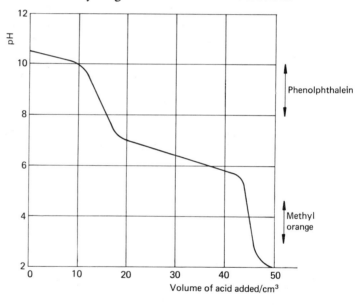

Fig. 3.1    pH changes in Question 11

12. A household cleaner contains ammonia. A 25.37 g sample of the cleaner is dissolved in water and made up to 250 cm³. A 25.0 cm³ portion of this solution requires 37.3 cm³ of 0.360 mol dm⁻³ sulphuric acid for neutralisation. What is the percentage by mass of ammonia in the cleaner?

13. A fertiliser contains ammonium sulphate and potassium sulphate. A sample of 0.225 g of fertiliser was warmed with sodium hydroxide solution. The ammonia evolved required 15.7 cm³ of 0.100 mol dm⁻³ hydrochloric acid for neutralisation. Calculate the percentage of ammonium sulphate in the sample.

14. Calculate the number of carboxyl groups in the compound $C_6H_8O_6$, given that $0.440\,g$ of it neutralised $37.5\,cm^3$ of sodium hydroxide of concentration $0.200\,mol\,dm^{-3}$.

15. $0.500\,g$ of a mixture of sodium hydrogencarbonate and anhydrous sodium carbonate was dissolved in water. Phenolphthalein was used as indicator when the solution was titrated against hydrochloric acid of concentration $0.100\,mol\,dm^{-3}$, $15.0\,cm^3$ of acid were required. Calculate the percentage by mass of sodium hydrogencarbonate in the mixture.

16. Sodium carbonate crystals ($27.823\,0\,g$) were dissolved in water and made up to $1.00\,dm^3$. $25.0\,cm^3$ of the solution were neutralised by $48.8\,cm^3$ of hydrochloric acid of concentration $0.100\,mol\,dm^{-3}$. Find $n$ in the formula $Na_2CO_3 \cdot nH_2O$.

# OXIDATION-REDUCTION REACTIONS

Oxidation–reduction (or 'redox') reactions involve a transfer of electrons. The oxidising agent accepts electrons, and the reducing agent gives electrons. In working out the equation for a redox reaction, a good method is to work out the 'half-reaction equation' for the oxidising agent and the 'half-reaction equation' for the reducing agent, and then add them together.

## Examples of half-reaction equations

a) Iron(III) salts are reduced to iron(II) salts. The equation is

$$Fe^{3+} \longrightarrow Fe^{2+}$$

For the equation to balance, the charge on the right-hand side (RHS) must equal the charge on the left-hand side (LHS). This can be accomplished by inserting an electron on the LHS:

$$Fe^{3+} + e^- \longrightarrow Fe^{2+}$$

b) When chlorine acts as an oxidising agent, it is reduced to chloride ions:

$$Cl_2 \longrightarrow 2Cl^-$$

To obtain a balanced half-reaction equation, $2e^-$ must be inserted on the LHS:

$$Cl_2 + 2e^- \longrightarrow 2Cl^-$$

c) Sulphites can be oxidised to sulphates:

$$SO_3^{2-} \longrightarrow SO_4^{2-}$$

To balance the equation with respect to mass, an extra oxygen atom is needed on the LHS. If $H_2O$ is introduced on the LHS to supply this oxygen, the equation becomes

$$SO_3^{2-} + H_2O \longrightarrow SO_4^{2-} + 2H^+$$

To balance the equation with respect to charge, $2e^-$ are needed on the RHS:

$$SO_3^{2-} + H_2O \longrightarrow SO_4^{2-} + 2H^+ + 2e^-$$

d) Potassium manganate(VII) is an oxidising agent. In acid solution, it is reduced to a manganese(II) salt:

$$MnO_4^- + H^+ \longrightarrow Mn^{2+}$$

To balance the equation with respect to mass, $8H^+$ are needed to combine with 4 oxygen atoms:

$$MnO_4^- + 8H^+ \longrightarrow Mn^{2+} + 4H_2O$$

To balance the equation with respect to charge, $5e^-$ are needed on the LHS:

$$MnO_4^- + 8H^+ + 5e^- \longrightarrow Mn^{2+} + 4H_2O$$

It is a good idea to make a final check. Charge on LHS $= -1 + 8 - 5 = +2$. Charge on RHS $= +2$. The equation is balanced.

e) Potassium dichromate(VI) is an oxidising agent in acid solution, being reduced to a chromium(III) salt:

$$Cr_2O_7^{2-} + H^+ \longrightarrow Cr^{3+}$$

To balance the equation for mass, $14H^+$ are needed:

$$Cr_2O_7^{2-} + 14H^+ \longrightarrow 2Cr^{3+} + 7H_2O$$

To balance the equation for charge, $6e^-$ are needed on the LHS:

$$Cr_2O_7^{2-} + 14H^+ + 6e^- \longrightarrow 2Cr^{3+} + 7H_2O$$

A final check shows that the charge on the LHS $= -2 + 14 - 6 = +6$.

Charge on RHS $= 2(+3) = +6$. The equation is balanced.

You may like to practise with the half-reaction equations on p. 54.

## Using half-reaction equations to obtain the equation for a redox reaction

a) In the reaction between iodine and thiosulphate ions, the two half-reaction equations are

$$I_2 + 2e^- \longrightarrow 2I^- \tag{1}$$

$$2S_2O_3^{2-} \longrightarrow S_4O_6^{2-} + 2e^- \tag{2}$$

Adding [1] and [2] gives

$$I_2 + 2e^- + 2S_2O_3^{2-} \longrightarrow 2I^- + S_4O_6^{2-} + 2e^-$$

Deleting the $2e^-$ term from both sides of the equation gives

$$I_2 + 2S_2O_3^{2-} \longrightarrow 2I^- + S_4O_6^{2-}$$

A check shows that the charges on the LHS and the RHS are both $-4$.


b) When potassium manganate(VII) oxidises an iron(II) salt to an iron (III) salt, the equations for the half-reactions are

$$MnO_4^- + 8H^+ + 5e^- \longrightarrow Mn^{2+} + 4H_2O \qquad [3]$$

$$Fe^{2+} \longrightarrow Fe^{3+} + e^- \qquad [4]$$

One manganate(VII) ion needs 5 electrons, and one iron(II) ion gives only one. Equation [4] must therefore be multiplied by 5:

$$5Fe^{2+} \longrightarrow 5Fe^{3+} + 5e^- \qquad [5]$$

Equations [3] and [5] can now be added to give

$$MnO_4^- + 8H^+ + 5Fe^{2+} \longrightarrow Mn^{2+} + 4H_2O + 5Fe^{3+}$$


c) When potassium manganate(VII) oxidises sodium ethanedioate, the equation for the manganate(VII) half-reaction is [3] as in Example 2, and the equation for the reduction of ethanediaote is

$$C_2O_4^{2-} \longrightarrow 2CO_2 + 2e^- \qquad [6]$$

d) One manganate(VII) ion needs $5e^-$, and one ethanedioate ion gives $2e^-$. Multiplying equation [3] by 2 and equation [6] by 5 and adding gives

$$2MnO_4^- + 16H^+ + 5C_2O_4^{2-} \longrightarrow 2Mn^{2+} + 10CO_2 + 8H_2O$$


e) Potassium dichromate(VI) oxidises iron(II) salt to iron(III) salts. The equations for the two half-reactions are

$$Cr_2O_7^{2-} + 14H^+ + 6e^- \longrightarrow 2Cr^{3+} + 7H_2O \quad [7]$$

$$Fe^{2+} \longrightarrow Fe^{3+} + e^- \qquad [8]$$

One dichromate ion will oxidise six iron(II) ions:

$$Cr_2O_7^{2-} + 14H^+ + 6Fe^{2+} \longrightarrow 2Cr^{3+} + 6Fe^{3+} + 7H_2O$$

You may like to try the problems on balancing equations on p. 54 before going on to tackle the numerical problems.

There is another method of balancing redox equations. It is explained in the following section on oxidation numbers.

## Oxidation numbers

It is helpful to discuss oxidation–reduction reactions in terms of the change in the *oxidation number* of each reactant. In the reaction

$$Cu(s) \ + \ \tfrac{1}{2}O_2(g) \longrightarrow Cu^{2+}O^{2-}(s)$$

copper is oxidised and oxygen is reduced. It is said that the oxidation number of copper increases from zero to $+2$, and the oxidation number of oxygen decreases from zero to $-2$. The following rules are followed in assigning oxidation numbers:

a) The oxidation number of an uncombined element is zero.

b) In ionic compounds, the oxidation number of each element is the charge on its ion. In NaCl, the oxidation number of Na $= +1$, and that of Cl $= -1$.

c) The sum of the oxidation numbers of all the elements in a compound is zero. In $AlCl_3$, the oxidation numbers are: Al $= +3$; Cl $= -1$, so that the sum of the oxidation numbers is $+3 + 3(-1) = 0$.

d) The sum of the oxidation numbers of all the elements in an ion is equal to the charge on the ion. In $SO_4^{2-}$, the oxidation numbers are S $= +6$, O $= -2$. The sum of the oxidation numbers for all the atoms is $+6 + 4(-2) = -2$, the same as the charge of the $SO_4^{2-}$ ion.

e) In a covalent compound, one element must be given a positive oxidation number and the other a negative oxidation number, such that the sum of the oxidation numbers for all the atoms is zero. The following elements always have the same oxidation numbers in all their compounds. A knowledge of their oxidation numbers helps one to assign oxidation numbers to the other elements combined with them:

Na, K $+1$    H $+1$, except in metal hydrides

Mg, Ca $+2$    F $-1$
Al    $+3$    Cl $-1$, except in compounds with O and F

O $-2$, except in peroxides and compounds with F

## The oxidation number method

A consideration of the changes in oxidation numbers which occur during a redox reaction helps you to decide which reactants have been oxidised and which have been reduced. It can also be very helpful when you need to balance the equation for the reaction. The following two points cover what is involved when you use the oxidation number method to balance the equation for a redox reaction:

a) When an element is oxidised, its oxidation number increases; when an element is reduced, its oxidation number decreases. If $x$ atoms (or ions) of an element A react with $y$ atoms (or ions) of an element B, i.e.

$$x\,A + y\,B \longrightarrow$$

then, if the oxidation number of A changes by $a$ units, and the oxidation number of B changes by $b$ units, you can see that

$$xa = yb$$

For example, in the reaction between tin(II) and iron(III) ions,

$$Sn^{2+}(aq) + 2Fe^{3+}(aq) \longrightarrow Sn^{4+}(aq) + 2Fe^{2+}$$

For Sn, no. of ions = 1, change in ox. no. = 2, and product = 2

For Fe, no. of ions = 2, change in ox. no. = 1, and product = 2

b) In a balanced equation

LHS sum of ox. nos. of elements = RHS sum of ox. nos. of elements

In the reaction

$$KIO_3(aq) + 2Na_2SO_3(aq) \longrightarrow KIO(aq) + 2Na_2SO_4(aq)$$

the elements K, Na and O keep the same oxidation states during the reaction, while I and S change.

Ox. no. of I in $KIO_3$ = +5; in $KIO$ = +1

Ox. no. of S in $Na_2SO_3$ = +4; in $Na_2SO_4$ = +6

Sum of ox. nos. on LHS = $(+5) + 2(+4)$ = +13

Sum of ox. nos. on RHS = $(+1) + 2(+6)$ = +13

---

When applying the oxidation number method to a reaction between A and B, remember:

$$\left( \begin{array}{c} \text{No. of atoms of A} \times \text{Change} \\ \text{in oxidation number of A} \end{array} \right) = \left( \begin{array}{c} \text{No. of atoms of B} \times \text{Change} \\ \text{in oxidation number of B} \end{array} \right)$$

Sum of ox. nos. on LHS = Sum of ox. nos. on RHS

---

**EXAMPLE 1**  What is the oxidation number of germanium in $GeCl_4$?

**METHOD**  Chlorine is one of the elements with a constant oxidation number of $-1$.

(Oxidation number of Ge) + $4(-1)$ = 0.

**ANSWER**  Oxidation number of Ge = +4.

**EXAMPLE 2**  What is the oxidation number of manganese in $Mn_2O_7$?

**METHOD**    Oxygen is one of the elements with a constant oxidation number of $-2$.

2(Oxidation number of Mn) + 7(−2) = 0.

**ANSWER**    Oxidation number of Mn = +7.


**EXAMPLE 3**    What is the oxidation number of iron in $Fe(CN)_6^{3-}$?

**METHOD**    Since the cyanide ion is $CN^-$, it has an oxidation number of −1.

(Oxidation number of Fe) + 6(−1) = −3.

**ANSWER**    Oxidation number of Fe = +3.


**EXAMPLE 4**    Use the oxidation number method to balance the equation

$$MnO_4^-(aq) + H^+(aq) + Fe^{2+}(aq) \longrightarrow Mn^{2+}(aq) + Fe^{3+}(aq) + H_2O(l)$$

**METHOD**    Hydrogen and oxygen have the same oxidation numbers on both sides of the equation; only manganese and iron need be considered.

In $MnO_4^-$, the oxidation number of Mn = +7

In $Mn^{2+}$, the oxidation number of Mn = +2

Thus, manganese decreases its oxidation number by 5 units, and iron must increase its oxidation number by 5 units.

From $Fe^{2+}$ to $Fe^{3+}$ is an increase of 1 unit; therefore the equation needs $5Fe^{2+} \longrightarrow 5Fe^{3+}$. This makes the equation

$$MnO_4^-(aq) + H^+(aq) + 5Fe^{2+}(aq) \longrightarrow Mn^{2+}(aq) + 5Fe^{3+}(aq) + H_2O(l)$$

To combine with 4 oxygen atoms, $8H^+$ are needed:

$$MnO_4^-(aq) + 8H^+(aq) + 5Fe^{2+}(aq) \longrightarrow Mn^{2+}(aq) + 5Fe^{3+}(aq) + 4H_2O(l)$$


# EXERCISE 11    Problems on Oxidation Numbers

1. What is the oxidation number of the named element in the following compounds?

a) Ba in $BaCl_2$

b) Fe in $Fe(CN)_6^{4-}$

c) Cl in $Cl_2$

d) Li in $Li_2O$

e) Fe in $Fe(CN)_6^{3-}$

f) Cl in $ClO^-$

g) P in $P_2O_3$

h) Br in $BrO_3^-$

i) Cl in $ClO_3^-$

j) C in $CCl_4$

k) I in $I_2$

l) Cl in $Cl_2O_7$

m) C in CO

n) I in $I^-$

o) Cl in $Cl_2O_3$

p) Cr in $CrO_3$

q) I in $IO_3^-$

r) O in $H_2O_2$

s) Cr in $CrO_4^{2-}$

t) N in $NO_2$

u) H in LiH

v) Cr in $Cr_2O_7^{2-}$

w) N in $N_2O_4$

x) H in HBr

y) S in $SO_3^{2-}$

z) P in $PO_4^{3-}$

2. a) Calculate the oxidation numbers of tin and lead on each side of the equation

$$PbO_2(s) + 4H^+(aq) + Sn^{2+}(aq) \longrightarrow Pb^{2+}(aq) + Sn^{4+}(aq) + 2H_2O(l)$$

and state which element has been oxidised and which has been reduced.

b) In the redox reaction

$$2Mn^{2+}(aq) + 5BiO_3^-(aq) + 14H^+(aq) \longrightarrow 2MnO_4^-(aq) + 5Bi^{3+}(aq) + 7H_2O(l)$$

calculate the oxidation numbers of all the elements, and state which have been oxidised and which have been reduced.

c) Calculate the oxidation numbers of arsenic and manganese in each of the species in the reaction:

$$5As_2O_3(s) + 4MnO_4^-(aq) + 12H^+(aq) \longrightarrow 5As_2O_5(s) + 4Mn^{2+}(aq) + 6H_2O(l)$$

State which element has been oxidised and which has been reduced.

3. In each of the following equations, one element is underlined. Calculate its oxidation number in each species, and state whether an oxidation or a reduction has occurred.

a) $2\underline{F}_2(g) + 2OH^-(aq) \longrightarrow \underline{F}_2O(g) + 2\underline{F}^-(aq) + H_2O(l)$

b) $3\underline{Cl}_2(g) + 6OH^-(aq) \longrightarrow \underline{Cl}O_3^-(aq) + 5\underline{Cl}^-(aq) + 3H_2O(l)$

c) $\underline{N}H_4^+\underline{N}O_3^-(s) \longrightarrow \underline{N}_2O(g) + 2H_2O(l)$

d) $\underline{Cr}_2O_7^{2-}(aq) + 14H^+(aq) + 6e^- \longrightarrow 2\underline{Cr}^{3+}(aq) + 7H_2O(l)$

e) $\underline{C}_2O_4^{2-}(aq) \longrightarrow 2\underline{C}O_2(g) + 2e^-$

4. a) Only N and I alter in oxidation number in the reaction

$$N_2H_6O(aq) + IO_3^-(aq) + 2H^+(aq) + Cl^-(aq) \longrightarrow N_2(g) + ICl(aq) + 4H_2O(l)$$

Calculate the oxidation number of N in $N_2H_6O$.

b) In the reaction below, only S and Br change in oxidation number.

$$Na_2H_{10}S_2O_8(aq) + 4Br_2(aq) \longrightarrow 2H_2SO_4(aq) + 2NaBr(aq) + 6HBr(aq)$$

Calculate the oxidation number of S in $Na_2H_{10}S_2O_8$.

5. Use the oxidation number method to balance the equations

a) $IO_4^-(aq) + I^-(aq) + H^+(aq) \longrightarrow I_2(aq) + 4H_2O(l)$

b) $BrO_3^-(aq) + I^-(aq) + H^+(aq) \longrightarrow Br^-(aq) + I_2(aq) + H_2O(l)$

c) $V^{3+}(aq) + H_2O_2(aq) \longrightarrow VO^{2+}(aq) + H^+(aq)$

d) $SO_2(g) + H_2O(l) + Br_2(aq) \longrightarrow H^+(aq) + SO_4^{2-}(aq) + Br^-(aq)$

e) $NH_3(g) + O_2(g) \longrightarrow N_2(g) + H_2O(g)$

f) $NH_3(g) + O_2(g) \longrightarrow N_2O(g) + H_2O(g)$

g) $NH_3(g) + O_2(g) \longrightarrow NO(g) + H_2O(g)$

h) $Fe^{2+}C_2O_4^{2-}(aq) + Ce^{3+}(aq) \longrightarrow CO_2(g) + Ce^{2+}(aq) + Fe^{3+}(aq)$

6. When potassium dichromate solution reacts with acidified potassium iodide solution, titration shows that 1 mole of potassium dichromate produces 3 moles of iodine. Use the oxidation number method to complete and balance the equation

$$Cr_2O_7^{2-}(aq) + I^-(aq) + H^+(aq) \longrightarrow 3I_2(aq)$$

## POTASSIUM MANGANATE(VII) TITRATIONS

When potassium manganate(VII) acts as an oxidising agent in acid solution, it is reduced to a manganese(II) salt:

$$MnO_4^-(aq) + 8H^+(aq) + 5e^- \longrightarrow Mn^{2+}(aq) + 4H_2O(l)$$

Potassium manganate(VII) is not sufficiently pure to be used as a primary standard, and solutions of the oxidant are standardised by titration against a primary standard such as sodium ethanedioate. This reductant can be obtained in a high state of purity as crystals of formula $Na_2C_2O_4 \cdot 2H_2O$, which are neither deliquescent nor efflorescent, and can be weighed out exactly to make a standard solution.

Once it has been standardised, a solution of potassium manganate(VII) can be used to estimate reducing agents such as iron(II) salts. No indicator is needed as the oxidant changes from purple to colourless at the end point.

**EXAMPLE 1**  *Standardising potassium manganate(VII) against the primary standard, sodium ethanedioate*
A 25.0 cm$^3$ portion of sodium ethanedioate solution of concentration 0.200 mol dm$^{-3}$ is warmed and titrated against a solution of potassium manganate(VII). If 17.2 cm$^3$ of potassium manganate(VII) are required, what is its concentration?

**METHOD**    Let $M$ = concentration of the manganate(VII) solution.
Amount (mol) of ethanedioate = $25.0 \times 10^{-3} \times 0.200$ mol
Amount (mol) of manganate(VII) = $17.2 \times 10^{-3} \times M$ mol

The equations for the half-reactions are

$$MnO_4^-(aq) + 8H^+(aq) + 5e^- \longrightarrow Mn^{2+}(aq) + 4H_2O(l) \quad [1]$$
$$C_2O_4^{2-}(aq) \longrightarrow 2CO_2(g) + 2e^- \quad [2]$$

Multiplying [1] by 2 and [2] by 5, and adding the two equations gives

$$2MnO_4^-(aq) + 16H^+(aq) + 5C_2O_4^{2-}(aq) \longrightarrow 2Mn^{2+}(aq) + 8H_2O(l) + 10CO_2(g)$$

No. of moles of $MnO_4^-$ = $\frac{2}{5} \times$ No. of moles of $C_2O_4^{2-}$.

$$\therefore \quad 17.2 \times 10^{-3} \times M = \tfrac{2}{5} \times 25.0 \times 10^{-3} \times 0.200$$

$$M = \frac{2 \times 25.0 \times 10^{-3} \times 0.200}{5 \times 17.2 \times 10^{-3}} = 0.116 \, \text{mol dm}^{-3}$$

**ANSWER** The potassium manganate(VII) solution has a concentration of $0.116 \, \text{mol dm}^{-3}$.

**EXAMPLE 2** *Oxidising iron(II) compounds*

Ammonium iron(II) sulphate crystals have the following formula: $(NH_4)_2SO_4 \cdot FeSO_4 \cdot nH_2O$. In an experiment to determine $n$, 8.492 g of the salt were dissolved and made up to 250 cm³ of solution with distilled water and dilute sulphuric acid. A 25.0 cm³ portion of the solution was further acidified and titrated against potassium manganate(VII) solution of concentration $0.0150 \, \text{mol dm}^{-3}$. A volume of 22.5 cm³ was required.

**METHOD** The equations for the two half-reactions are

$$MnO_4^-(aq) + 8H^+(aq) + 5e^- \longrightarrow Mn^{2+}(aq) + 4H_2O(l) \quad [1]$$
$$Fe^{2+}(aq) \longrightarrow Fe^{3+}(aq) + e^- \quad [2]$$

Multiplying [2] by 5 and then adding it to [1] gives

$$MnO_4^-(aq) + 8H^+(aq) + 5Fe^{2+}(aq) \longrightarrow Mn^{2+}(aq) + 5Fe^{3+}(aq) + 4H_2O(l)$$

Amount (mol) of manganate(VII) $= 22.5 \times 10^{-3} \times 0.0150$

$$= 0.338 \times 10^{-3} \, \text{mol}$$

No. of moles of iron(II) $= 5 \times$ No. of moles of manganate(VII)

$$= 5 \times 0.338 \times 10^{-3} = 1.69 \times 10^{-3} \, \text{mol}$$

Concn of iron(II) $= \dfrac{1.69 \times 10^{-3}}{25.0 \times 10^{-3}} = 0.0674 \, \text{mol dm}^{-3}$

Concn of $(NH_4)_2SO_4 \cdot FeSO_4 \cdot nH_2O = \dfrac{\text{Mass in 1 dm}^3 \text{ of solution}}{\text{Molar mass}}$

$$= \frac{4 \times 8.492}{\text{Molar mass}}$$

$$0.0674 = \frac{4 \times 8.492}{\text{Molar mass}}$$

$$\text{Molar mass} = 503.9 \, \text{g mol}^{-1}$$

Molar mass of $(NH_4)_2SO_4 \cdot FeSO_4 \cdot nH_2O = 284 + 18n = 504 \, \text{g mol}^{-1}$

$$\therefore \quad n = 12$$

**ANSWER** The formula of the crystals is $(NH_4)_2SO_4 \cdot 12H_2O$

**EXAMPLE 3**    *Oxidising hydrogen peroxide*

A solution of hydrogen peroxide was diluted 20.0 times. A $25.0 \, cm^3$ portion of the diluted solution was acidified and titrated against $0.0150 \, mol \, dm^{-3}$ potassium manganate(VII) solution. $45.7 \, cm^3$ of the oxidant were required. Calculate the concentration of the hydrogen peroxide solution a) in $mol \, dm^{-3}$ and b) the 'volume concentration'. (This means the number of volumes of oxygen obtained from one volume of the solution.)

**METHOD**    The equations for the half-reactions are

$$MnO_4^-(aq) + 8H^+(aq) + 5e^- \longrightarrow Mn^{2+}(aq) + 4H_2O(l) \quad [1]$$

$$H_2O_2(aq) \longrightarrow O_2(g) + 2H^+(aq) + 2e^- \quad [2]$$

Multiplying [1] by 2 and [2] by 5, and adding the two equations gives

$$2MnO_4^-(aq) + 6H^+(aq) + 5H_2O_2(aq) \longrightarrow 2Mn^{2+}(aq) + 8H_2O(l) + 5O_2(g)$$

Amount (mol) of $MnO_4^-(aq)$ = $45.7 \times 10^{-3} \times 0.0150$

$$= 0.685 \times 10^{-3} \, mol$$

No. of moles of $H_2O_2$ = $\frac{5}{2} \times$ No. of moles of $MnO_4^-$

$$= \frac{5}{2} \times 0.685 \times 10^{-3} = 1.71 \times 10^{-3} \, mol$$

Concn of $H_2O_2$ = $(1.71 \times 10^{-3})/(25.0 \times 10^{-3})$ = $0.0684 \, mol \, dm^{-3}$

Concn of original solution = $20.0 \times 0.0684$ = $1.37 \, mol \, dm^{-3}$.

When hydrogen peroxide decomposes,

$$2H_2O_2(aq) \longrightarrow 2H_2O(l) + O_2(g)$$

2 moles of hydrogen peroxide form 1 mole of oxygen. Therefore a solution of hydrogen peroxide of concentration $2 \, mol \, dm^{-3}$ is a 22.4 volume solution (the volume of 1 mole of oxygen).

A solution of $H_2O_2$ of concentration $1.37 \, mol \, dm^{-3}$ is a $22.4 \times 1.37/2$ = 15.4 volume solution.

**ANSWER**    The concentration of hydrogen peroxide is: a) $1.37 \, mol \, dm^{-3}$, and b) 15.4 volume.

**EXAMPLE 4**    *Finding the percentage of iron in ammonium iron(III) sulphate*

Iron(III) ions can be estimated by first reducing them to iron(II) ions, and then, after destroying the excess of reducing agent, oxidising them to iron(III) ions with a standard solution of potassium manganate(VII). Zinc amalgam and sulphuric acid are used as the reducing agent. Note that hydrochloric acid cannot be used, and the reducing agent tin(II) chloride cannot be used as potassium manganate(VII) oxidises chloride ions to chlorine.

7.418 g of ammonium iron(III) sulphate are dissolved and made up to 250 cm$^3$ after the addition of dilute sulphuric acid, 25.0 cm$^3$ of the solution are pipetted into a bottle containing zinc amalgam, and shaken until a drop of the solution gives no colour when tested with a solution of a thiocyanate (which turns deep red in the presence of iron(III) ions). The aqueous solution is then separated by decantation from the zinc amalgam. On addition of more dilute sulphuric acid and titration against standard potassium manganate(VII) solution, 18.7 cm$^3$ of 0.0165 mol dm$^{-3}$ solution are required. Calculate the percentage of iron in ammonium iron(III) sulphate.

**METHOD**

Amount (mol) of manganate(VII) in volume used $= 18.7 \times 10^{-3} \times 0.0165$

$$= 0.0309 \times 10^{-3} \, mol$$

From the equation

$$MnO_4^-(aq) + 8H^+(aq) + 5Fe^{2+}(aq) \longrightarrow Mn^{2+}(aq) + 5Fe^{3+}(aq) + 4H_2O(l)$$

Amount (mol) of $Fe^{2+}$(aq) in 25.0 cm$^3$ $= 5 \times 0.309 \times 10^{-3}$

$$= 1.55 \times 10^{-3} \, mol$$

Amount (mol) of $Fe^{2+}$(aq) in whole solution $= 1.55 \times 10^{-2} \, mol$

Mass of iron in sample $=$ Amount (mol) $\times$ Relative atomic mass

$$= 1.55 \times 10^{-2} \times 55.8 = 0.865 \, g$$

**ANSWER**   Percentage of iron $= \dfrac{0.865}{7.418} \times 100 = 11.7\%$.

## POTASSIUM DICHROMATE(VI) TITRATIONS

Potassium dichromate(VI) can be obtained in a high state of purity, and its aqueous solutions are stable. It is used as a primary standard. The colour change when chromium(VI) changes to chromium(III) in the reaction

$$Cr_2O_7^{2-}(aq) + 14H^+(aq) + 6e^- \longrightarrow 2Cr^{3+}(aq) + 7H_2O(l)$$

is from orange to green. As it is not possible to see a sharp change in colour, an indicator is used. Barium N-phenylphenylamine-4-sulphonate gives a sharp colour change, from blue-green to violet, when a slight excess of potassium dichromate has been added. Phosphoric(V) acid must be present to form a complex with the $Fe^{3+}$ ions formed during the oxidation reaction; otherwise $Fe^{3+}$ ions affect the colour change of the indicator.

Since dichromate(VI) has a slightly lower redox potential than manganate(VII), it can be used in the presence of chloride ions, without oxidising them to chlorine.

**EXAMPLE**    *Determination of the percentage of iron in iron wire*
A piece of iron wire of mass 2.225 g was put into a conical flask containing dilute sulphuric acid. The flask was fitted with a bung carrying a Bunsen valve, to allow the hydrogen generated to escape but prevent air from entering. The mixture was warmed to speed up reaction. When all the iron had reacted, the solution was cooled to room temperature and made up to 250 cm³ in a graduated flask. With all these precautions, iron is converted to $Fe^{2+}$ ions only, and no $Fe^{3+}$ ions are formed. 25.0 cm³ of the solution were acidified and titrated against a 0.0185 mol dm⁻³ solution of potassium dichromate(VI). The volume required was 31.0 cm³. Calculate the percentage of iron in the iron wire.

**METHOD**    Amount (mol) of $Cr_2O_7{}^{2-}(aq)$ used $= 31.0 \times 10^{-3} \times 0.0185$

$$= 0.574 \times 10^{-3} \, mol$$

The equations for the two half-reactions are

$$Cr_2O_7{}^{2-}(aq) + 14H^+(aq) + 6e^- \longrightarrow 2Cr^{3+}(aq) + 7H_2O(l) \quad [1]$$
$$Fe^{2+}(aq) \longrightarrow Fe^{3+}(aq) + e^- \quad [2]$$

Multiplying [2] by 6 and adding gives

$$Cr_2O_7{}^{2-}(aq) + 14H^+(aq) + 6Fe^{2+}(aq) \longrightarrow 2Cr^{3+}(aq) + 6Fe^{3+}(aq) + 7H_2O(l)$$

Amount (mol) of $Fe^{2+}$ in 25.0 cm³ $= 6 \times 0.574 \times 10^{-3}$

$$= 3.45 \times 10^{-3} \, mol$$

Amount (mol) of $Fe^{2+}$ in the whole solution $= 3.45 \times 10^{-2} \, mol$

Mass of Fe in the whole solution $= 3.45 \times 10^{-2} \times 55.8 = 1.93 \, g$

Percentage of Fe in wire $= \dfrac{1.93}{2.225} \times 100 = 86.7\%$

**ANSWER**    The wire is 86.7% iron.

## SODIUM THIOSULPHATE TITRATIONS

Sodium thiosulphate reduces iodine to iodide ions, and forms sodium tetrathionate, $Na_2S_4O_6$:

$$2S_2O_3{}^{2-}(aq) + I_2(aq) \longrightarrow 2I^-(aq) + S_4O_6{}^{2-}(aq)$$

Sodium thiosulphate, $Na_2S_2O_3 \cdot 5H_2O$, is not used as a primary standard as the water content of the crystals is variable. A solution of sodium thiosulphate can be standardised against a solution of iodine, or a solution of potassium iodate(V) or potassium dichromate or potassium manganate(VII).

**EXAMPLE 1**   *Standardisation of a sodium thiosulphate solution, using iodine*
Iodine has a limited solubility in water. It dissolves in a solution of potassium iodide because it forms the very soluble complex ion, $I_3^-$.

$$I_2(s) + I^-(aq) \rightleftharpoons I_3^-(aq)$$

An equilibrium is set up between iodine and tri-iodide ions, and if iodine molecules are removed from solution by a reaction, tri-iodide ions dissociate to form more iodine molecules. A solution of iodine in potassium iodide can thus be titrated as though it were a solution of iodine in water.

When sufficient of a solution of thiosulphate is added to a solution of iodine, the colour of iodine fades to a pale yellow. Then $2\,cm^3$ of starch solution are added to give a blue colour with the iodine. Addition of thiosulphate is continued drop by drop, until the blue colour disappears.

$2.835\,g$ of iodine and $6\,g$ of potassium iodide are dissolved in distilled water and made up to $250\,cm^3$. A $25.0\,cm^3$ portion titrated against sodium thiosulphate solution required $17.7\,cm^3$ of the solution. Calculate the concentration of the thiosulphate solution.

**METHOD**   Molar mass of iodine $= 2 \times 127 = 254\,g\,mol^{-1}$
Concn of iodine solution $= 2.835 \times 4/254 = 0.0446\,mol\,dm^{-3}$
Amount (mol) of $I_2$ in $25.0\,cm^3 = 25.0 \times 10^{-3} \times 0.0446$
$$= 1.115 \times 10^{-3}\,mol$$

From the equation

$$2S_2O_3^{2-}(aq) + I_2(aq) \longrightarrow 2I^-(aq) + S_4O_6^{2-}(aq)$$

No. of moles of 'thio' $= 2 \times$ No. of moles of $I_2$
Amount (mol) of 'thio' in volume used $= 2.23 \times 10^{-3}\,mol$

$$\text{Concn of 'thio'} = \frac{2.23 \times 10^{-3}}{17.7 \times 10^{-3}} = 0.126\,mol\,dm^{-3}$$

**ANSWER**   The concentration of the thiosulphate solution is $0.126\,mol\,dm^{-3}$.

**EXAMPLE 2**   *Standardisation of thiosulphate against potassium iodate (V)*
Potassium iodate(V) is a primary standard. It reacts with iodide ions in the presence of acid to form iodine:

$$IO_3^-(aq) + 5I^-(aq) + 6H^+(aq) \longrightarrow 3I_2(aq) + 3H_2O(l)$$

A standard solution of iodine can be prepared by weighing out the necessary quantity of potassium iodate(V) and making up to a known volume of solution. When a portion of this solution is added to an excess of potassium iodide in acid solution, a calculated amount of iodine is liberated.

1.015 g of potassium iodate(V) are dissolved and made up to 250 cm³. To a 25.0 cm³ portion are added an excess of potassium iodide and dilute sulphuric acid. The solution is titrated with a solution of sodium thiosulphate, starch solution being added near the end-point. 29.8 cm³ of thiosulphate solution are required. Calculate the concentration of the thiosulphate solution.

**METHOD**     Molar mass of $KIO_3$ = $39.1 + 127 + (3 \times 16.0)$ = $214\,g\,mol^{-1}$

Concn of $KIO_3$ solution = $1.015 \times 4/214$ = $0.0189\,mol\,dm^{-3}$

Amount (mol) of $KIO_3$ in 25 cm³ = $25.0 \times 10^{-3} \times 0.0189$

$$= 0.473 \times 10^{-3}\,mol$$

Since

$$IO_3^-(aq) + 5I^-(aq) + 6H^+(aq) \longrightarrow 3I_2(aq) + 3H_2O(l)$$

and                    $$2S_2O_3^{2-}(aq) + I_2(aq) \longrightarrow 2I^-(aq) + S_4O_6^{2-}(aq)$$

No. of moles of 'thio' = $6 \times$ No. of moles of $IO_3^-$

$$= 6 \times 0.473 \times 10^{-3} = 2.84 \times 10^{-3}\,mol$$

Concn of 'thio' = $(2.84 \times 10^{-3})/(29.8 \times 10^{-3})$ = $0.0950\,mol\,dm^{-3}$

**ANSWER**     The sodium thiosulphate solution has a concentration $0.0950\,mol\,dm^{-3}$.

**EXAMPLE 3**  *Standardisation of thiosulphate solution with potassium dichromate(VI)*
A standard solution is made by dissolving 1.015 g of potassium dichromate(VI) and making up to 250 cm³. A 25.0 cm³ portion is added to an excess of potassium iodide and dilute sulphuric acid, and the iodine liberated is titrated with sodium thiosulphate solution. 19.2 cm³ of this solution are needed. Find the concentration of the thiosulphate solution.

**METHOD**     Molar mass of $K_2Cr_2O_7$ = $294\,g\,mol^{-1}$

Concn of dichromate solution = $1.015 \times 4/294$ = $0.0138\,mol\,dm^{-3}$

Amount (mol) of dichromate in 25 cm³ = $25.0 \times 10^{-3} \times 0.138\,mol$

$$= 0.345 \times 10^{-3}\,mol$$

The equations for the half-reactions are

$$Cr_2O_7^{2-}(aq) + 14H^+(aq) + 6e^- \longrightarrow 2Cr^{3+}(aq) + 7H_2O(l) \quad [1]$$

$$2I^-(aq) \longrightarrow I_2(aq) + 2e^- \quad [2]$$

Multiplying [2] by 3, and adding to [1] gives the equation

$$Cr_2O_7^{2-}(aq) + 14H^+(aq) + 6I^-(aq) \longrightarrow 2Cr^{3+}(aq) + 7H_2O(l) + 3I_2(aq)$$

No. of moles of $I_2$ = $3 \times$ No. of moles of $Cr_2O_7^{2-}$

Amount (mol) of $I_2$ in 25 cm³ $= 3 \times 0.345 \times 10^{-3}$
$$= 1.035 \times 10^{-3} \, mol$$
No. of moles of 'thio' $= 2 \times$ No. of moles of $I_2$ (see Example 1)
Amount (mol) of 'thio' in volume used $= 2.07 \times 10^{-3} \, mol$
Concn. of 'thio' $= (2.07 \times 10^{-3})/(19.2 \times 10^{-3}) = 0.108 \, mol \, dm^{-3}$

**ANSWER**    The concentration of the thiosulphate solution is $0.108 \, mol \, dm^{-3}$.

**EXAMPLE 4** *Estimation of chlorine*
Chlorine displaces iodine from iodides. The iodine formed can be determined by titration with a standard thiosulphate solution. Chlorate(I) solutions are often used as a source of chlorine as they liberate chlorine readily on reaction with acid:

$$ClO^-(aq) + 2H^+(aq) + Cl^-(aq) \longrightarrow Cl_2(aq) + H_2O(l)$$

The amount of chlorine available in a domestic bleach which contains sodium chlorate(I) can be found by allowing the bleach to react with an iodide solution to form iodine, and then titrating with thiosulphate solution:

$$ClO^-(aq) + 2H^+(aq) + 2I^-(aq) \longrightarrow I_2(aq) + Cl^-(aq) + H_2O(l)$$

A domestic bleach in solution is diluted by pipetting 10.0 cm³ and making this volume up to 250 cm³. A 25.0 cm³ portion of the solution is added to an excess of potassium iodide and ethanoic acid and titrated against sodium thiosulphate solution of concentration $0.0950 \, mol \, dm^{-3}$, using starch as an indicator. The volume required is 21.3 cm³. Calculate the percentage of available chlorine in the bleach.

**METHOD**    Amount (mol) of 'thio' $= 21.3 \times 10^{-3} \times 0.0950 = 2.03 \times 10^{-3} \, mol$

Since $2S_2O_3^{2-}(aq) + I_2(aq) \longrightarrow S_4O_6^{2-}(aq) + 2I^-(aq)$

Amount (mol) of $I_2 = 1.015 \, mol$

Since iodine is produced in the reaction

$$ClO^-(aq) + 2I^-(aq) + 2H^+(aq) \longrightarrow I_2(aq) + Cl^-(aq) + H_2O(l)$$

Amount (mol) of $ClO^-$ in 25 cm³ of solution $= 1.015 \times 10^{-3} \, mol$

Since chlorate(I) liberates chlorine in the reaction

$$ClO^-(aq) + 2H^+(aq) + Cl^-(aq) \longrightarrow Cl_2(aq) + H_2O(l)$$

No. of moles of $Cl_2 =$ No. of moles of chlorate(I)
$$= 1.015 \times 10^{-3} \, mol$$

Mass of chlorine $= 71.0 \times 1.015 \times 10^{-3} = 0.0720 \, g$

This is the mass of chlorine available in 25 cm³ of solution

$$\text{Percentage of available Cl}_2 = \frac{\text{Mass of Cl}_2 \text{ from 250 cm}^3 \text{ solution}}{\text{Mass of bleach solution used}} \times 100$$

$$= \frac{0.0720 \times 10}{10} \times 100 = 7.2\%$$

**ANSWER**    The percentage of available chlorine in bleach is 7.2%.

**EXAMPLE 5**  *Estimation of copper(II) salts*
Copper(II) ions oxidise iodide ions to iodine. The iodine produced can be titrated with standard thiosulphate solution, and, from the amount of iodine produced, the concentration of copper(II) ions in the solution can be calculated.

A sample of 4.256 g of copper(II) sulphate-5-water is dissolved and made up to 250 cm³. A 25.0 cm³ portion is added to an excess of potassium iodide. The iodine formed required 18.0 cm³ of a 0.0950 mol dm⁻³ solution of sodium thiosulphate for reduction. Calculate the percentage of copper in the crystals.

**METHOD**    Amount (mol) of 'thio' $= 18.0 \times 10^{-3} \times 0.0950 = 1.71 \times 10^{-3}$ mol
No. of moles of $I_2 = \frac{1}{2} \times$ No. of moles of 'thio' $= 0.855 \times 10^{-3}$ mol

Since    $2Cu^{2+}(aq) + 4I^-(aq) \longrightarrow Cu_2I_2(s) + I_2(aq)$

No. of moles of Cu $= 2 \times$ No. of moles of $I_2 = 1.71 \times 10^{-3}$ mol
Mass of Cu $= 63.5 \times 1.71 \times 10^{-3} = 0.109$ g
Mass of Cu in whole solution $= 1.09$ g

$$\text{Percentage of Cu} = \frac{1.09}{4.256} \times 100 = 25.6\%.$$

**ANSWER**    The percentage of copper in the crystals is 25.6%.

**\*EXAMPLE 6**  *Deriving an equation for the reaction between bromine and thiosulphate ions*
A solution of bromine was prepared and two titrations were performed:

a) 25.0 cm³ of the solution were added to an excess of potassium iodide. The iodine liberated required 19.5 cm³ of a 0.120 mol dm⁻³ solution of sodium thiosulphate.

b) 25.0 cm³ of the bromine solution were titrated directly against the thiosulphate solution. 2.45 cm³ of thiosulphate solution were required.

c) The solution from titration b) was tested for the presence of various anions. Sulphate ions were detected.

Derive an equation for the reaction.

**METHOD**     In titration a)

Amount (mol) of 'thio' $= 19.5 \times 10^{-3} \times 0.120 = 2.35 \times 10^{-3}$ mol

No. of moles of $I_2 = \frac{1}{2} \times$ No. of moles of 'thio' $= 1.18 \times 10^{-3}$ mol

Since the reaction is

$$Br_2(aq) + 2I^-(aq) \longrightarrow 2Br^-(aq) + I_2(aq)$$

No. of moles of $Br_2 =$ No. of moles of $I_2$

Amount (mol) of $Br_2$ in 25.0 cm$^3 = 1.18 \times 10^{-3}$ mol

In titration b)

Amount (mol) of 'thio' in volume used $= 2.45 \times 10^{-3} \times 0.120$

$$= 0.294 \text{ mol}$$

Moles of $Br_2$/Moles of $S_2O_3^{2-} = 1.18 \times 10^{-3}/0.294 \times 10^{-3} = 4/1$

Bromine is reduced to bromide ions:

$$4Br_2(aq) + 8e^- \longrightarrow 8Br^-(aq)$$

Thiosulphate ions form sulphate ions:

$$S_2O_3^{2-}(aq) \longrightarrow 2SO_4^{2-}(aq)$$

To balance the equation, $H_2O$ is needed to supply the extra oxygen:

$$S_2O_3^{2-}(aq) + 5H_2O(l) \longrightarrow 2SO_4^{2-}(aq) + 10H^+(aq) + 8e^-$$

Putting the two half-reactions together gives

**ANSWER**     $4Br_2(aq) + S_2O_3^{2-}(aq) + 5H_2O(l) \longrightarrow 8Br^-(aq) + 2SO_4^{2-}(aq) + 10H^+(aq)$

## EXERCISE 12     Problems on Redox Reactions

1. Write balanced half-reaction equations for the oxidation of each of the following:
   a) $NO_2^-$ to $NO_3^-$     b) $AsO_3^{3-}$ to $AsO_4^{3-}$   c) $Hg_2^{2+}$ to $Hg^{2+}$
   d) $H_2O_2$ to $O_2$       e) $V^{3+}$ to $VO^{2+}$

2. Write balanced half-reaction equations for the reduction of each of the following in acid solution:
   a) $NO_3^-$ to $NO_2$      b) $NO_3^-$ to $NO$      c) $NO_3^-$ to $NH_4^+$
   d) $BrO_3^-$ to $Br_2$     e) $PbO_2$ to $Pb^{2+}$

3. Complete and balance the following ionic equations:
   a) $MnO_4^-(aq) + H_2O_2(aq) + H^+(aq) \longrightarrow$
   b) $MnO_2(s) + H^+(aq) + Cl^-(aq) \longrightarrow$
   c) $MnO_4^-(aq) + C_2O_4^{2-}(aq) + H^+ \longrightarrow$
   d) $Cr_2O_7^{2-}(aq) + C_2O_4^{2-}(aq) + H^+(aq) \longrightarrow$
   e) $Cr_2O_7^{2-}(aq) + I^-(aq) + H^+(aq) \longrightarrow$
   f) $H_2O_2(aq) + NO_2^-(aq) \longrightarrow$

4. How many moles of the following reductants will be oxidised by $3.0 \times 10^{-3}$ mol of potassium manganate(VII) in acid solution?

    a) $Fe^{2+}$    b) $Sn^{2+}$    c) $(CO_2^-)_2$    d) $H_2O_2$    e) $I^-$

5. How many moles of the following will be oxidised by $1.0 \times 10^{-4}$ mol of potassium dichromate(VI)?

    a) $Fe^{2+}$    b) $SO_3^{2-}$    c) $Br^-$    d) $(CO_2^-)_2$    e) $Hg_2^{2+}$?

6. How many moles of the following will be reduced by $2.0 \times 10^{-3}$ moles of $Sn^{2+}$?

    a) $Fe(CN)_6^{3-}$    b) $Cl_2$    c) $Mn^{4+}$ (to $Mn^{2+}$)
    d) $Ce^{4+}$ (to $Ce^{3+}$)    e) $BrO_3^-$ (to $Br^-$)?

7. What volumes of the following solutions will be oxidised by $25.0$ cm$^3$ of $0.0200$ mol dm$^{-3}$ potassium manganate(VII) in acid solution?

    a) $0.0200$ mol dm$^{-3}$ tin(II) nitrate
    b) $0.0100$ mol dm$^{-3}$ iron(II) sulphate
    c) $0.250$ mol dm$^{-3}$ hydrogen peroxide
    d) $0.200$ mol dm$^{-3}$ chromium(II) nitrate
    e) $0.150$ mol dm$^{-3}$ sodium ethanedioate

8. What volumes of the following solutions will be oxidised by $20.0$ cm$^3$ of $0.0150$ mol dm$^{-3}$ potassium dichromate(VI) in acid solution?

    a) $0.0200$ mol dm$^{-3}$ tin(II) chloride
    b) $0.150$ mol dm$^{-3}$ iron(II) chloride
    c) $0.125$ mol dm$^{-3}$ sodium ethanedioate
    d) $0.300$ mol dm$^{-3}$ sodium sulphite (sulphate(IV))
    e) $0.100$ mol dm$^{-3}$ mercury(I) nitrate, $Hg_2(NO_3)_2$

9. $25.0$ cm$^3$ of a sodium sulphite solution require $45.0$ cm$^3$ of $0.0200$ mol dm$^{-3}$ potassium manganate(VII) solution for oxidation. What is the concentration of the sodium sulphite solution?

10. $35.0$ cm$^3$ of potassium manganate(VII) solution are required to oxidise a $0.2145$ g sample of ethanedioic acid-2-water, $H_2C_2O_4 \cdot 2H_2O$. What is the concentration of the potassium manganate(VII) solution?

11. $37.5$ cm$^3$ of cerium(IV) sulphate solution are required to titrate a $0.2245$ g sample of sodium ethanedioate, $Na_2C_2O_4$. What is the concentration of the cerium(IV) sulphate solution?

12. A piece of iron wire weighs $0.2756$ g. It is dissolved in acid, reduced to the $Fe^{2+}$ state, and titrated with $40.8$ cm$^3$ of $0.0200$ mol dm$^{-3}$ potassium dichromate solution. What is the percentage purity of the iron wire?

13. A piece of limestone weighing $0.1965$ g was allowed to react with an excess of hydrochloric acid. The calcium in it was precipitated as calcium ethanedioate. The precipitate was dissolved in sulphuric acid, and the ethanedioate in the solution needed $35.6$ cm$^3$ of a $0.0200$ mol dm$^{-3}$ solution of potassium manganate(VII) for titration. Calculate the percentage of $CaCO_3$ in the limestone.

14. A solution of potassium dichromate is standardised by titration with sodium ethanedioate solution. If $47.0\,cm^3$ of the dichromate solution were needed to oxidise $25.0\,cm^3$ of ethanedioate solution of concentration $0.0925\,mol\,dm^{-3}$, what is the concentration of the potassium dichromate solution?

15. $2.4680\,g$ of sodium ethanedioate are dissolved in water and made up to $250\,cm^3$ of solution. When a $25.0\,cm^3$ portion of the solution is titrated against cerium(IV) sulphate, $35.7\,cm^3$ of the cerium(IV) sulphate solution are required. What is its concentration?

16. A $25.0\,cm^3$ aliquot of a solution containing $Fe^{2+}$ ions and $Fe^{3+}$ ions was acidified and titrated against potassium manganate(VII) solution. $15.0\,cm^3$ of a $0.0200\,mol\,dm^{-3}$ solution of potassium manganate(VII) were required. A second $25.0\,cm^3$ aliquot was reduced with zinc and titrated against the same manganate(VII) solution. $19.0\,cm^3$ of the oxidant solution were required. Calculate the concentrations of
a) $Fe^{2+}$, and b) $Fe^{3+}$ in the solution.

17. a) What volume of acidified potassium manganate(VII) of concentration $0.0200\,mol\,dm^{-3}$ is decolourised by $100\,cm^3$ of hydrogen peroxide of concentration $0.0100\,mol\,dm^{-3}$?
b) What volume of oxygen is evolved at s.t.p.?

18. A $0.6125\,g$ sample of potassium iodate(V), $KIO_3$, is dissolved in water and made up to $250\,cm^3$. A $25.0\,cm^3$ aliquot of the solution is added to an excess of potassium iodide in acid solution. The iodine formed requires $22.5\,cm^3$ of sodium thiosulphate solution for titration. What is the concentration of the thiosulphate solution?

19. $25.0\,cm^3$ of a solution of $X_2O_5$ of concentration $0.100\,mol\,dm^{-3}$ is reduced by sulphur dioxide to a lower oxidation state. To reoxidise X to its original oxidation number required $50.0\,cm^3$ of $0.0200\,mol\,dm^{-3}$ potassium manganate(VII) solution. To what oxidation number was X reduced by sulphur dioxide?

20. Manganese(II) sulphate is oxidised to manganese(IV) oxide by potassium manganate(VII) in acid solution. A flocculant is added to settle the solid $MnO_2$ so that it does not obscure the colour of the manganate(VII). If $25.0\,cm^3$ of manganese(II) sulphate solution require $22.5\,cm^3$ of $0.0200\,mol\,dm^{-3}$ potassium manganate(VII) solution, what is the concentration of $MnSO_4$?

*21. A solution of hydroxylamine hydrochloride contains $0.1240\,g$ of $NH_2OH \cdot HCl$. On boiling, it is oxidised by an excess of acidified iron(III) sulphate. The iron formed is titrated against potassium manganate(VII) solution of concentration $0.0160\,mol\,dm^{-3}$. A volume of $44.6\,cm^3$ of the oxidant is required.
a) Find the ratio of moles $NH_2OH$ : moles $Fe^{3+}$.
b) State the change in oxidation number of Fe.
c) State the oxidation number of N in $NH_2OH$.

d) Deduce the oxidation number of N in the product of the reaction.
e) Decide what compound of nitrogen in this oxidation state is likely to be formed in the reaction.
f) Write the equation for the reaction.

22. A piece of impure copper was allowed to react with dilute nitric acid. The copper(II) nitrate solution formed liberated iodine from an excess of potassium iodide solution. The iodine was estimated by titration with a solution of sodium thiosulphate. If a 0.877 g sample of copper was used, and the volume required was 23.7 cm$^3$ of 0.480 mol dm$^{-3}$ thiosulphate solution, what is the percentage of copper in the sample?

23. A household bleach contains sodium chlorate(I), NaOCl. The chlorate(I) ion will react with potassium iodide to give iodine, which can be estimated with a standard thiosulphate solution.

a) Write the equations for the reaction of ClO$^-$ and I$^-$ to give I$_2$ and for the reaction of iodine and thiosulphate ions.

b) A 25.0 cm$^3$ sample of household bleach is diluted to 250 cm$^3$. A 25.0 cm$^3$ portion of the solution is added to an excess of potassium iodide solution and titrated against 0.200 mol dm$^{-3}$ sodium thiosulphate solution. The volume required is 18.5 cm$^3$. What is the concentration of sodium chlorate(I) in the bleach?

# COMPLEXOMETRIC TITRATIONS

The complexes formed by a number of metal ions with

bis[bis(carboxymethyl)amino]ethane,

(HO$_2$CCH$_2$)$_2$NCH$_2$CH$_2$N(CH$_2$CO$_2$H)$_2$,

which is usually referred to as edta (short for its old name) are very stable, and can be used for the estimation of metal ions by titration. The end-point in the titration is shown by an indicator which forms a coloured complex with the metal ion being titrated. If Eriochrome Black T is used as indicator, the metal-indicator colour of red is seen at the beginning of the titration. As the titrant is added, the metal ions are removed from the indicator and complex with edta. At the end-point, the colour of the free indicator, blue, is seen:

Metal-indicator (red) + edta $\longrightarrow$ Metal-edta + Indicator (blue)

**EXAMPLE**     *Determination of the hardness of tap water*
Hardness in water is caused by the presence of calcium ions and magnesium ions. Both these ions complex strongly with edta. The amounts of temporary hardness and permanent hardness can be determined separately by performing complexometric titrations on tap water and boiled tap water. 100 cm$^3$ of tap water are measured into a flask. An alkaline buffer and Eriochrome Black T are added, and the solution is titrated against 0.100 mol dm$^{-3}$ edta solution. The volume required is 2.10 cm$^3$.

A second 100 cm³ of tap water are measured into a 250 cm³ beaker, and boiled for 30 minutes. After cooling, the water is filtered into a 100 cm³ graduated flask, and made up to the mark by the addition of distilled water. On titration as before, the volume of edta needed is 1.25 cm³. Calculate the concentration of calcium and magnesium present as permanent hardness and the concentration of calcium and magnesium present as temporary hardness.

**METHOD**

| Total hardness requires | 2.10 cm³ of 0.100 mol dm⁻³ edta |
|---|---|

Total hardness requires    $2.10 \text{ cm}^3$ of $0.100 \text{ mol dm}^{-3}$ edta

Permanent hardness requires    $1.25 \text{ cm}^3$ of $0.100 \text{ mol dm}^{-3}$ edta

Temporary hardness requires    $0.85 \text{ cm}^3$ of $0.100 \text{ mol dm}^{-3}$ edta

Amount (mol) of metal as permanent hardness

$$= 1.25 \times 10^{-3} \times 0.100 \text{ mol}$$

$$= 0.125 \times 10^{-3} \text{ mol in } 100 \text{ cm}^3 \text{ water}$$

$$= 1.25 \times 10^{-3} \text{ mol dm}^{-3}$$

Amount (mol) of metal as temporary hardness

$$= 0.85 \times 10^{-3} \times 0.100 \text{ mol}$$

$$= 0.085 \times 10^{-3} \text{ mol in } 100 \text{ cm}^3 \text{ water}$$

$$= 0.85 \times 10^{-3} \text{ mol dm}^{-3}$$

**ANSWER** The concentration of calcium and magnesium present as temporary hardness is $8.5 \times 10^{-4} \text{ mol dm}^{-3}$; the concentration of calcium and magnesium present as permanent hardness is $1.25 \times 10^{-3} \text{ mol dm}^{-3}$.

## EXERCISE 13    Problems on Complexometric Titrations

1. Calculate the concentration of a solution of zinc sulphate from the following data. 25.0 cm³ of the solution, when added to an alkaline buffer and Eriochrome Black T indicator, required 22.3 cm³ of a $1.05 \times 10^{-2} \text{ mol dm}^{-3}$ solution of edta for titration. The equation for the reaction can be represented as

$$Zn^{2+}(aq) + edta^{4-}(aq) \longrightarrow Znedta^{2-}(aq)$$

2. To a 50.0 cm³ sample of tap water were added a buffer and a few drops of Eriochrome Black T. On titration against a 0.0100 mol dm⁻³ solution of edta, the indicator turned blue after the addition of 9.80 cm³ of the titrant. Calculate the hardness of water in parts per million of calcium, assuming that the hardness is entirely due to the presence of calcium salts. (1 p.p.m. = 1 g in 10⁶ g water.)

3. A 0.2500 g sample of a mixture of magnesium oxide and calcium oxide was dissolved in dilute nitric acid and made up to 1.00 dm³ of solution with distilled water. A 50.0 cm³ portion was buffered and, after addition of indicator, was titrated against 0.0100 mol dm⁻³ edta solution. 25.8 cm³ of the titrant were required. Find the percentage by mass of calcium oxide and magnesium oxide in the mixture.

4. Find $n$ in the formula $Al_2(SO_4)_3 \cdot nH_2O$ from the following analysis. 2.000 g of aluminium sulphate hydrate were weighed out and made up to $250 \, cm^3$. A $25.0 \, cm^3$ portion was allowed to complex with edta by being boiled with $50.0 \, cm^3$ of edta solution of concentration $1.00 \times 10^{-2} \, mol \, dm^{-3}$. The excess of edta was determined by adding Eriochrome Black T and titrating against a solution of $1.115 \times 10^{-2}$ $mol \, dm^{-3}$ solution of zinc sulphate. $17.9 \, cm^3$ of zinc sulphate solution were required to turn the indicator from blue to red. The reactions taking place are

$$Al^{3+}(aq) \; + \; edta^{4-}(aq) \; \longrightarrow \; Aledta^{-}(aq)$$
$$Zn^{2+}(aq) \; + \; edta^{4-}(aq) \; \longrightarrow \; Znedta^{2-}(aq)$$

## PRECIPITATION TITRATIONS

In a precipitation titration, the two solutions react to form a precipitate of an insoluble salt. A solution of a chloride can be estimated by finding the volume of a standard solution of silver nitrate that will precipitate all the chloride ions as insoluble silver chloride:

$$Ag^{+}(aq) \; + \; Cl^{-}(aq) \; \longrightarrow \; AgCl(s)$$

Bromides and iodides and thiocyanates can be titrated in the same way. There are various ways of finding out when the end-point has been reached.

EXAMPLE 1  *Determination of chlorides*
$25.0 \, cm^3$ of a sodium chloride solution required $18.7 \, cm^3$ of $0.100 \, mol$ $dm^{-3}$ silver nitrate solution for complete precipitation. Calculate the concentration of the sodium chloride solution.

METHOD     Since           $Ag^{+}(aq) \; + \; Cl^{-}(aq) \; \longrightarrow \; AgCl(s)$

No. of moles of $AgNO_3$ = No. of moles of $Cl^-$

Amount (mol) of $AgNO_3$ in volume used = $18.7 \times 10^{-3} \times 0.100$
$$= 1.87 \times 10^{-3} \, mol$$

Amount (mol) of $Cl^-$ in $25.0 \, cm^3$ = $1.87 \times 10^{-3} \, mol$

Concn of $Cl^-$ = $(1.87 \times 10^{-3})/(25.0 \times 10^{-3}) = 7.50 \times 10^{-2} \, mol \, dm^{-3}$

ANSWER     The concentration of sodium chloride is $7.50 \times 10^{-2} \, mol \, dm^{-3}$.

EXAMPLE 2  *Determination of a mixture of halides*
2.95 g of a mixture of potassium chloride and potassium bromide is dissolved in water, and the solution is made up to $250 \, cm^3$. $25.0 \, cm^3$ of this solution required $31.5 \, cm^3$ of $0.100 \, mol \, dm^{-3}$ silver nitrate solution. Calculate the percentages of potassium chloride and potassium bromide in the mixture.

**METHOD**  Let $x$ grams = Mass of potassium chloride

Then $(2.95 - x)$ grams = Mass of potassium bromide

Amount (mol) of KCl in $25.0\,\text{cm}^3$ = $\dfrac{x}{74.5} \times \dfrac{25.0}{250} = \dfrac{0.1x}{74.5}\,\text{mol}$

The KCl in $25.0\,\text{cm}^3$ requires $\dfrac{x}{74.5}\,\text{dm}^3$ of $0.100\,\text{mol dm}^{-3}\,\text{AgNO}_3(\text{aq})$

Amount (mol) of KBr in $25.0\,\text{cm}^3$ = $\dfrac{(2.95 - x)}{119} \times \dfrac{25.0}{250}$

$= \dfrac{0.100(2.95 - x)}{119}$

The KBr in $25.0\,\text{cm}^3$ requires $\dfrac{(2.95 - x)}{119}\,\text{dm}^3$ of $0.100\,\text{mol dm}^{-3}\,\text{AgNO}_3(\text{aq})$

Therefore, $\dfrac{x}{74.5} + \dfrac{(2.95 - x)}{119} = 31.5 \times 10^{-5}$

$x = 1.34$

**ANSWER**  The mass of potassium chloride is $1.34\,\text{g}$; the mass of potassium bromide is $1.61\,\text{g}$.

## EXERCISE 14    Problems on Precipitation Reactions

1. A $25.0\,\text{cm}^3$ portion of a solution of potassium chloride required $18.5\,\text{cm}^3$ of a silver nitrate solution of concentration $0.0200\,\text{mol dm}^{-3}$ for titration. What is the concentration of the KCl solution?

2. A solid mixture contains sodium chloride and sodium nitrate. A $0.5800\,\text{g}$ sample of the mixture was dissolved in water and made up to $250\,\text{cm}^3$. A $25.0\,\text{cm}^3$ portion was titrated against silver nitrate solution of concentration $0.0180\,\text{mol dm}^{-3}$. The volume required was $27.4\,\text{cm}^3$. Calculate the percentage by mass of sodium chloride in the mixture.

3. $1.2400\,\text{g}$ of a mixture of sodium chloride and sodium bromide was dissolved and made up to $1.00\,\text{dm}^3$. A $25.0\,\text{cm}^3$ portion of this solution required $21.7\,\text{cm}^3$ of a silver nitrate solution of concentration $0.0175\,\text{mol dm}^{-3}$ for titration. Calculate the percentage composition by mass of the mixture.

4. $25.0\,\text{cm}^3$ of a solution of potassium cyanide required $19.8\,\text{cm}^3$ of a solution of silver nitrate of concentration $0.0350\,\text{mol dm}^{-3}$ for titration. Calculate the concentration of the KCN solution.

5. A solution contains sodium chloride and hydrochloric acid. A 25.0 cm³ aliquot required 38.2 cm³ of a 0.0325 mol dm⁻³ solution of silver nitrate for titration. A second 25.0 cm³ aliquot required 7.2 cm³ of a 0.0550 mol dm⁻³ solution of sodium hydroxide for neutralisation. Calculate the concentrations of a) sodium chloride, and b) hydrochloric acid in the solution.

6. Find the percentage by mass of silver in an alloy from the following information. A sample of 1.245 g of the alloy was dissolved in dilute nitric acid and made up to 250 cm³. A 25.0 cm³ portion required 29.8 cm³ of a 0.0214 mol dm⁻³ solution of potassium thiocyanate for titration.

## EXERCISE 15    Questions from A-level Papers

*1. The following is a method by which the reaction between iron(III) ions and hydroxylammonium chloride, $NH_3OH^+Cl^-$, may be investigated.

25.0 cm³ of a solution containing 3.60 g dm⁻³ of hydroxylammonium chloride is added to a solution containing an excess of $Fe^{3+}$ ions and about 25 cm³ of 1 M sulphuric acid, and the mixture boiled. It is then diluted with water, allowed to cool, and the $Fe^{2+}$ ions titrated with 0.02 M potassium manganate(VII) (potassium permanganate) of which 25.9 cm³ were required.

a) Calculate the molar ratio $Fe^{3+}/NH_3OH^+$ in the reaction, and hence determine the oxidation number of nitrogen in the product.

b) Using the oxidation number concept, or otherwise, deduce the equation for the reaction (H = 1, N = 14, O = 16, Cl = 35.5).

(L80, S)

*2. Sulphur dioxide was passed through 100 cm³ of solution A of potassium chlorate(V) until reduction was complete according to the equation

$$KClO_3 + 3SO_2 + 3H_2O \longrightarrow KCl + 3H_2SO_4$$

The solution was then boiled and treated with a slight excess of silver nitrate solution. The precipitate of silver chloride was collected and found to weigh 0.3010 g. Exactly 25 cm³ of a solution B of iron(II) sulphate in dilute sulphuric acid were titrated with a solution of 0.016 67 mol l⁻¹ potassium dichromate(VI); 44.00 cm³ of dichromate(VI) were required.

$$K_2Cr_2O_7 + 6FeSO_4 + 7H_2SO_4 \longrightarrow K_2SO_4 + Cr_2(SO_4)_3 + 3Fe_2(SO_4)_3 + 7H_2O$$

Calculate the concentrations in mol l⁻¹ of solutions A and B.

25 cm³ of solution A were then mixed with 25 cm³ of solution B and boiled for ten minutes.

To find the iron(II) sulphate remaining the mixture was then titrated with the $0.016.67 \text{ mol } l^{-1}$ potassium dichromate(VI) solution used above. $12.50 \text{ cm}^3$ were required.

Calculate the number of moles of $Fe^{2+}$ which have reacted with one mole of $ClO_3^-$.

Hence suggest a balanced ionic equation for the reaction of the type

$$ClO_3^- + xFe^{2+} + yH^+ \longrightarrow \qquad \text{(JMB80, S)}$$

3. When $0.20 \text{ g}$ of basic oxide MO was reacted with $125 \text{ cm}^3$ of $0.10 \text{ M}$ hydrochloric acid, the excess acid required $25 \text{ cm}^3$ of $0.10 \text{ M}$ sodium hydroxide solution for neutralisation. $(O = 16.)$ The relative atomic mass of M is

a 4          b 20          c 24

d 34          e 40                               (NI81)

4. $0.180 \text{ g}$ of a sample of sodium ethanedioate (oxalate), $Na_2C_2O_4$, which is contaminated with an inert substance, reacts at $333 \text{ K}$ with $25.0 \text{ cm}^3$ of $0.02 \text{ M}$ potassium tetraoxomanganate(VII) (permanganate) to which an excess of dilute sulphuric acid has been added. $(C = 12, O = 16, Na = 23.)$

$$MnO_4^- + 8H^+ + 5e^- \longrightarrow Mn^{2+} + 4H_2O$$

$$C_2O_4^{2-} \longrightarrow 2CO_2 + 2e^-$$

The percentage by weight of sodium ethanedioate in the sample is

a 14.9          b 37.2          c 61.1

d 74.4          e 93.1                             (NI81)

5. a) $4.41 \text{ g}$ of dark blue crystals of a copper/ammonia compound $A$ were dissolved in water and made up to $250.0 \text{ cm}^3$. $25.00 \text{ cm}^3$ of this solution, $B$, formed a brown mixture when added to approximately $10 \text{ cm}^3$ of a 10 per cent solution of potassium iodide.

$$Cu^{2+}(aq) + 2I^-(aq) \longrightarrow CuI(s) + \tfrac{1}{2}I_2(aq)$$

The iodine liberated was equivalent to $17.60 \text{ cm}^3$ of aqueous sodium thiosulphate $(Na_2S_2O_3)$ of concentration $0.100 \text{ mol dm}^{-3}$. A white precipitate remained after the removal of the iodine. Calculate the amount of copper, in moles, present in $100 \text{ g}$ of the crystals $A$; and give the name and formula of the white precipitate.

b) $1.00 \text{ g}$ of crystals $A$ were dissolved in water, 2 drops of methyl orange were added and the mixture was titrated with sulphuric acid of concentration $0.500 \text{ mol dm}^{-3}$ until the colour changed from green to pink/violet. $15.84 \text{ cm}^3$ of the acid were required.

Assuming that the ammonia in the crystals $A$ reacts with the acid,

$$NH_3(aq) + H^+(aq) \longrightarrow NH_4^+(aq)$$

calculate the amount of ammonia, in moles, in $100 \text{ g}$ of the crystals $A$.

c) Use the results from a) and b) to calculate the proportion of ammonia to copper in the crystals $A$.

d) Addition of aqueous barium chloride to a sample of $A$ dissolved in water give a white precipitate which remains on addition of hydrochloric acid.

Addition of concentrated hydrochloric acid to a sample of $A$ dissolved in water gave a yellow solution which turned turquoise on dilution.

From the results in c) and d) suggest a formula for $A$.

e) State three properties of copper in the $+2$ oxidation state which are characteristic of transition elements. Give a short explanation in each case. (AEB84)

6. This question concerns a hydrated double salt $A$,

$$Cu_w(NH_4)_x(SO_4)_y \cdot zH_2O,$$

whose formula may be determined from the following experimental data.

a) 2 g of salt $A$ was boiled with excess sodium hydroxide and the ammonia expelled collected by absorption in $40 \text{ cm}^3$ of 0.5 M hydrochloric acid in a cooled flask. Subsequently this solution required $20 \text{ cm}^3$ of 0.5 M sodium hydroxide for neutralisation.
   i) Calculate the number of moles of ammonium ion in 2 g of salt $A$.
   ii) Calculate the mass of ammonium ion present in 2 g of salt $A$.

b) A second sample of 2 g of salt $A$ was dissolved in water and treated with an excess of barium chloride solution. The mass of the precipitate formed, after drying, was found to be 2.33 g.
   i) Calculate the number of moles of sulphate ion in 2 g of salt $A$.
   ii) Calculate the mass of sulphate ion in 2 g of salt $A$.

c) A third sample of 10 g of salt $A$ was dissolved to give $250 \text{ cm}^3$ of solution. $25 \text{ cm}^3$ of this solution, treated with an excess of potassium iodide, gave iodine equivalent to $25 \text{ cm}^3$ of 0.1 M sodium thiosulphate solution.

$$2Cu^{2+} + 4I^- \longrightarrow 2CuI + I_2$$

$$I_2 + 2S_2O_3^{2-} \longrightarrow 2I^- + S_4O_6^{2-}$$

   i) Calculate the number of moles of copper(II) ion in 2 g of salt $A$.
   ii) Calculate the mass of copper(II) ion in 2 g of salt $A$.

d) i) Calculate the mass of water of crystallisation in 2 g of salt $A$.
   ii) Calculate the number of moles of water of crystallisation in 2 g of salt $A$.

e) Determine the formula of the hydrated salt $A$. (L84)

**7.** a) Show, by means of equations, how each of the following compounds can be prepared in the laboratory:
   i) iron(II)sulphate-7-water from iron
   ii) tin(IV)chloride from tin
   iii) potassium chromate(VI) from a chromium(III) salt.

   b) i) Write a balanced equation for the reaction between $Fe^{2+}$ and $Cr_2O_7^{2-}$ in acidic, aqueous solution.
   ii) A medicinal tablet for patients suffering from iron deficiency contains $FeSO_4 \cdot 7H_2O$ as the iron-containing ingredient. A tablet weighing 0.2000 g was dissolved in an excess of dilute sulphuric acid and the resulting solution titrated with 0.0100 M potassium dichromate(VI); 11.51 cm$^3$ of the potassium dichromate(VI) solution were required for complete oxidation of the iron(II).
   Calculate the percentage of $FeSO_4 \cdot 7H_2O$ in the tablet.(JMB83)

**8.** The following method was used to determine the solubility of oxygen in water at 288 K.

Approximately 700 cm$^3$ of distilled water were shaken with air at 288 K for some time. 500 cm$^3$ of this water were then placed in a 500 cm$^3$ graduated flask. Using a pipette 1.00 cm$^3$ of a concentrated aqueous solution of sodium hydroxide was delivered into the flask below the surface of the liquid, followed by 1.00 cm$^3$ of a concentrated aqueous solution of manganese(II)chloride in the same way. The contents were gently agitated for a few minutes, the precipitate allowed to settle, the stopper removed and crystals of solid potassium iodide added followed by about 5 cm$^3$ of concentrated hydrochloric acid. The contents of the flask were then transferred to a litre vessel and titrated with aqueous sodium thiosulphate of concentration 0.0200 mol dm$^{-3}$.

   a) The initial precipitate of manganese(II) hydroxide is considered to be oxidised by the dissolved oxygen to hydrated manganese(III) oxide. Using oxidation states, determine the number of moles of oxygen required to produce one mole of hydrated manganese(III) oxide.

   b) In acid solution manganese(III) oxide oxidises iodide ions to iodine and is reduced to manganese(II) ions. Write an equation for this reaction.

   c) The volume of aqueous sodium thiosulphate required in the above titration was 30.0 cm$^3$. Calculate the concentration in mol dm$^{-3}$ of dissolved oxygen in water at 288 K.

   d) State *one* method by which the concentration of dissolved oxygen is maintained in 'healthy' natural waters and *one* pollutant which may cause a reduction in the level of dissolved oxygen. (AEB85)

9. a) Compare ammonia, $NH_3$, phosphine, $PH_3$, and arsine, $AsH_3$, in terms of i) shapes of molecules and ii) thermal stability.

   b) $25.0 \, cm^3$ of aqueous potassium iodate(V) was mixed with excess potassium iodide and acidified with aqueous sulphuric acid. The mixture then required $25.0 \, cm^3$ of aqueous sodium thiosulphate, of concentration $0.100 \, mol \, dm^{-3}$, to reduce the liberated iodine.

   To $100 \, cm^3$ (an excess) of the same aqueous potassium iodate(V), $37.5 \, cm^3$ of aqueous hydrazine ($N_2H_4$) of concentration $0.0333$ $mol \, dm^{-3}$ were added. On subsequent addition of acidified potassium iodide, the liberated iodine required $50 \, cm^3$ of the same aqueous sodium thiosulphate.

   i) Calculate the concentration of the potassium iodate(V) in $mol \, dm^{-3}$.

   ii) Deduce the stoichiometry of the reaction between iodate(V) ions and hydrazine. Write an equation.

   iii) From your equation state the original and final oxidation states for each of the iodine and the nitrogen.          (AEB84S)

10. When $10.0 \, cm^3$ of a $0.10 \, mol \, dm^{-3}$ solution of an alkali metal salt, $MXO_3$, was reduced with an excess of acidified potassium iodide solution, the resulting iodine required $60.0 \, cm^3$ of $0.10 \, mol \, dm^{-3}$ sodium thiosulphate solution for its reduction. The anion could have been reduced to

   a $XO_2$          b $XO_2^-$          c $XO^-$

   d $XO$          e $X^-$          (O & C82)

11. Use the following experimental data to deduce an equation for the reaction between chlorine and sodium thiosulphate in aqueous solution, and explain your reasoning.

   $3.10 \, g$ of $Na_2S_2O_3 \cdot 5H_2O$ were dissolved in water to give $250 \, cm^3$ of solution, and chlorine was passed through until reaction was complete. Excess chlorine was then swept from the solution using gaseous nitrogen. $25.0 \, cm^3$ samples of the resulting solution were found

   a) to require $12.5 \, cm^3$ of $1.00 \, M$ KOH for neutralisation,

   b) to require $10.0 \, cm^3$ of $1.00 \, M$ $AgNO_3$ to complete the precipitation of chloride ions, and

   c) to give $0.583 \, g$ of white precipitate when treated with an excess of aqueous $BaCl_2$.

   Iodine reacts with sodium thiosulphate according to the equation

   $$I_2 + 2Na_2S_2O_3 \longrightarrow 2NaI + Na_2S_4O_6.$$

   Compare the changes in oxidation number of the chlorine and sulphur in the first reaction with those of iodine and sulphur in the second. What explanation can you offer for this difference between the two halogen elements?

   (Relative atomic masses: $H = 1$, $O = 16$, $Na = 23$, $S = 32$, $Ba = 137$.)          (L83)

12. Laminaria sea weed concentrates iodide ions derived from sea water. The iodine can be estimated:

    'The sea weed is burned in a limited supply of air. The ash is boiled with water, filtered, and the filtrate treated with an excess of acidified hydrogen peroxide solution. The liberated iodine is extracted in trichloromethane, the appropriate layer separated and titrated against standard sodium thiosulphate solution ($Na_2S_2O_3$).'

    a) Why is the sea weed burned to ash?

    b) In what form is the iodine liberated by boiling with water?

    c) Explain the *type* of change undergone by iodide ions in the reaction with hydrogen peroxide. Write the reaction equation.

    d) Why is it necessary to use an organic extraction rather than titrate the liberated iodine directly?

    e) In which layer will the iodine concentrate during the extraction?

    f) What will be the colour of the iodine in this layer?

    g) Where are the principal sources of iodine loss in the entire method of extraction?

    h) 1 kg of sea weed produces sufficient iodine to react with 100.0 cm$^3$ of 0.100 mol dm$^{-3}$ sodium thiosulphate solution. Calculate the mass of iodine extracted from 1 kg of sea weed.

    i) $^{131}_{53}I$ is used as a radioactive tracer in the diagnosis of thyroid gland malfunction. Explain the significance of the numbers quoted on the iodine symbol. (SUJB84)

13. a) State two ways in which copper and zinc resemble each other chemically and two ways in which they differ chemically.

    b) How do copper(II) ions react with iodide ions in aqueous solution?

    c) How do zinc(II) and copper(II) ions react with the hydrogen form of an ion-exchange resin?

    d) A sample of brass, containing only copper and zinc, was dissolved and a solution was prepared containing $Cu^{2+}$ and $Zn^{2+}$ as the only cations present. This solution was then treated in the following way.

    i) A measured portion of the solution was treated with an excess of potassium iodide. The iodine liberated required 15.0 cm$^3$ of 0.100 M sodium thiosulphate for titration.

    ii) A further sample, of the same volume as used in experiment i), was added to an ion-exchange resin in the H$^+$-form and after complete exchange the resulting solution required 40.0 cm$^3$ of 0.100 M sodium hydroxide solution for titration.

    What was the composition, in moles, of the brass sample?

    (The reaction between sodium thiosulphate and iodine is given by the equation

    $$2S_2O_3{}^{2-} + I_2 = S_4O_6{}^{2-} + 2I^-.$$

    M represents mol dm$^{-3}$.) (O & C83)

14. Discuss the compounds in which iodine exhibits positive oxidation states and give examples.

When iodic(V) acid ($HIO_3$) reacts with an aqueous solution of sulphur dioxide, sulphuric acid and iodine are formed. Write a balanced equation for this reaction.

$25.00 cm^3$ of a solution $X$ of iodic(V) acid required $23.80 cm^3$ of a $0.1125 M$ sodium hydroxide solution for neutralisation. Calculate the concentration in $mol\, l^{-1}$ of the iodic(V) acid solution $X$.

Another $25.00 cm^3$ portion of the iodic(V) acid solution $X$ was treated with an excess of sulphur dioxide solution. The iodine formed and the excess of sulphur dioxide were removed by boiling. The resulting solution of sulphuric acid was treated with a slight excess of barium chloride solution and the resulting precipitate of barium sulphate weighed $1.5623 g$ after washing and drying. Show that this result agrees with a quantitative reaction according to the equation you obtained. ($A_r$: Ba, 137.3; S, 32.06; O, 16.00.)     (JMB82,S)

15. A white crystalline compound $X$, of relative molecular mass 97, has the following composition by mass:

N, 14.4%; H, 3.1%; S, 33.0%; O, 49.5%.

An aqueous solution of $X$ had no immediate reaction with barium chloride solution but when $X$ was treated with nitrous acid in aqueous solution, $0.25 g$ of $X$ gave $57.7 cm^3$ of nitrogen, measured at s.t.p. The resulting solution, on treatment with barium chloride, gave $0.60 g$ of a white precipitate. $24.3 cm^3$ of a solution $X$ of concentration $10.0 g\, dm^{-3}$ was neutralised exactly by $25.0 cm^3$ of $0.10 M$ sodium hydroxide, it being noted that a wide range of indicators is suitable for the titration.

Deduce the structure of $X$ and show how the above information supports your deduction. (Relative atomic masses: H = 1, N = 14, O = 16, S = 32, Ba = 137; molar volume of gas = $22.4 dm^3$ at s.t.p.)     (L84)

16. a) Explain in detail the choice of indicator used (if any) when carrying out the following volumetric estimations in aqueous solution:
    i) sodium chloride using silver nitrate,
    ii) ethanoic (*acetic*) acid using sodium hydroxide,
    iii) ethanedioic (*oxalic*) acid using acidified potassium manganate(VII).
    b) $25.0 cm^3$ of a solution containing only sulphuric acid and ethanedioic acid required $37.5 cm^3$ of $0.1 M$ NaOH for complete neutralisation. Excess dilute sulphuric acid was then added to the resulting solution and it was found that $12.5 cm^3$ of $0.02 M\, KMnO_4$ were required for oxidation. Determine the concentration of the sulphuric acid and the ethanedioic acid in the original solution.     (JMB82)

**17.** This question involves a practical analysis of a compound of copper. The compound consisted of light blue crystals containing 27.04% by mass of water of crystallisation.

The addition of dilute hydrochloric acid and barium chloride solution to a solution of the compound gave a dense white precipitate.

When sodium hydroxide solution was added to the compound, a gas with a distinctive odour was evolved which turned red litmus blue and gave white fumes when a rod, previously dipped in concentrated hydrochloric acid, was brought close to the tube.

Which ions, other than copper, were present?

a)  10.00 g of the copper compound were dissolved in water and made up to 250.0 cm$^3$. This solution was labelled $A$.

To 25.00 cm$^3$ of the solution $A$, excess barium chloride solution was added. The dense white precipitate was carefully filtered off and dried and it was found to weigh 1.1672 g (0.6868 g of the precipitate was due to barium).

Calculate the percentage, by mass, of the anion in the copper compound.

b)  To determine the copper present in the compound, 25.00 cm$^3$ of solution $A$ were placed in a conical flask. Excess (1 g) of solid potassium iodide was added and the iodine liberated was titrated with a standard solution, containing 6.100 g of sodium thiosulphate $Na_2S_2O_3 \cdot 5H_2O$ ($M_r = 248.2$) crystals in 250.0 cm$^3$ of water. The following burette readings were obtained.

| Final burette reading (cm$^3$) | 25.70 | 26.70 | 33.15 | 26.43 |
|---|---|---|---|---|
| Initial burette reading (cm$^3$) | 0.15 | 1.30 | 7.70 | 1.00 |
| Titre (cm$^3$) | | | | |

i)  Enter the titres in the appropriate spaces.

ii)  Give the mean accurate titre.

iii)  State which titres you used to calculate the mean accurate titre.

iv)  Calculate the concentration (mol dm$^{-3}$) of the standard sodium thiosulphate solution.

v)  Show, by giving appropriate equations, that $1\,Cu^{2+} \equiv 1\,S_2O_3^{2-}$

vi)  Calculate the concentration (mol dm$^{-3}$) of copper(II) ions in solution $A$.

vii)  Calculate the percentage, by mass, of copper in the compound.

c) Deduce the formula of the compound by completing the table below, making any appropriate calculations.

| Constituent | Copper | Water | Anion | Residual |
|---|---|---|---|---|
| $A_r$ or $M_r$ of constituent | | | | |
| Percentage by mass in the compound | | 27.04 | | |
| Mole ratios in the compound | | | | |

($A_r(H) = 1.01; A_r(C) = 12.01; A_r(N) = 14.01; A_r(O) = 16.00;$ $A_r(Na) = 22.99; A_r(S) = 32.06; A_r(Cu) = 63.55;$ $A_r(Ba) = 137.34.$)

(WJEC84,p)

# 4  The Atom

## MASS SPECTROMETRY

In a mass spectrometer, an element or compound is vaporised and then ionised. The ions are accelerated, collimated into a beam and deflected by a magnetic field. The amount of the deflection depends on the ratio of mass/charge of the ions, as well as the values of the accelerating voltage and the magnetic field. The magnetic field is kept constant while the accelerating voltage is varied continuously to focus each species in turn into the ion detector. The detector records each species as a peak on a trace. From the value of the voltage associated with a particular peak the ratio of mass/charge for that ionic species can be found. Since each ion has a charge of $+1$, the ratio mass/charge is equal to the mass of the ion. The mass spectrometer can be calibrated to read out ionic masses directly. The heights of the peaks are proportional to the relative abundance of the different ions.

**EXAMPLE 1** The mass spectrum of boron shows two peaks, one at $10.0\,m_u$, and the other at $11.0 m_u$. The heights of the peaks are in the ratio $18.7\% : 81.3\%$. Calculate the relative atomic mass of boron.

**METHOD** The relative heights of the peaks show that the relative abundance of $^{10}B$ and $^{11}B$ is $18.7\%\ ^{10}B : 81.3\%\ ^{11}B$.

In 1000 atoms, there are 187 of mass $10.0 m_u = 1870 m_u$

and 813 of mass $11.0 m_u = 8943 m_u$

The mass of 1000 atoms $= 10\,813 m_u$

The average atomic mass $= 10.8 m_u$

**ANSWER** The relative atomic mass of boron is 10.8.

**EXAMPLE 2** The mass spectrum of neon shows three peaks, corresponding to masses of 22, 21 and $20 m_u$. The heights of the peaks are in the ratio $11.2 : 0.2 : 114$. Calculate the average atomic mass of neon.

**METHOD** Multiplying the relative abundance (the height of the peak) by the mass to find the total mass of each isotope present gives

Mass of neon-22 $= 11.2 \times 22.0 = 246.4 m_u$

Mass of neon-21 $= 0.2 \times 21.0 = 4.2 m_u$

Mass of neon-20 $= 114 \times 20.0 = 2280 m_u$

Totals are $125.4 = 2530.6 m_u$

Average mass of neon atom $= 2530.6/125.4 = 20.18 m_u$.

**ANSWER** The average atomic mass of neon is $20.2 m_u$.

**EXERCISE 16**    Problems on Mass Spectrometry

1. The mass spectrum of rubidium consists of a peak at mass 85 and a peak at mass $87\,m_u$. The relative abundance of the isotopes is 72 : 28. Calculate the mean atomic mass of rubidium.

2. If $^{69}Ga$ and $^{71}Ga$ occur in the proportions 60 : 40, calculate the average atomic mass of gallium.

3. Fig. 4.1 shows the mass spectrum of magnesium. The heights of the three peaks and the mass numbers of the isotopes are shown in Fig. 4.1. Calculate the relative atomic mass of magnesium.

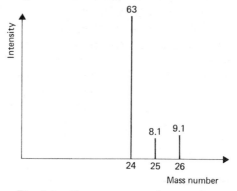

Fig. 4.1    Mass spectrum of magnesium

4. The mass spectrum of chlorine shows peaks at masses 70, 72 and $74\,m_u$. The heights of the peaks are in the ratio of 9 to 6 to 1. What is the relative abundance of $^{35}Cl$ and $^{37}Cl$? What is the average atomic mass of chlorine?

5. Calculate the relative atomic mass of lithium, which consists of 7.4% of $^{6}Li$ and 92.6% of $^{7}Li$.

6. A sample of water containing $^{1}H$, $^{2}H$ and $^{16}O$ was analysed in a mass spectrometer. The trace showed peaks at mass numbers 1, 2, 3, 4, 17, 18, 19 and 20. Suggest which ions are responsible for these peaks.

7. Calculate the average atomic mass of potassium, which consists of 93% $^{39}K$ and 7.0% $^{41}K$.

8. Fig. 4.2 shows a mass spectrometer trace for copper nitrate. Each of the eight peaks is produced by a different species of ion. Suggest what these ions are.

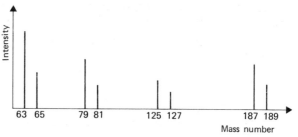

Fig. 4.2    The mass spectrum of copper nitrate

**EXERCISE 17**     Questions from A-level Papers

1. The mass spectrum of neon consists of two major peaks, one at mass 20 and the other at mass 22, of relative abundance 10 : 1. Show how these data can be used to determine the relative atomic mass of neon.

   Interpret as fully as you can the mass spectrum of sulphur vapour shown in the figure below.                    (C80, p)

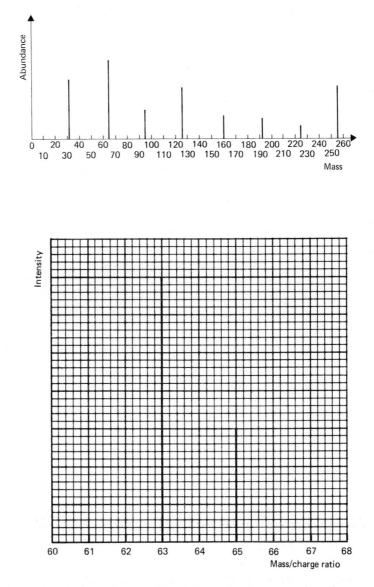

2.

The mass spectrum of a metal is shown above. Assuming that each ion has unit charge, the relative atomic mass of the metal is

a  63.2          b  63.4          c  63.6
d  63.8          e  64.0                    (NI82)

3. a) Discuss what is meant by the terms *isotopes* and *isotopic abundance.*

  b) Describe in outline how in the mass spectrometer:
    i) positive ions are formed from organic molecules;
    ii) the ions are separated according to mass and charge.

  c) Separate samples of *ethene gas* were reacted with bromine vapour, aqueous bromine (bromine water), and aqueous bromine containing some dissolved sodium chloride. The organic products were examined using a mass spectrometer, peaks being observed as follows:

| Experiment | Reagent | Mass/charge ratio | Information on peaks |
|---|---|---|---|
| 1 | Bromine vapour | 186, 188, 190 | Peak height in ratio 1 : 2 : 1 |
| 2 | Aqueous bromine | 124, 126 186, 188, 190 | All five are principal peaks |
| 3 | Aqueous bromine with dissolved sodium chloride | 124, 126 142, 144, 146 186, 188, 190 | All eight are principal peaks |

Assuming that natural samples of carbon, hydrogen and oxygen contain only one isotope whilst chlorine contains $^{35}Cl$ and $^{37}Cl$:
    i) use the results of experiment 1 to determine the isotopic composition of bromine,
    ii) give the names and structural formulae of the compounds responsible for the additional peaks in experiments 2 and 3.
    iii) show how the mechanism of the reaction accounts for the formation of these compounds.             (AEB85)

4. a) State the meaning of *each* of the following:
    i) atomic number;     ii) mass number;     iii) isotopes.

  b) The relative atomic mass of an element may be defined as the weighted mean of the masses of all the atoms in a normal isotopic sample of the element on the scale where the mass of one atom of carbon-12 has the value of twelve exactly.

  The element copper has a relative atomic mass of 63.55 and contains atoms with mass numbers of 63 and 65.

  Calculate the percentage composition of a normal isotopic sample of copper.

  c) Chlorine contains $^{35}_{17}Cl$ and $^{37}_{17}Cl$ in a 3 : 1 ratio and bromine contains $^{79}_{35}Br$ and $^{81}_{35}Br$ in a 1 : 1 ratio. Assuming that carbon contains atoms of mass number 12 only and hydrogen atoms of mass number 1 only, calculate the possible values for the molecular ion peaks in the mass spectrum of a compound of formula $C_2H_3Br_2Cl$ and the relative heights of these peaks.

  d) Explain what is meant by the *mass deficit* of a nucleus.     (AEB85)

**5.** a) Define:
   i) mass number;
   ii) mass deficit.

   b) The mass spectra of methylbenzene and butanone are shown below.

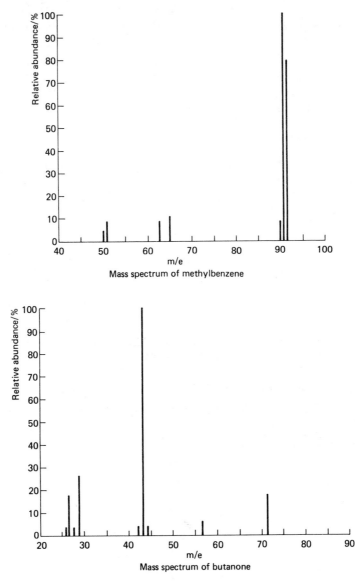

Mass spectrum of methylbenzene

Mass spectrum of butanone

   i) Draw structural formulae for methylbenzene and butanone.
   ii) Identify the peak in each spectrum corresponding to the parent molecule ion.
   iii) Deduce from the spectra the main fragmentation process that occurs in the mass spectrometer for molecules of methylbenzene and for molecules of butanone.

## NUCLEAR REACTIONS

In a nuclear reaction, a rearrangement of the protons and neutrons in the nuclei of the atoms takes place, and new elements are formed.

The atomic number or proton number, Z, of an element is the number of protons in the nucleus of an atom of the element. The mass number or nucleon number, A, is the number of protons and neutrons in the nucleus of an atom. Isotopes of an element differ in mass number but have the same atomic number. Isotopes are represented as $^A_Z$ Symbol, e.g. $^{12}_6C$. Protons are represented as $^1_1H$, electrons ($\beta$-particles) as $^0_{-1}e$, neutrons as $^1_0n$, and $\alpha$-particles as $^4_2He$. In the equation for a nuclear reaction, the sum of the mass numbers is the same on both sides, and the sum of the atomic numbers is the same on both sides of the equation.

For practice in balancing nuclear equations, study the following examples.

**EXAMPLE 1**  Complete the equation

$$^{16}_7N \longrightarrow {}^a_bO + {}^0_{-1}e$$

**METHOD**  Consider mass numbers:  $16 = a + 0 \quad \therefore a = 16$

Consider atomic numbers:  $7 = b + (-1) \quad \therefore b = 8$

**ANSWER**  $^{16}_7N \longrightarrow {}^{16}_8O + {}^0_{-1}e$

**EXAMPLE 2**  Find the values of $a$ and $b$ in the equation

$$^{27}_{13}Al + {}^1_0n \longrightarrow {}^4_2He + {}^a_bX$$

**METHOD**  Consider mass numbers:  $27 + 1 = 4 + a \quad \therefore a = 24$

Consider atomic numbers:  $13 + 0 = 2 + b \quad \therefore b = 11$

**ANSWER**  $a = 24$ and $b = 11$.

## EXERCISE 18    Problems on Nuclear Reactions

Complete the following equations, supplying values for the missing mass numbers (nucleon numbers) and atomic numbers (proton numbers).

1.  a) $^9_4Be + \gamma \longrightarrow {}^8_4Be + {}^a_bX$

b) $^{14}_7N + {}^4_2He \longrightarrow {}^1_1H + {}^a_bY$

c) $^9_4Be + {}^1_1H \longrightarrow {}^6_3Li + {}^a_bZ$

d) $^{209}_{83}Bi + {}^2_1D \longrightarrow {}^a_bX + {}^1_1H$

e) $^{16}_8O + {}^1_0n \longrightarrow {}^{13}_6C + {}^a_bY$

f) $^{10}_5B + {}^a_bY \longrightarrow {}^{13}_7N + {}^1_0n$

g) $^{14}_7N + {}^1_0n \longrightarrow {}^a_bQ + {}^1_1H$

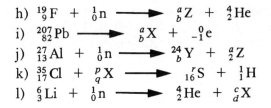

h) $^{19}_{9}F + ^{1}_{0}n \longrightarrow ^{a}_{b}Z + ^{4}_{2}He$

i) $^{207}_{82}Pb \longrightarrow ^{a}_{b}X + ^{0}_{-1}e$

j) $^{27}_{13}Al + ^{1}_{0}n \longrightarrow ^{24}_{b}Y + ^{a}_{2}Z$

k) $^{35}_{17}Cl + ^{p}_{q}X \longrightarrow ^{r}_{16}S + ^{1}_{1}H$

l) $^{6}_{3}Li + ^{1}_{0}n \longrightarrow ^{4}_{2}He + ^{c}_{d}X$

## EXERCISE 19     Questions from A-level Papers

1. Describe the extraction of magnesium by an electrolytic process. Explain what is meant by the statement that magnesium has a relative atomic mass of 24.31.

   An element $X$ consists of a single isotope. Its compound with magnesium has a formula $MgX_2$ and contains 61.00 per cent by mass of X. Deduce

   a) the relative atomic mass of X,

   b) the mass number of the isotope,

   c) which group of the periodic table contains X, and

   d) the number of neutrons in the nucleus of an atom of X.

   Describe and explain the acidic properties of aqueous solutions of HX.
   (WJEC80)

2. a)  i) Balance the following equation in terms of the numbers of electrons and neutrons produced. The total relative atomic masses of the reactants is 236.0526 and that of the products is 235.8332. Discuss the commercial use of such processes.

   $$^{235}_{92}U + ^{1}_{0}n = ^{95}_{42}Mo + ^{139}_{57}La + \qquad +$$

   For nuclear reactors, explain the use of graphite and describe how control is achieved.

   ii) Give the essential nuclear reactions which occur in fast breeder reactors.

   b) Describe the basic chemistry and design of a hydrogen-oxygen fuel cell, outlining the advantages of such a cell.

   (Velocity of light $= 2.998 \times 10^8 \, m \, s^{-1}$.)     (WJEC84)

3. Astatine, the heaviest of the halogens, exists only as radioactive isotopes. Which one of these is produced by the bombardment of $^{209}_{83}Bi$ with α-particles ? Because of the very short half-life and the intense radioactivity, work on astatine can only be done at concentrations of about $10^{-14} \, mol \, dm^{-3}$. What sort of technique can be used to compare the solution chemistry of astatine with that of the other halogens?
   (O85,S,p)

4. Under the conditions prevailing in the centre of a star like the sun, where the temperature is about $1.6 \times 10^7 \, K$, hydrogen $^1_1H$ undergoes nuclear changes according to the following sequence of

$$^1_1H \ + \ ^1_1H \ \longrightarrow \ X \ + \ ^0_{+1}e \quad\quad + \quad\quad v$$

a positron                  a neutrino
(or positive electron)   (which has no mass)

$$^2_1H \ + \ Y \ \longrightarrow \ ^3_2He \ + \ \gamma$$

gamma rays

$$^3_2He \ + \ Z \ \longrightarrow \ ^4_2He \ + \ ^1_1H \ + \ ^1_1H$$

The positrons are annihilated by meeting electrons, with the consequent production of gamma rays.

$$^0_{+1}e \ + \ ^0_{-1}e \ \longrightarrow \ \gamma$$

a) Identify the nuclei $X$, $Y$ and $Z$, giving their atomic symbols, atomic numbers and mass numbers.

b) What new element has been formed from hydrogen?         (C84,p)

# 5   **Gases**

## THE GAS LAWS

The behaviour of gases is described by the Gas Laws: Boyle's law, Charles' law, the equation of state for an ideal gas, Graham's law of diffusion, Gay-Lussac's law, Avogadro's law, Dalton's law of partial pressures, Henry's law of the solubility of gases and the ideal gas equation. We look at each of these in turn.

## BOYLE'S LAW

Boyle's law states that the pressure of a fixed mass of gas at a constant temperature is inversely proportional to its volume:

$$PV = \text{Constant}$$

where $P$ = pressure, $V$ = volume.

## CHARLES' LAW

Charles' law states that the volume of a fixed mass of gas at constant pressure is directly proportional to its temperature on the Kelvin scale:

$$\frac{V}{T} = \text{Constant}$$

where $T$ = temperature in kelvins.

Temperature on the Kelvin scale (called the absolute temperature) is obtained by adding 273 to the temperature on the Celsius scale.

$$\text{Temperature (K)} = \text{Temperature (}^{\circ}\text{C)} + 273$$

$$273\,\text{K} = 0\,^{\circ}\text{C}$$

## THE EQUATION OF STATE FOR AN IDEAL GAS

Gases which obey Boyle's law and Charles' law are called *ideal* gases. By combining these two laws, the following equation can be obtained. It is called the *equation of state for an ideal gas*:

$$\boxed{\frac{P_1 V_1}{T_1} = \frac{P_2 V_2}{T_2}}$$

A gas has a volume of $V_1$ at a temperature $T_1$ and pressure $P_1$. If the conditions are changed to a pressure $P_2$ and a temperature $T_2$, the new volume can be calculated from the equation. It is usual to compare gas volumes at $0\,^\circ C$ and 1 atmosphere (abbreviated as 1 atm). These conditions are referred to as standard temperature and pressure (s.t.p.) or normal temperature and pressure (n.t.p.). Some authors calculate volumes at room temperature ($20\,^\circ C$) and 1 atm. The SI unit of pressure is the pascal.

$$1\,atm = 1.01 \times 10^5 \text{ pascals (Pa)} = 101 \text{ kilopascals (kPa)}$$

$$= 1.01 \times 10^5 \text{ newtons per square metre (N m}^{-2}\text{)}$$

$$= 760 \text{ mm mercury}$$

Volumes can be measured in the SI unit, the cubic metre ($m^3$) or in cubic decimetres ($dm^3$) or cubic centimetres ($cm^3$).

$$10^3\,cm^3 = 1\,dm^3 = 10^{-3}\,m^3$$

Temperatures must be in kelvins.

**EXAMPLE**    A volume of gas, 265 cm³, is collected at $70\,^\circ C$ and $1.05 \times 10^5\,N\,m^{-2}$. What volume would the gas occupy at s.t.p.?

**METHOD 1**    The experimental conditions are

$$P_1 = 1.05 \times 10^5\,N\,m^{-2}$$

$$T_1 = 273 + 70 = 343\,K$$

$$V_1 = 265\,cm^3$$

Standard conditions are

$$P_2 = 1.01 \times 10^5\,N\,m^{-2}$$

$$T_2 = 273\,K$$

$$\frac{P_1 V_1}{T_1} = \frac{P_2 V_2}{T_2}$$

$$\frac{1.05 \times 10^5 \times 265}{343} = \frac{1.01 \times 10^5 \times V_2}{273}$$

$$V_2 = 219\,cm^3$$

**ANSWER**    The volume of gas at s.t.p. would be 219 cm³.

(*Note* that the pressure and volume are in the same units on both sides of the equation.)

**METHOD 2**    Some students prefer to tackle this type of calculation in a slightly different manner.

First, consider the effect of the change in pressure.

Since the pressure decreases from $1.05 \times 10^5 \, \mathrm{N\,m^{-2}}$ to $1.01 \times 10^5 \, \mathrm{N\,m^{-2}}$, the volume will increase in the same ratio.

$$\text{New volume} = 265 \times \frac{1.05 \times 10^5}{1.01 \times 10^5}$$

Now consider the effect of the change in temperature.

Since the temperature drops from 343 to 273, the volume will decrease by the same ratio.

$$\text{New volume} = 265 \times \frac{1.05 \times 10^5}{1.01 \times 10^5} \times \frac{273}{343} = 219 \, \mathrm{cm^3} \quad \text{(as before)}.$$

## EXERCISE 20    Problems on Gas Volumes

1. Correct the following gas volumes to s.t.p.:
   a) $205 \, \mathrm{cm^3}$ at $27\,^\circ\mathrm{C}$ and $1.01 \times 10^5 \, \mathrm{N\,m^{-2}}$
   b) $355 \, \mathrm{cm^3}$ at $310 \, \mathrm{K}$ and $1.25 \times 10^5 \, \mathrm{N\,m^{-2}}$
   c) $5.60 \, \mathrm{dm^3}$ at $425 \, \mathrm{K}$ and $1.75 \times 10^5 \, \mathrm{N\,m^{-2}}$
   d) $750 \, \mathrm{cm^3}$ at $308 \, \mathrm{K}$ and $2.00 \times 10^4 \, \mathrm{N\,m^{-2}}$
   e) $1.25 \, \mathrm{dm^3}$ at $25\,^\circ\mathrm{C}$ and $2.14 \times 10^5 \, \mathrm{N\,m^{-2}}$

2. A certain mass of an ideal gas has a volume of $3.25 \, \mathrm{dm^3}$ at $25\,^\circ\mathrm{C}$ and $1.01 \times 10^5 \, \mathrm{N\,m^{-2}}$. What pressure is required to compress it to $1.88 \, \mathrm{dm^3}$ at the same temperature?

3. An ideal gas occupies a volume $2.00 \, \mathrm{dm^3}$ at $25\,^\circ\mathrm{C}$ and $1.01 \times 10^5 \, \mathrm{N\,m^{-2}}$. What will the volume of gas become at $40\,^\circ\mathrm{C}$ and $2.25 \times 10^5 \, \mathrm{N\,m^{-2}}$?

4. An ideal gas occupies $2.75 \, \mathrm{dm^3}$ at $290 \, \mathrm{K}$ and $8.70 \times 10^4 \, \mathrm{N\,m^{-2}}$. At what temperature will it occupy $3.95 \, \mathrm{dm^3}$ at $1.01 \times 10^5 \, \mathrm{N\,m^{-2}}$?

5. An ideal gas occupies $365 \, \mathrm{cm^3}$ at $298 \, \mathrm{K}$ and $1.56 \times 10^5 \, \mathrm{N\,m^{-2}}$. What will be its volume at $310 \, \mathrm{K}$ and $1.01 \times 10^5 \, \mathrm{N\,m^{-2}}$?

6. Correct the following gas volumes to s.t.p.:
   a) $256 \, \mathrm{cm^3}$ of an ideal gas measured at $50\,^\circ\mathrm{C}$ and $650 \, \mathrm{mm\,Hg}$
   b) $47.2 \, \mathrm{cm^3}$ of an ideal gas measured at $62\,^\circ\mathrm{C}$ and $726 \, \mathrm{mm\,Hg}$
   c) $10.0 \, \mathrm{dm^3}$ of an ideal gas measured at $200\,^\circ\mathrm{C}$ and $850 \, \mathrm{mm\,Hg}$
   d) $4.25 \, \mathrm{dm^3}$ of an ideal gas measured at $370\,^\circ\mathrm{C}$ and $2.12 \, \mathrm{atm}$
   e) $600 \, \mathrm{cm^3}$ of an ideal gas measured at $95\,^\circ\mathrm{C}$ and $0.98 \, \mathrm{atm}$

## GRAHAM'S LAW OF DIFFUSION

At constant temperature and pressure, the rate of diffusion of a gas is inversely proportional to the square root of its density:

$$r \propto \frac{1}{\sqrt{\rho}}$$

where $r$ = rate of diffusion and $\rho$ = density.

Comparing the rates of diffusion of two gases A and B gives

$$\frac{r_A}{r_B} = \sqrt{\frac{\rho_B}{\rho_A}}$$

This expression applies to rates of effusion (passage through a small aperture) as well as to diffusion (passage from a region of high concentration to a region of low concentration). It provides a method of measuring molar masses. The molar mass of a gas is proportional to its density (see p. 81: Density = Molar mass/Gas molar volume).

Graham's law can therefore be written as

$$\frac{r_A}{r_B} = \sqrt{\frac{M_B}{M_A}}$$

where $M_A$ and $M_B$ are the molar masses of A and B.

**EXAMPLE 1**   A gas, A, diffuses through a porous plug at a rate of $1.43 \text{ cm}^3\text{s}^{-1}$. Carbon dioxide diffuses through the plug at a rate of $0.43 \text{ cm}^3\text{s}^{-1}$. Calculate the molar mass of A.

**METHOD**   Molar mass of carbon dioxide $= 44.0 \text{ g mol}^{-1}$

$$\frac{r_{CO_2}}{r_A} = \sqrt{\frac{M_A}{M_{CO_2}}}$$

$$\frac{0.43}{1.43} = \sqrt{\frac{M_A}{44.0}}$$

$$M_A = 4.0$$

**ANSWER**   The molar mass of A is $4.0 \text{ g mol}^{-1}$.

**EXAMPLE 2**   It takes 54.4 seconds for $100 \text{ cm}^3$ of a gas, X, to effuse through an aperture, and 36.5 seconds for $100 \text{ cm}^3$ of oxygen to effuse through the same aperture. What is the molar mass of X?

**METHOD**   Since

$$\frac{r_{O_2}}{r_X} = \sqrt{\frac{M_X}{M_{O_2}}}$$

$$\frac{100/36.5}{100/54.4} = \sqrt{\frac{M_X}{32}}$$

$$M_X = 71$$

**ANSWER**   The molar mass of X is $71 \text{ g mol}^{-1}$.

## EXERCISE 21     Problems on Diffusion and Effusion

1. A certain volume of hydrogen takes 2 min 10 s to diffuse through a porous plug, and an oxide of nitrogen takes 10 min 23 s. What is:
   a) the molar mass, b) the formula of the oxide of nitrogen?

2. Plugs of cotton wool, one soaked in concentrated ammonia solution and the other soaked in concentrated hydrochloric acid, are inserted into opposite ends of a horizontal glass tube. A disc of solid ammonium chloride forms in the tube. If the tube is 1 m long, how far from the ammonia plug is the solid deposit?

3. A certain volume of sulphur dioxide diffuses through a porous plug in 10.0 min, and the same volume of a second gas takes 15.8 min. Calculate the relative molecular mass of the second gas.

4. Nickel forms a carbonyl, $Ni(CO)_n$. Deduce the value of $n$ from the fact that carbon monoxide diffuses 2.46 times faster than the carbonyl compound.

5. A certain volume of oxygen diffuses through an apparatus in 60.0 seconds. The same volumes of gases A and B, in the same apparatus under the same conditions, diffuse in 15.0 and 73.5 seconds respectively. Gas A is flammable and gas B turns starch-iodide paper blue. Identify A and B.

6. $25 \text{ cm}^3$ of ethane effuses through a small aperture in 40 s. What time is taken by $25 \text{ cm}^3$ of carbon dioxide?

7. $100 \text{ cm}^3$ of oxygen effused in 42 s through a small hole. $100 \text{ cm}^3$ of nitrogen dioxide took 60 s. Calculate the molar mass of nitrogen dioxide. How does your answer compare with the molar mass calculated from the formula? Explain the difference.

8. $200 \text{ cm}^3$ of chlorine diffuse out of a porous container in 2 min 14 s. How long will it take for the same volume of argon to diffuse?

9. Xenon diffuses through a pin-hole at a rate of $2.00 \text{ cm}^3 \text{min}^{-1}$. At what rate will hydrogen effuse through the same hole at the same temperature and pressure?

10. In 3.00 minutes, $7.50 \text{ cm}^3$ of carbon dioxide effuse through a pinhole. What volume of helium would effuse through the same hole under the same conditions in the same time?

11. In 5.00 minutes, $12.6 \text{ cm}^3$ of a gas, X, diffuse through a porous partition. In the same time, $14.8 \text{ cm}^3$ of oxygen diffuse through the same partition. Calculate the molar mass of X.

12. A mixture of carbon monoxide and carbon dioxide diffuses through a porous diaphragm in one half of the time taken for the same volume of bromine vapour. What is the composition by volume of the mixture?

13. In 4.00 minutes, $16.2 \text{ cm}^3$ of water vapour effuse through a small hole. In the same time, $8.1 \text{ cm}^3$ of a mixture of $NO_2$ and $N_2O_4$ effuse through the same hole. Calculate the percentage by volume of $NO_2$ in the mixture.

## GAY-LUSSAC'S LAW

Gay-Lussac's law states that in a reaction between gases, the volumes of the reacting gases, measured at the same temperature and pressure, are in simple ratio to one another and to the volumes of any gaseous products.

The law was stated as a result of observations such as the fact that one volume of oxygen reacts with twice its volume of hydrogen to form two volumes of steam. An explanation of Gay-Lussac's observations was provided by Avogadro.

## AVOGADRO'S LAW

Avogadro's law states that equal volumes of gases, measured at the same temperature and pressure, contain the same number of molecules.

Experimental results on reacting volumes of gases can be interpreted using Avogadro's law to give the molecular formulae of gases.

Since

$$\begin{pmatrix} 1 \text{ volume of} \\ \text{hydrogen} \end{pmatrix} + \begin{pmatrix} 1 \text{ volume of} \\ \text{chlorine} \end{pmatrix} \longrightarrow \begin{pmatrix} 2 \text{ volumes of} \\ \text{hydrogen chloride} \end{pmatrix}$$

then

$$\begin{pmatrix} 1 \text{ molecule of} \\ \text{hydrogen} \end{pmatrix} + \begin{pmatrix} 1 \text{ molecule of} \\ \text{chlorine} \end{pmatrix} \longrightarrow \begin{pmatrix} 2 \text{ molecules of} \\ \text{hydrogen chloride} \end{pmatrix}$$

The formation of one molecule of hydrogen chloride can occur if the molecules of hydrogen and chlorine consist of two atoms, so that

$$\tfrac{1}{2}(H_2) + \tfrac{1}{2}(Cl_2) \longrightarrow 1 \text{ molecule of hydrogen chloride}$$

Hydrogen chloride must therefore have the formula $HCl$.

**EXAMPLE**  Cyanogen is a compound of carbon and nitrogen. On combustion in excess oxygen, $250 \text{ cm}^3$ of cyanogen form $500 \text{ cm}^3$ of carbon dioxide and $250 \text{ cm}^3$ of nitrogen (measured at the same temperature and pressure). What is the formula of cyanogen?

**METHOD**  Let the formula of cyanogen be $C_xN_y$.

Then

$$\begin{array}{ccccccc}
C_xN_y & + & O_2 & \longrightarrow & N_2 & + & CO_2 \\
250 \text{ cm}^3 & & & & 250 \text{ cm}^3 & & 500 \text{ cm}^3 \\
1 \text{ volume} & & & & 1 \text{ volume} & & 2 \text{ volumes}
\end{array}$$

Therefore,

$$C_xN_y + nO_2 \longrightarrow N_2 + 2CO_2$$

Balancing the C atoms gives $x = 2$, and balancing the N atoms gives $y = 2$.

ANSWER    The formula is $C_2N_2$.

More examples and problems are given in the section on Reacting Volumes of Gases in Chapter 2, on p. 19 onwards.

## THE GAS MOLAR VOLUME

Avogadro's law states that equal volumes of gases, measured at the same temperature and pressure, contain equal numbers of molecules. It follows that the volume occupied by a mole of gas is the same for all gases. It is called the *gas molar volume* and measures $22.4 \, dm^3$ at s.t.p. ($24.0 \, dm^3$ at $20 \, °C$ and 1 atm).

If the volume occupied by a known mass of gas is known, the molar mass of the gas can be calculated.

EXAMPLE    $11.0 \, g$ of a gas occupy $5.60 \, dm^3$ at s.t.p. What is the molar mass of the gas?

METHOD    Mass of $5.60 \, dm^3$ of gas $= 11.0 \, g$
Mass of $22.4 \, dm^3$ of gas $= 11.0 \times 22.4/5.60 = 44.0 \, g$

ANSWER    The molar mass of the gas is $44.0 \, g \, mol^{-1}$.

## EXERCISE 22    Problems on Gas Molar Volume

Use $R = 8.314 \, J \, K^{-1} \, mol^{-1}$; GMV $= 22.41 \, dm^3$ at s.t.p.

1. Calculate the molar mass of a gas which has a density of $1.798 \, g \, dm^{-3}$ at 298 K and $101 \, kN \, m^{-2}$.

2. At 273 K and $1.01 \times 10^5 \, N \, m^{-2}$, $2.965 \, g$ of argon occupy $1.67 \, dm^3$. Calculate the molar mass of the gas.

3. Calculate the volume occupied by $0.250 \, mol$ of an ideal gas at $1.01 \times 10^5 \, N \, m^{-2}$ and $20 \, °C$.

4. A volume, $500 \, cm^3$ of krypton, measured at $0 \, °C$ and $9.8 \times 10^4 \, N \, m^{-2}$, has a mass of $1.809 \, g$. Calculate the molar mass of krypton.

5. What amount (number of moles) of an ideal gas occupies $5.80 \, dm^3$ at $2.50 \times 10^5 \, N \, m^{-2}$ and 300 K?

6. Propane has a density of $1.655 \, g \, dm^{-3}$ at 323 K and $1.01 \times 10^5 \, N \, m^{-2}$. Calculate its molar mass.

7. What volume is occupied by $0.250 \, mole$ of an ideal gas at 373 K and $1.25 \times 10^5 \, N \, m^{-2}$?

8. An ideal gas occupies $1.50\,dm^3$ at $300\,K$ and $1.25 \times 10^5\,N\,m^{-2}$. What is the amount (in moles) of gas present?

## DALTON'S LAW OF PARTIAL PRESSURES

In a mixture of gases, the total pressure is the sum of the pressures that each of the gases would exert if it alone occupied the same volume as the mixture. The contribution that each gas makes to the total pressure is called the *partial pressure.*

EXAMPLE     $3.0\,dm^3$ of carbon dioxide at a pressure of $200\,kPa$ and $1.0\,dm^3$ of nitrogen at a pressure of $300\,kPa$ are introduced into a $1.5\,dm^3$ vessel. What is the total pressure in the vessel?

METHOD     When the carbon dioxide contracts from $3.0\,dm^3$ to $1.5\,dm^3$, the pressure increases from 200 to $200 \times \dfrac{3.0}{1.5}\,kPa$, i.e. $400\,kPa$. The partial pressure of carbon dioxide in the vessel is $400\,kPa$.

When the nitrogen expands from $1.0\,dm^3$ to $1.5\,dm^3$, the pressure decreases from 300 to $300 \times 1.0/1.5 = 200\,kPa$. The partial pressure of nitrogen is $200\,kPa$.

$$\text{Total pressure} = P_{CO_2} + P_{N_2}$$
$$= 400 + 200 = 600\,kPa$$

ANSWER     The total pressure is $600\,kPa$.

## EXERCISE 23     Problems on Partial Pressures of Gases

1. Use the following values of the vapour pressure of water at various temperatures.

| Temperature | Vapour Pressure/$N\,m^{-2}$ |
|---|---|
| $15\,°C$ | $1.70 \times 10^3$ |
| $20\,°C$ | $2.33 \times 10^3$ |
| $25\,°C$ | $3.16 \times 10^3$ |
| $30\,°C$ | $4.23 \times 10^3$ |

a) $200\,cm^3$ of oxygen are collected over water at an atmospheric pressure of $9.80 \times 10^4\,N\,m^{-2}$ and a temperature of $20\,°C$. What is the partial pressure of the oxygen? What will be its volume at s.t.p.?

b) $250\,cm^3$ of gas are collected over water at an atmospheric pressure of $9.70 \times 10^4\,N\,m^{-2}$ and a temperature of $30\,°C$. What is the partial pressure of the gas? Correct its volume to s.t.p.

c) What is the volume of 1 mole of nitrogen measured over water at an atmospheric pressure of $9.70 \times 10^4\,N\,m^{-2}$ and a temperature of $25\,°C$?

2. $2.00\,dm^3$ of nitrogen at a pressure of $1.01 \times 10^5\,N\,m^{-2}$ and $5.00\,dm^3$ of hydrogen at a pressure of $5.05 \times 10^5\,N\,m^{-2}$ are injected into a $10.0\,dm^3$ vessel. What is the pressure of the mixture of gases?

3. A mixture of gases at a pressure of $1.01 \times 10^5\,N\,m^{-2}$ contains 25.0% by volume of oxygen. What is the partial pressure of oxygen in the mixture?

4. Into a $5.00\,dm^3$ vessel are introduced $2.50\,dm^3$ of methane at a pressure of $1.01 \times 10^5\,N\,m^{-2}$, $7.50\,dm^3$ of ethane at a pressure of $2.525 \times 10^5\,N\,m^{-2}$ and $0.500\,dm^3$ of propane at a pressure of $2.02 \times 10^5\,N\,m^{-2}$. What is the resulting pressure of the mixture?

5. A mixture of gases at a pressure $7.50 \times 10^4\,N\,m^{-2}$ has the volume composition 40% $N_2$; 35% $O_2$; 25% $CO_2$.
   a) What is the partial pressure of each gas?
   b) What will the partial pressures of nitrogen and oxygen be if the carbon dioxide is removed by the introduction of some sodium hydroxide pellets?

6. A mixture of gases at $1.50 \times 10^5\,N\,m^{-2}$ has the composition 40% $NH_3$; 25% $H_2$; 35% $N_2$ by volume.
   a) What is the partial pressure of each gas?
   b) What will the partial pressures of the other gases become if the ammonia is removed by the addition of some solid phosphorus(V) oxide?

# HENRY'S LAW

Henry's law is concerned with the solubility of gases in liquids. It states that the mass of gas dissolved at constant temperature per unit volume of solvent is directly proportional to the partial pressure of the gas. This law may be expressed as

$$m_s = kp$$

where $m_s$ is the mass of gas dissolved, $p$ is its partial pressure, and $k$ is a constant.

The *solubility* of a gas is the volume that will dissolve in unit volume of the solvent under stated conditions of temperature and pressure.

The *absorption coefficient* is the volume of gas (at s.t.p.) that will dissolve in unit volume of liquid at a stated temperature, under a pressure of 1 atmosphere.

Henry's law does not apply to gases which combine chemically with the solvent, for example, hydrogen chloride in solution in water. It applies only to solution as a physical process.

**EXAMPLE**    Taking the composition by volume of air as 80% nitrogen, 20% oxygen, calculate the volume composition of the air which is dissolved in water at 298 K. The absorption coefficients at 298 K are: nitrogen, 0.016; oxygen, 0.031.

**METHOD**    Air is composed of 80% by volume of nitrogen and 20% by volume of oxygen.

The mole fractions are therefore: nitrogen, 0.80; oxygen, 0.20.

The partial pressure of each gas is proportional to its mole fraction.

The absorption coefficient must be multiplied by the mole fraction to give the absorption of the gas at its partial pressure.

Vol. of $N_2$ at s.t.p. dissolving in 100 $cm^2$ water

$$= 100 \times 0.016 \times 0.80 = 1.28 \ cm^3$$

Vol. of $O_2$ at s.t.p. dissolving in 100 $cm^3$ water

$$= 100 \times 0.031 \times 0.20 = 0.62 \ cm^3$$

**ANSWER**    The dissolved air has the volume composition 1.28 $cm^3 \ N_2$, 0.62 $cm^3 \ O_2$

$$= 67\% \ nitrogen: \ 33\% \ oxygen$$

## EXERCISE 24    Problems on Solubility of Gases

Use the following room-temperature coefficients of solubility:

$N_2$ = 0.0150        CO = 0.0380        $O_2$ = 0.0300
$Cl_2$ = 2.26          $CO_2$ = 1.00

1. If 1 $dm^3$ of water is shaken with a gaseous mixture of 75% $N_2$, 20% $O_2$, 5% $CO_2$, calculate a) the volume of the dissolved gas, and b) its percentage by volume composition.

2. A mixture consisting of 79.0% $N_2$, 20.9% $O_2$ and 0.030% $CO_2$ was shaken with 1 $dm^3$ of water. Calculate: a) the volume of gas dissolved, and b) the percentage composition by volume.

3. A mixture of CO and $CO_2$ which has a 50:50 by volume composition is shaken with 1 $dm^3$ of water. What is the percentage by volume composition of the dissolved gas?

4. A mixture of 20% chlorine and 80% by volume of oxygen is shaken with water. What is the composition of the dissolved gas?

## THE IDEAL GAS EQUATION

Gases which obey Boyle's law and Charles' law are called *ideal gases*. Combining these two laws gives the equation:

$$\frac{P \times V}{T} = \text{Constant for a given mass of gas}$$

It follows from Avogadro's law that, if a mole of gas is considered, the constant will be the same for all gases. It is called the universal gas constant, and given the symbol $R$, so that the equation becomes

$$PV = RT$$

This equation is called the *ideal gas equation*. For $n$ moles of gas, the equation becomes

$$\boxed{PV = nRT}$$

The value of the constant $R$ can be calculated. Consider 1 mole of gas at s.t.p. Its volume is $22.414\,dm^3$. Inserting values of $P$, $V$ and $T$ in SI units into the ideal gas equation.

$P = 1.0132 \times 10^5\,N\,m^{-2}$     $T = 273.15\,K$
$V = 22.414 \times 10^{-3}\,m^3$     $n = 1\,mol$

gives     $1.0132 \times 10^5 \times 22.414 \times 10^{-3} = 1 \times 273.15 \times R$

$$R = 8.314$$

The units of $R$ are $PV/nT$, i.e.

$$\frac{N\,m^{-2}\,m^3}{mol\,K} = N\,m\,mol^{-1}K^{-1} = J\,K^{-1}mol^{-1}$$

Thus, $R = 8.314\,J\,K^{-1}\,mol^{-1}$ (joules per kelvin per mole).

## THE KINETIC THEORY OF GASES

To explain the gas laws, the kinetic theory of gases was put forward. The kinetic theory considers that the molecules of gas are in constant motion in straight lines. The pressure which the gas exerts results from the bombardment of the walls of the container by the molecules.

The kinetic energy of a molecule $= \frac{1}{2}mc^2$ ($m$ = mass, $c$ = velocity).

The kinetic energy of the gas $= \frac{1}{2}mN\overline{c^2}$ ($N$ = number of molecules, $\overline{c^2}$ = average value of the square of the velocity for all the molecules; $\sqrt{\overline{c^2}}$ = root mean square velocity).

From the kinetic theory can be derived the equation

$$PV = \frac{1}{3}mN\overline{c^2}$$

Since the kinetic energy of the molecules is proportional to $T$ (kelvins)

$$PV = \text{Constant} \times T$$

This is the ideal gas equation. The agreement between theory and experimental results is good support for the kinetic theory.

The kinetic theory can be used to calculate the root mean square velocity of gas molecules.

**EXAMPLE**   Calculate the root mean square velocity of hydrogen molecules at s.t.p.

**METHOD 1**   Use $M(H_2) = 2.02 \text{ g mol}^{-1}$. In the equation $PV = \frac{1}{3}mN\overline{c^2}$, substitute $PV = RT$ for 1 mole of gas, and $mN = M$, the molar mass of gas in kg. Substituting $mN = 2.02 \times 10^{-3} \text{ kg mol}^{-1}$ in $\frac{1}{3}mN\overline{c^2} = RT$, gives

$$\overline{c^2} = 3 \times 8.31 \times 273/(2.02 \times 10^{-3})$$

$$\sqrt{\overline{c^2}} = 1.84 \times 10^3 \text{ m s}^{-1}$$

**ANSWER**   The root mean square velocity of hydrogen molecules at s.t.p. is $1.84 \times 10^3 \text{ m s}^{-1}$.

**METHOD 2**   Use the density of hydrogen $(9.00 \times 10^{-2} \text{ kg m}^{-3}$ at s.t.p.). Since $mN/V = \rho$, the density of the gas, substituting in $P = \frac{1}{3}\rho\overline{c^2}$ gives

$$\sqrt{\overline{c^2}} = \sqrt{\frac{3 \times 1.01 \times 10^5}{9.00 \times 10^{-2}}} = 1.84 \times 10^3 \text{ m s}^{-1}$$

**ANSWER**   As before, the root mean square velocity is $1.84 \times 10^3 \text{ m s}^{-1}$.

## EXERCISE 25    Problems on the Kinetic Theory and the Ideal Gas Equation

Use $R = 8.314 \text{ J K}^{-1} \text{mol}^{-1}$.

1. Krypton has a density of $3.44 \text{ g dm}^{-3}$ at $25\,^{\circ}C$ and $1.01 \times 10^5 \text{ N m}^{-2}$. Calculate its molar mass.

2. The density of hydrogen at 273 K and $1.01 \times 10^5 \text{ N m}^{-2}$ is $8.96 \times 10^{-2} \text{ g dm}^{-3}$. Calculate the root mean square velocity of the hydrogen molecules under these conditions.

3. Using the equation $PV = \frac{1}{3}mN\overline{c^2}$ calculate the kinetic energy of the molecules in one mole of an ideal gas at $0\,^{\circ}C$.

4. Calculate the root mean square velocity for argon at s.t.p. $(M_r(Ar) = 40.0)$.

5. A volume of $1.00 \text{ dm}^3$ is occupied by 1.798 g of a gas at 298 K and 101 kPa. Calculate the molar mass of the gas.

6. Calculate the ratio of the root mean square velocities of oxygen and xenon molecules at $27\,^{\circ}C$. $(A_r(O) = 16.0, A_r(Xe) = 131.)$

7. Calculate the root mean square velocity of hydrogen iodide molecules at $27\,^{\circ}C$. $(A_r(I) = 127.)$

8. a) Calculate the ratio of the root mean square velocity of hydrogen molecules to the root mean square velocity of argon molecules at the same temperature.

   b) At what temperature will argon molecules have the same root mean square velocity as hydrogen molecules at $0\,^{\circ}C$? $(A_r(Ar) = 40.0.)$

**EXERCISE 26**    Questions from A-level Papers

1. a) State Graham's law.

   b) Two gases X and Y effuse separately from a container under identical conditions of constant temperature and constant pressure difference. The total volumes effused (measured at s.t.p.) were recorded and the results are plotted in Fig. 5.1.

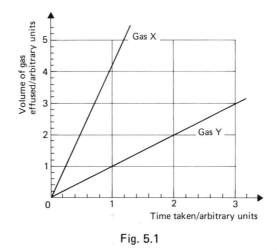

Fig. 5.1

   i) State, giving reasons, which gas has the larger relative molecular mass.

   ii) Given that gas X is deuterium, $^2_1H_2$, deduce the nature of *element* Y.

   c) Without carrying out any further calculations, show qualitatively on a tracing of Fig. 5.2 the form of the graphs you would expect to obtain if, under the same conditions as in b), effusion experiments were carried out using i) hydrogen, $^1_1H_2$, ii) an equimolecular mixture of gases X and Y. Label the graphs 'H₂' and 'X + Y' respectively.

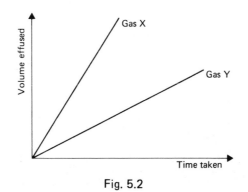

Fig. 5.2                                              (C80)

2. An evacuated glass bulb of volume $1\,dm^3$ was weighed. It was then filled with hydrogen at a pressure of one atmosphere ($1 \times 10^5\,N\,m^{-2}$) and temperature of 293 K, and re-weighed. What change of mass was observed? (H $= 1.008$; molar volume at s.t.p. $= 22.4\,dm^3\,mol^{-1}$.)

   a  decrease of 0.0839 g          b  no change
   c  increase of 0.0419 g          d  increase of 0.0839 g
   e  increase of 0.0900 g

(NI81)

3. The time taken for a certain volume of methane (relative molecular mass 16) to effuse through an orifice was 60 s. The time, in s, taken for an equal volume of sulphur dioxide (relative molecular mass 64) to effuse through the same orifice under the same conditions of temperature and pressure would be

   a  15                b  30                c  60
   d  120               e  240

(AEB83)

4. a) Give a simple description of the metallic bond. Describe *four* physical properties of metals, relating them as far as possible to your description of the bonding.

   b) Describe *four* characteristic chemical properties of the transition metals, giving *one* example of *each* property.

   c) When 0.0958 g of a metal X was heated in chlorine, 0.3794 g of its higher chloride $XCl_n$ was formed, the vapour of this chloride occupying a volume of $82.07\,cm^3$ at 1 atm pressure and 500 K. Calculate
      i) the number of moles of the chloride formed,
      ii) the relative atomic mass of the metal, and
      iii) the formula of the chloride.

   Indicate where in the Periodic Table you would expect X to be found.

   (1 atm $= 101325\,Nm^{-2}$; $1\,cm^3 = 10^{-6}\,m^3$; $R = 8.314\,J\,mol^{-1}\,K^{-1}$; $A_r(Cl) = 35.45$.)

(WJEC85)

5. a) Discuss briefly, but critically, *three* basic postulates underlying the kinetic theory of gases and explain the use of $\overline{c^2}$, the mean square velocity, in the equation

$$pV = \tfrac{1}{3}mN\overline{c^2}$$

   where $N$ molecules of ideal (perfect) gas, each of mass $m$, occupy a volume $V$ at a pressure $p$.

   b) By sketching graphs of
      i) $pV$ plotted against $p$,
      ii) $p$ against $V$,
      for an ideal gas and a real gas (of your own choice) over a suitable temperature range, relate what is seen to the earlier discussion.

c) Calculate:
   i) in terms of $R/J\,mol^{-1}K^{-1}$, the universal gas constant, the kinetic energy $/J\,mol^{-1}$ of the molecules in one mole of ideal gas at 27 °C;
   ii) in terms of $R/J\,mol^{-1}K^{-1}$, the universal gas constant, the *root* mean square velocity $/m\,s^{-1}$ of sulphur dioxide molecules at 27 °C in the gaseous phase, assuming ideal behaviour;
   iii) the ratio of the *root* mean square velocities of methane and sulphur dioxide, assuming ideal behaviour, at 27 °C;
   iv) the time expected to be taken for a given volume of methane at 27 °C to effuse (or diffuse) through a pin-hole if the same volume of sulphur dioxide under the same conditions takes 80 s. Relate what is calculated to the appropriate gas law.
   (H = 1, C = 12, O = 16, S = 32; 0 °C = 273 K.)

(SUJB83)

6. 33 cm³ of chlorine reacted with excess aqueous ammonia according to the equation

$$3Cl_2(g) + 8NH_3(aq) \longrightarrow N_2(g) + 6NH_4Cl(aq)$$

The volume of nitrogen gas, in cm³, formed (measured at the original temperature and pressure) was

a 11                    b 33                    c 99
d 198                   e 264                                   (AEB83)

7. a) Give *two* assumptions made in the kinetic theory about the behaviour of ideal gas molecules.

   b) 0.71 g of a gas at 227 °C when contained in a vessel of volume 0.821 litre exerted a pressure of 0.5 atm. Calculate the relative molecular mass of the gas.

   c) Suppose you had a sample of this gas in a sealed vessel at 20 °C and the pressure in the vessel was 0.8 atm. If the temperature was raised to 200 °C what would be the pressure in the vessel?

   d) If the root mean square speed of the molecules of a gas is $100\,m\,s^{-1}$ at 0 °C, what will be their root mean square speed at 200 °C?

(JMB85)

8. a) State
      i) Dalton's law of partial pressures,
      ii) Avogadro's law.

   b) A mixture of gases $X$, consisted of 20 cm³ methane, 40 cm³ ethene and 90 cm³ butane. This mixture was exploded with 850 cm³ oxygen and the products were cooled to room temperature. All volumes were measured at room temperature and a pressure of 101.3 kPa.
      i) Calculate the partial pressure of each gas in the mixture, $X$.
      ii) How could the presence of ethene be detected in the mixture?

iii) Calculate the volume of oxygen needed for the complete combustion of mixture $X$.

iv) Calculate the partial pressure of carbon dioxide in the final mixture.                                                  (AEB82,p)

*9. The following data refer to two of the chlorides, $A$ and $B$, of a transition element $X$.

a) 0.500 g of chloride $A$ occupies a volume of 100 cm³ at 177 °C and 740 mm Hg. Calculate the relative molecular mass of chloride $A$.

b) Treatment of 0.660 g of chloride $A$ with silver nitrate under suitable conditions gave 2.00 g of silver chloride. Calculate the percentage of chlorine in $A$ and hence the relative atomic mass of $X$.

c) Reduction of chloride $A$ to chloride $B$ gave a solution in which the concentration of $X$ was 6.00 g dm⁻³. 25.0 cm³ of this solution were treated with an excess of iron(III) sulphate and the mixture titrated with 0.100 M cerium(IV) sulphate, of which 31.3 cm³ were required.

Cerium(IV) and iron(II) ions react together as follows:

$$Ce^{4+} + Fe^{2+} \longrightarrow Ce^{3+} + Fe^{3+}$$

Determine the formulae of the chlorides $A$ and $B$.

(Molar volume of a gas at s.t.p. is 22.4 dm³. Relative atomic masses: Ag = 108, Cl = 35.5.)                                       (L83,S)

10. Using the data given, explain the following information concerning carbon monoxide and carbon dioxide.

a) 20.0 cm³ of carbon monoxide diffused through a porous partition in the same time as 15.95 cm³ of carbon dioxide at the same temperature and pressure.

b) Carbon dioxide can be liquefied at a room temperature of 298 K whereas carbon monoxide cannot.

c) 250 cm³ of carbon dioxide at a temperature of 27 °C and under a total pressure of 202.6 kPa contains the same number of molecules as 0.568 g of carbon monoxide.                                      (AEB81,p)

# 6 Liquids

## DETERMINATION OF MOLAR MASS

### The gas syringe method

The gas syringe method can be used to find the molar mass of a liquid with a low boiling point. A small weighed quantity of liquid is injected into a gas syringe. The volume of vapour formed is measured, and its temperature and pressure are noted. From the values of mass and volume, the molar mass can be calculated.

EXAMPLE 1    A gas syringe contains $18.4 \, cm^3$ of air at $57 \, °C$. $0.187 \, g$ of a volatile liquid is injected into the syringe. The volume of gas in the syringe is then $54.6 \, cm^3$ at $57 \, °C$ and $1.01 \times 10^5 \, Pa$. Calculate the molar mass of the liquid.

METHOD    Using the values

$$P = 1.01 \times 10^5 \, Pa$$
$$V = 36.2 \, cm^3 = 36.2 \times 10^{-6} \, m^3$$
$$T = 273 + 57 = 330 \, K$$
$$R = 8.314 \, J \, K^{-1} mol^{-1}$$

in the equation $PV = \dfrac{m}{M} RT$ gives

$$1.01 \times 10^5 \times 36.2 \times 10^{-6} = \frac{0.187}{M} \times 8.314 \times 330$$

$$M = 140$$

ANSWER    The molar mass is $140 \, g \, mol^{-1}$.

The values of molar mass obtained by this method are not very accurate. A knowledge of the empirical formula enables the value to be corrected. For example, if the compound has the empirical formula $CH_2O$ and an experimental value of $57 \, g \, mol^{-1}$ for the molar mass, one can see that $C_2H_4O_2$ is the molecular formula, and $60 \, g \, mol^{-1}$ is the correct molar mass.

## ANOMALOUS RESULTS FROM MEASUREMENTS OF MOLAR MASS

Sometimes, an unexpectedly low result for molar mass is obtained. This happens when the molecules of the vapour on which measurements are being made dissociate, causing an increase in the actual number of particles present. If 1 mole of molecules of XY dissociate partially into X and Y, and $\alpha$ is the degree of dissociation, then

Species:                    XY $\rightleftharpoons$  X + Y

Number of moles:  $(1 - \alpha)$              $\alpha$    $\alpha$    Total = $(1 + \alpha)$

$(1 - \alpha)$ moles of XY remain, and $\alpha$ moles of X and $\alpha$ moles of Y are formed.

Thus          $\dfrac{\text{Actual number of moles}}{\text{Expected number of moles}} = \dfrac{1 + \alpha}{1}$

Since the volume occupied by a gas is proportional to the number of moles of gas,

$$\frac{\text{Actual volume of gas}}{\text{Expected volume of gas}} = \frac{1 + \alpha}{1}$$

Since we are finding molar mass from the equation, given on p. 80,

$$PV = nRT = \frac{m}{M}RT$$

where $m$ = mass of substance, and $M$ = its molar mass, if the volume, $V$, is greater than expected, $M$, the molar mass, is less than expected.

Thus

$$\frac{\text{Actual volume}}{\text{Expected volume}} = \frac{\text{Molar mass calculated from formula}}{\text{Measured molar mass}} = \frac{1 + \alpha}{1}$$

If the volume is kept constant, the pressure increases instead of the gas expanding and

$$\frac{\text{Calculated molar mass}}{\text{Measured molar mass}} = \frac{\text{Measured pressure}}{\text{Calculated pressure}} = \frac{1 + \alpha}{1}$$

If one molecule dissociates into $n$ particles, the expression becomes:

$$\frac{\text{Calculated molar mass}}{\text{Measured molar mass}} = \frac{\text{Measured pressure}}{\text{Calculated pressure}} = \frac{1 + (n - 1)\alpha}{1}$$

**EXAMPLE 1**  The molar mass of phosphorus(V) chloride at 140 °C is 166. Calculate the degree of dissociation.

**METHOD**  The molar mass of $PCl_5$ = $31.0 + (5 \times 31.5)$ = $208.5\,\text{g mol}^{-1}$

The dissociation which occurs is

$$PCl_5(g) \rightleftharpoons PCl_3(g) + Cl_2(g)$$

$$\frac{\text{Calculated molar mass}}{\text{Measured molar mass}} = \frac{1 + \alpha}{1}$$

Thus, $n = 2$, and

$$\frac{\text{Calculated molar mass}}{\text{Measured molar mass}} = \frac{208.5}{166} = \frac{1 + \alpha}{1}$$

Therefore $\alpha = 0.26$ (26%).

**ANSWER**  The degree of dissociation is 0.26.

**EXAMPLE 2**   When 1.00 g of iodine is heated at 1200°C in a 500 cm³ vessel a pressure of $1.51 \times 10^2$ kPa develops. Calculate the degree of dissociation.

**METHOD**   Since

$$PV = nRT,$$

$$P = \frac{1.00}{254} \times \frac{8.314 \times 1473}{500 \times 10^{-6}}$$

$$= 9.64 \times 10^4 \, \text{Pa}$$

$$\frac{\text{Observed pressure}}{\text{Calculated pressure}} = \frac{1.51 \times 10^5}{9.64 \times 10^4} = \frac{1 + (n-1)\alpha}{1}$$

Since the dissociation is

$$I_2(g) \rightleftharpoons 2I(g)$$

$$n = 2 \quad \text{and} \quad \frac{1.51 \times 10^5}{9.60 \times 10^4} = 1 + \alpha$$

Solving this equation gives $\alpha = 0.58$ (or 58%).

**ANSWER**   The degree of dissociation is 0.58.

A measurement of molar mass higher than the value calculated from the formula is a sign that molecules are associated. In 1 mole of A, if 2 molecules of A form a dimer, and if the degree of dimerisation is $\alpha$,

Species:           $2A \rightleftharpoons A_2$

No. of moles:  $(1 - \alpha)$           $\alpha/2$     Total $= (1 - \alpha/2)$

$$\frac{\text{Actual no. of moles}}{\text{Expected no. of moles}} = \frac{1 - \alpha/2}{1}$$

$$\frac{\text{Actual volume}}{\text{Calculated volume}} = 1 - \alpha/2$$

$$\frac{\text{Calculated molar mass}}{\text{Measured molar mass}} = 1 - \alpha/2$$

In general, if $n$ molecules associate,

$$\frac{\text{Calculated molar mass}}{\text{Measured molar mass}} = 1 - \frac{(n-1)\alpha}{n}$$

**EXAMPLE 3**   A value of 200 is obtained for the molar mass of aluminium chloride. Calculate the degree of dimerisation of aluminium chloride at the temperature at which the measurement was made.

**METHOD**

$$\text{Calculated molar mass} = 133.5 \, \text{g mol}^{-1}$$

$$\frac{\text{Calculated molar mass}}{\text{Measured molar mass}} = 1 - \frac{\alpha}{2}$$

$$133.5/200 = 1 - \frac{\alpha}{2}$$

$$\alpha = 0.67$$

**ANSWER**    The degree of association is 0.67.

## EXERCISE 27    Problems on Molar Masses of Volatile Liquids

The gas constant, $R = 8.314 \, \text{J K}^{-1} \, \text{mol}^{-1}$.

$1 \, \text{atm} = 1.01 \times 10^5 \, \text{N m}^{-2} = 1.01 \times 10^5 \, \text{Pa} = 101 \, \text{kPa}$.

1. Calculate the molar mass of a liquid B, given that 0.850 g of B produced 55.5 cm³ of vapour (corrected to s.t.p.).

2. A compound of phosphorus and fluorine contains 24.6% by mass of phosphorus. 1.000 g of this compound has a volume of 178 cm³ at s.t.p. Deduce the molecular formula of the compound.

3. 0.110 g of a liquid produced 42.0 cm³ of vapour, measured at 147 °C and $1.01 \times 10^5 \, \text{N m}^{-2}$. What is the molar mass of the liquid?

4. 0.228 g of liquid was injected into a gas syringe. The volume of vapour formed was 84.0 cm³ at 17 °C and $1.01 \times 10^5 \, \text{N m}^{-2}$. Calculate the molar mass of the substance.

5. 0.452 g of a volatile solid displaced 82.0 cm³ of air, collected at 20 °C and $1.023 \times 10^5 \, \text{N m}^{-2}$. If the saturated vapour pressure of water at 20 °C is $2.39 \times 10^3 \, \text{N m}^{-2}$, calculate the molar mass of the solid.

6. Fig. 6.1 shows the results of gas syringe measurements on ethanol ($\circ$), propanone ($\square$) and ethoxyethane ($\bullet$), all at 80 °C and 1 atm. For each liquid, several measurements of the volume of vapour formed after the injection of a known mass of liquid were made.

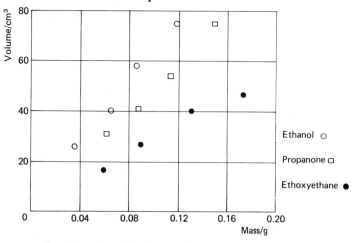

Fig. 6.1    Results of gas syringe measurements

Trace the results on to a piece of paper, and plot the best straight line through the points for each vapour. Find the slope of each line.

The reciprocal of the slope, mass/volume, is the density of the vapour, $\rho$. From the equation

$$PV = \frac{m}{M} RT$$

$$\left( \text{since} \quad \frac{m}{V} = \rho \right)$$

$$M = \rho \frac{RT}{P}$$

a) Insert the value you have obtained for the density of ethanol vapour into the equation, and find the molar mass of ethanol. Do the same for b) propanone, and c) ethoxyethane.

## EXERCISE 28    Problems on Association and Dissociation

1. 20.85 g of phosphorus(V) chloride are allowed to vaporise in a 5.00 dm³ vessel at 175 °C. A pressure of $1.04 \times 10^5 \, \mathrm{N \, m^{-2}}$ develops. Calculate the degree of dissociation of $PCl_5$ into $PCl_3$ and $Cl_2$.

2. 10.32 g of aluminium chloride are allowed to vaporise in a 1.00 dm³ vessel at 80 °C. A pressure of $1.70 \times 10^5 \, \mathrm{N \, m^{-2}}$ develops. What is the degree of association of $AlCl_3$ into $Al_2Cl_6$ molecules?

3. Nitrogen dioxide exists in an equilibrium mixture:

$$N_2O_4(g) \rightleftharpoons 2NO_2(g)$$

The relative molar mass of nitrogen dioxide at 25 °C is 80.0. What percentage of the molecules in the mixture is $N_2O_4$?

4. A sample of iodine of mass 25.4 g is vaporised in a 2.00 dm³ vessel at 800 K. A pressure of $4.32 \times 10^5 \, \mathrm{N \, m^{-2}}$ develops. Calculate the degree of dissociation of iodine molecules into atoms.

5. The molar mass of iron(III) chloride measured at 900 K is 246 g mol⁻¹. Calculate the degree of dimerisation of $FeCl_3$ molecules.

## VAPOUR PRESSURE

In a liquid, the molecules are in constant motion. Some molecules, those with energy considerably above average, will have enough energy to escape from the liquid into the vapour state. If a liquid is introduced into a closed container, some of the liquid will evaporate. The molecules in the vapour state will exert a pressure. When equilibrium is reached between the liquid state and the vapour state, the pressure exerted by the vapour is called the *vapour pressure* of the liquid. To be correct, one should call it the *saturated vapour pressure* or the *equilibrium vapour pressure*. The magnitude of the vapour pressure depends on the identity of the liquid and on the temperature: it does not depend on the amount of liquid present.

**EXAMPLE**    The saturated vapour pressure of water at $65°C$ is $25.05\,kN\,m^{-2}$. What mass of water will be present in the vapour phase if $10.0\,cm^3$ of water are injected into a $1.000\,dm^3$ vessel?

**METHOD**    Use the ideal gas equation, $PV = nRT$, and substitute

$$P = 25.05 \times 10^3\,N\,m^{-2} \qquad R = 8.314\,J\,K^{-1}\,mol^{-1}$$

$$T = 338\,K \qquad V = 1.000\,dm^3 = 1.000 \times 10^{-3}\,m^3$$

giving    $25.05 \times 10^{-3} \times 1.000 \times 10^{-3} = n \times 8.314 \times 338$

Amount (mol) of water,    $n = 8.92 \times 10^{-3}\,mol$

Mass of water    $= 18.0 \times 8.92 \times 10^{-3} = 0.161\,g$

**ANSWER**    The mass of water that evaporates is $0.161\,g$.

## EXERCISE 29    Problems on Vapour Pressures of Liquids

1. $10.0\,cm^3$ of ethyl ethanoate are introduced into an evacuated $10.0\,dm^3$ vessel at $25\,°C$. What mass of ethyl ethanoate will vaporise? The saturated vapour pressure of ethyl ethanoate at $25\,°C$ is $9.55 \times 10^3\,N\,m^{-2}$.

2. At $95\,°C$, the saturated vapour pressure of bromobenzene is $1.54 \times 10^4\,N\,m^{-2}$. What mass of bromobenzene will vaporise when a small amount of liquid bromobenzene is introduced into a $2.50\,dm^3$ flask at $95\,°C$?

3. At $0\,°C$, the saturated vapour pressure of water is $6.10 \times 10^2\,N\,m^{-2}$. How many molecules of water vapour will be present in each $cm^3$ of air in a vessel containing ice at $0\,°C$?

4. If analysis shows that $0.0230\,g$ of water are present in $1.00\,dm^3$ of air at $25\,°C$, what is the saturated vapour pressure of water at $25\,°C$?

## SOLUTIONS OF LIQUIDS IN LIQUIDS

How do you express the composition of a liquid–liquid mixture? One way is by stating the mole fraction of each constituent:

$$\text{Mole fraction of A in A–B mixture} = \frac{\text{No. of moles of A}}{\text{Total no. of moles}}$$

$$= \frac{n_A}{n_A + n_B}$$

The vapour above a mixture of the liquids A and B will contain both A and B.

*Raoult's law* states that the saturated vapour pressure of each component in the mixture is equal to the product of the mole fraction of that component and the saturated vapour pressure of that component when pure, at the same temperature.

If $\qquad p_A$ = vapour pressure of A,

$\qquad\qquad p_A^0$ = saturated vapour pressure of pure A,

and $\qquad x_A$ = mole fraction of A, then

$$p_A = x_A \times p_A^0 \qquad p_B = x_B \times p_B^0$$

Raoult's law is obeyed by mixtures of similar compounds. They are said to form *ideal solutions*. The vapour above a mixture of liquids does not have the same composition as the mixture. It is richer in the more volatile component. The mole fractions of A and B in the vapour phase are in the ratio of their mole fractions in the liquid phase multiplied by the ratio of the saturated vapour pressures of the two liquids. If $x_A'$ and $x_B'$ are the mole fractions of A and B in the vapour phase,

$$\frac{x_A'}{x_B'} = \frac{x_A}{x_B} \times \frac{p_A^0}{p_B^0}$$

**EXAMPLE 1** Calculate the vapour pressure of a solution containing 50.0 g heptane and 38.0 g octane at 20 °C. The vapour pressures of the pure liquids at 20 °C are heptane 473 Pa; octane 140 Pa.

**METHOD** Amount (mol) of heptane = 50/100 = 0.50 mol

Amount (mol) of octane = 38/114 = 0.33 mol

Mole fraction of heptane = 0.50/0.83

Mole fraction of octane = 0.33/0.83

$$p\,(\text{heptane}) = p^0(\text{heptane}) \times x(\text{heptane})$$
$$= 473 \times 0.50/0.83 = 284.0$$
$$p\,(\text{octane}) = p^0(\text{octane}) \times x(\text{octane})$$
$$= 140 \times 0.33/0.83 = 55.9$$

**ANSWER** Total vapour pressure = 284.0 + 55.9 = 340 Pa.

**EXAMPLE 2** Two pure liquids A and B have vapour pressures $1.50 \times 10^4\,\text{N m}^{-2}$ and $3.50 \times 10^4\,\text{N m}^{-2}$ at 20 °C. If a mixture of A and B obeys Raoult's law, calculate the mole fraction of A in a mixture of A and B which has a total vapour pressure of $2.90 \times 10^4\,\text{N m}^{-2}$ at 20 °C.

**METHOD** If $n_A$ is the mole fraction of A, $(1 - n_A)$ is the mole fraction of B.

Then, $(n_A \times 1.50 \times 10^4) + (1 - n_A)(3.50 \times 10^4) = 2.90 \times 10^4$

$$1.50 n_A + 3.50 - 3.50 n_A = 2.90$$
$$2.00 n_A = 0.60$$
$$n_A = 0.30$$

**ANSWER** The mole fraction of A is 0.30.

**EXERCISE 30**    Problems on Vapour Pressures of Solutions of Two Liquids

1. Two pure liquids A and B have vapour pressures of respectively 17000 and 35000 N m$^{-2}$ at 25 °C. An equimolar mixture of A and B has a vapour pressure of 26 000 N m$^{-2}$ at 25 °C. Calculate the vapour pressure of a mixture containing four moles of A and one mole of B at 25 °C.

2. Hexane and heptane are totally miscible and form an ideal two component system. If the vapour pressures of the pure liquids are 56000 and 24000 N m$^{-2}$ at 50 °C calculate:

   a) the total vapour pressure, and

   b) the mole fraction of heptane in the vapour above an equimolar mixture of hexane and heptane.

3. The vapour pressure of water at 298 K is 3.19 × 10$^3$ Pa. What are the partial vapour pressures of water in mixtures of:

   a) 27 g water and 69 g ethanol

   b) 9.0 g of water and 92 g of ethanol

   at this temperature?

4. A and B are two miscible liquids which form an ideal solution. The vapour pressures at 20 °C are: A, 40 kPa, B, 32 kPa. Calculate the total pressure of the vapour in equilibrium with mixtures of:

   a) 3 moles of A and 1 mole of B at 20 °C

   b) 1 mole of A and 4 moles of B at 20 °C.

## IMMISCIBLE LIQUIDS: SUM OF VAPOUR PRESSURES

### Steam distillation

In a system of immiscible liquids, each liquid exerts its own vapour pressure independently of the other. The vapour pressure of the system is equal to the sum of the vapour pressures of the pure components. This is the basis for steam distillation. Phenylamine will distil over in steam at 98 °C, although its boiling point is 184 °C. At 98 °C, the sum of the vapour pressures of phenylamine and water is equal to atmospheric pressure. The ratio of the amounts of the two liquids in the distillate is equal to the ratio of their vapour pressures:

$$\frac{n_A}{n_W} = \frac{p_A}{p_W}$$

where $n_A$ and $n_W$ are the amounts of phenylamine and water in the distillate, and $p_A$ and $p_W$ are the vapour pressures of phenylamine and water at 98 °C.

Since $n = m/M$ (where $m$ = mass, $M$ = molar mass)

$$\frac{m_A}{M_A} \times \frac{M_W}{m_W} = \frac{p_A}{p_W}$$

This equation can be used to find $m_A/m_W$, the ratio of masses of amine and water in the distillate. Steam distillation has been used as a method of determining molar masses. In this case, the masses of the liquid and water in the distillate must be measured and inserted into the equation to give the unknown molar mass.

**EXAMPLE 1**  Bromobenzene distils in steam at 95 °C. The vapour pressures of bromobenzene and water at 95 °C are $1.59 \times 10^4\,\mathrm{N\,m^{-2}}$ and $8.50 \times 10^4\,\mathrm{N\,m^{-2}}$. Calculate the percentage by mass of bromobenzene in the distillate.

**METHOD**  Let the percentage of bromobenzene $= y$.

In the equation $\quad \dfrac{n_{C_6H_5Br}}{n_{H_2O}} = \dfrac{p_{C_6H_5Br}}{p_{H_2O}}$

$$\frac{y/157}{(100-y)/18} = \frac{1.59 \times 10^4}{8.50 \times 10^4}$$

$$y = 62.0$$

**ANSWER**  The distillate contains 62.0% by mass of bromobenzene.

**EXAMPLE 2**  An organic liquid which does not mix with water distils in steam at 96 °C under a pressure of $1.01 \times 10^5\,\mathrm{N\,m^{-2}}$. The vapour pressure of water at 96 °C is $8.77 \times 10^4\,\mathrm{N\,m^{-2}}$. The distillate contains 51.0% by mass of the organic liquid. Calculate its molar mass.

**METHOD**  Let $M$ be the molar mass of the organic compound, $n_A$ and $n_W$ be the amounts (mol) of organic compound and water in the distillate, and $p_A$ and $p_W$ be the vapour pressures of the organic compound and water.

Then $\quad n_A/n_W = p_A/p_W$

$$\frac{51.0/M}{49.0/18.0} = \frac{(1.01 \times 10^5) - (8.77 \times 10^4)}{8.77 \times 10^4}$$

$$M = 123\,\mathrm{g\,mol^{-1}}$$

**ANSWER**  The molar mass of the organic compound is $123\,\mathrm{g\,mol^{-1}}$.

**EXERCISE 31**    Problems on Steam Distillation

1.  An organic liquid distils in steam. The partial pressures of the two liquids at the boiling point are X, 5.30 kPa; $H_2O$, 96.0 kPa. The distillate contains the liquids in a ratio of 0.480 g X to 1.00 g water. What is the molar mass of X?

2.  The liquid A distils in steam. At the boiling point, the partial pressures of the two liquids are $A = 6.59 \times 10^3 \, N \, m^{-2}$; $H_2O = 9.44 \times 10^4 \, N \, m^{-2}$. If the molar mass of A is $95 \, g \, mol^{-1}$, what is the percentage by mass of A in the distillate?

3.  Compound B distils in steam at 96 °C and $1.01 \times 10^5 \, N \, m^{-2}$. The vapour pressure of water at this temperature is $7.24 \times 10^4 \, N \, m^{-2}$. The distillate contains 80.0% by mass of B. Calculate the molar mass of B.

4.  Phenylamine, $C_6H_5NH_2$, distils in steam at 98 °C and $1.01 \times 10^5 \, N \, m^{-2}$. If the saturation vapour pressure of water is $9.40 \times 10^4 \, N \, m^{-2}$, what is the percentage by mass of phenylamine in the distillate?

5.  Naphthalene, $C_{10}H_8$, distils in steam at 98 °C and $1.01 \times 10^5 \, N \, m^{-2}$. If the vapour pressure of water is $9.50 \times 10^4 \, N \, m^{-2}$, calculate the mass of distillate that contains 10.0 g of naphthalene.

6.  A substance X distils in steam at 98 °C and 97 kPa. The percentage of X by mass in the distillate is 15.8%. The saturated vapour pressure of water at 98 °C is 94.7 kPa. Calculate the molar mass of X.

## DISTRIBUTION OF A SOLUTE BETWEEN TWO IMMISCIBLE SOLVENTS

Consider a solid which is appreciably soluble in both of a pair of immiscible liquids. When the solid is shaken with the two liquids, it distributes itself between the two layers. It is found that the ratio of the solute concentrations in the two layers is always the same. If $c_U$ and $c_L$ are the concentrations in the upper and lower layers, then

$$c_U/c_L = k$$

The constant $k$ is called the *partition coefficient* or *distribution coefficient*, and is constant for a particular temperature.

**EXAMPLE 1**  The partition coefficient for iodine between water and carbon disulphide at 20 °C is $2.43 \times 10^{-3}$. A 100 cm³ sample of a solution of iodine in water, of concentration $1.00 \times 10^{-3} \, mol \, I_2 \, dm^{-3}$ is shaken with 10.0 cm³ of carbon disulphide. What fraction of the iodine is extracted by carbon disulphide?

**METHOD**  Let $x$ be the number of moles of $I_2$ extracted by $CS_2$.

No. of moles of $I_2$ (total) $= 100 \times 10^{-3} \times 10^{-3} = 1.00 \times 10^{-4}$

Use $c_U/c_L = k$   ($c_U$ for water layer, $c_L$ for $CS_2$ layer):

Then,
$$\frac{(1.00 \times 10^{-4})-x}{100} \bigg/ \frac{x}{10.0} = 2.43 \times 10^{-3}$$

$$(1.00 \times 10^{-4})-x = 2.43 \times 10^{-2}x$$

giving
$$x = 9.76 \times 10^{-5}$$

**ANSWER**  Fraction of iodine extracted $= (9.76 \times 10^{-5})/(1.00 \times 10^{-4})$
$$= 0.98.$$

**EXAMPLE 2**  The partition coefficient of X between ether and water is 25.0 at 20 °C. Calculate the mass of X extracted from a solution containing 10.0 g of X in 1.00 dm$^3$ of water by a) 100 cm$^3$ of ether, b) two successive portions of 50.0 cm$^3$ of ether.

**METHOD**  Let the mass of X extracted by 100 cm$^3$ of ether be $m_1$.

Use $c_U/c_L = k$   ($c_U$ for ether layer, $c_L$ for water layer)

Concn of X in ether $= m_1/100 \text{ g cm}^{-3}$

Concn of X in water $= (10.0 - m_1)/1000 \text{ g cm}^{-3}$

$$\frac{m_1}{100} \bigg/ \frac{(10.0 - m_1)}{1000} = 25.0$$

giving $m_1 = 7.14$ g.

Let $m_2 =$ mass of X extracted by the first 50.0 cm$^3$ of ether, and $m_3 =$ mass of X extracted by the second 50.0 cm$^3$ of ether.

Then
$$\frac{m_2}{50.0} \bigg/ \frac{(10.0 - m_2)}{1000} = 25.0$$

giving $m_2 = 5.55$ g.

If 5.55 g of X are extracted by ether, 4.45 g remain in the aqueous solution.

$\therefore$
$$\frac{m_3}{50} \bigg/ \frac{(4.45 - m_3)}{1000} = 25.0$$

giving $m_3 = 2.47$ g.

**ANSWER**  Total mass of X extracted by ether in two portions $= 5.55 \text{ g} + 2.47 \text{ g}$
$$= 8.02 \text{ g}.$$

(*Note* that this is greater than the value of 7.14 g calculated for the mass of X extracted by using all the ether at once.)

Partition can be used to investigate an equilibrium in aqueous solution between a covalent species and an ionic species, for example, the equilibrium

$$I_2(aq) \; + \; I^-(aq) \; \rightleftharpoons \; I_3^-(aq)$$

Only the covalent $I_2$ molecules will dissolve in an organic solvent. If an aqueous solution of iodine in iodide ions is shaken with an organic solvent, the concentration of iodine in the solvent can be measured and divided by the partition coefficient to give the concentration of iodine molecules in the aqueous layer. The concentration of iodine combined as $I_3^-$ ions is obtained by subtracting the free iodine from the total iodine concentration. The concentration of $I^-$ ions is obtained by subtracting $[I_3^-]$ from the original concentration of $I^-$ ions.

**EXAMPLE 3**   Iodine is dissolved in water containing $0.160 \, mol \, dm^{-3}$ of potassium iodide, and the solution is shaken with tetrachloromethane. The concentration of iodine in the aqueous layer was found to be $0.080 \, mol \, dm^{-3}$; that in the organic layer $0.100 \, mol \, dm^{-3}$. The partition coefficient for iodine between tetrachloromethane and water is 85. Calculate the equilibrium constant for the reaction:

$$I_2(aq) \; + \; I^-(aq) \; \rightleftharpoons \; I_3^-(aq).$$

**METHOD**   Since     $[I_2]$ in $CCl_4$ = $0.100 \, mol \, dm^{-3}$

$[I_2]$ free in water = $0.100/85$ = $0.001\,18 \, mol \, dm^{-3}$

$[I_2]$ total = $0.080 \, mol \, dm^{-3}$

$[I_2]$ combined as $I_3^-$ = $0.080 - 0.001\,18$ = $0.0788 \, mol \, dm^{-3}$

$[I^-]$ total = $0.160 \, mol \, dm^{-3}$

$[I^-]$ free = $0.160 - 0.0788$ = $0.0812 \, mol \, dm^{-3}$

Putting these values into the expression

$$\frac{[I_3^-]}{[I_2][I^-]} = K$$

gives           $K = \dfrac{0.0788}{0.001\,18 \times 0.0812}$ = $822 \, mol^{-1} \, dm^3$

**ANSWER**   The equilibrium constant is $820 \, mol^{-1} \, dm^3$.

**EXAMPLE 4**   $100 \, cm^3$ of aqueous copper(II) sulphate solution of concentration $0.100 \, mol \, dm^{-3}$ were mixed with $100 \, cm^3$ of ammonia solution of concentration $1.12 \, mol \, dm^{-3}$. The mixture was shaken with $100 \, cm^3$ of trichloromethane. The trichloromethane layer required $12.0 \, cm^3$ of hydrochloric acid of concentration $1.00 \, mol \, dm^{-3}$ for neutralisation.

Assume that the partitition coefficient

$$\frac{\text{Concn of free ammonia in water}}{\text{Concn of free ammonia in trichloromethane}} = 25.0$$

Deduce the formula for the complex formed by ammonia and copper(II) ions.

**METHOD**    Amount of $NH_3$ in $CHCl_3$ = $12.0 \times 10^{-3} \times 1.00$ = $1.20 \times 10^{-2}$ mol
$[NH_3]$ in $CHCl_3$ = $1.20 \times 10^{-3}/0.100$ = $0.0120$ mol dm$^{-3}$
$[NH_3]$ free in water = $25.0 \times 0.0120$ = $0.300$ mol dm$^{-3}$
Amount of $NH_3$ free in water = $0.300 \times 0.200$ = $0.0600$ mol
Total amount of $NH_3$ = $100 \times 10^{-3} \times 1.12$ = $0.112$ mol
Amount of combined $NH_3$ in water = $0.112 - 0.0600 - 0.0120$ mol
$$= 0.0400 \text{ mol}$$
Amount of $Cu^{2+}$ ions in water = $100 \times 10^{-3} \times 0.100$ = $0.0100$ mol
Ratio (moles of $Cu^{2+}$)/(moles of $NH_3$) = $0.0100/0.0400$ = $1/4$

**ANSWER**    The formula of the complex is $[Cu(NH_3)_4]^{2+}$.

## *Distribution law for association and dissociation

If the solute has a different molecular formula in the two solvents, then the distribution law has to be modified. For example, ethanoic acid is dimerised in benzene. When ethanoic acid is partitioned between benzene and water, the equilibrium is

$$(CH_3CO_2H)_2 \text{ (benzene)} \rightleftharpoons 2CH_3CO_2H \text{ (water)}$$

and    $\dfrac{[CH_3CO_2H \text{ (benzene)}]}{[CH_3CO_2H \text{ (water)}]^2} = C,$    where $C$ is a constant

∴    $lg [CH_3CO_2H \text{ (benzene)}] = lg C + 2 lg [CH_3CO_2H \text{ (water)}]$

In general, if a solute is monomeric in solvent 1 and polymeric in solvent 2:

$$nA \text{ (solvent 1)} \rightleftharpoons A_n \text{ (solvent 2)}$$

then    $lg [A \text{ (solvent 2)}] = n lg [A \text{ (solvent 1)}] + lg C$

A plot of $lg [A \text{ (solvent 2)}]$ against $lg [A \text{ (solvent 1)}]$ gives a straight line of gradient $n$ and intercept $lg C$ (see Fig. 6.2).

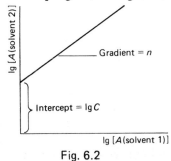

Fig. 6.2

## EXERCISE 32    Problems on Partition

1. X is 12.0 times more soluble in trichloromethane than in water. What mass of X will be extracted from $1.00\,dm^3$ of an aqueous solution containing 25.0 g by shaking with $100\,cm^3$ of trichloromethane?

2. The partition coefficient of Y between ethoxyethane (ether) and water is 80. If $200\,cm^3$ of an aqueous solution containing 5.00 g of Y is shaken with $50.0\,cm^3$ of ethoxyethane, what mass of Y is extracted from the solution?

3. Z is allowed to reach an equilibrium distribution between the liquids ethoxyethane and water. The ether layer is $50.0\,cm^3$ in volume and contains 4.00 g of Z. The aqueous layer is $250\,cm^3$ in volume and contains 1.00 g of Z. What is the partition coefficient of Z between ethoxyethane and water?

4. $500\,cm^3$ of an aqueous solution of concentration $0.120\,mol\,dm^{-3}$ is shaken with $50.0\,cm^3$ of ethoxyethane. The partition coefficient of the solute between ethoxyethane and water is 60.0. Calculate the amount (in mol) of solute which will be extracted by ethoxyethane.

5. The distribution coefficient of A between ethoxyethane and water is 90. An aqueous solution of A with a volume of $500\,cm^3$ contains 5.00 g. What mass of A will be extracted by:

   a) $100\,cm^3$ of ethoxyethane, and

   b) two successive portions of $50.0\,cm^3$ of ethoxyethane?

6. An organic acid is allowed to reach an equilibrium distribution in a separating funnel containing $50.0\,cm^3$ of ethoxyethane and $500\,cm^3$ of water. On titration, $25.0\,cm^3$ of the ethoxyethane layer required $22.5\,cm^3$ of $1.00\,mol\,dm^{-3}$ sodium hydroxide solution, and $25.0\,cm^3$ of the aqueous layer required $9.0\,cm^3$ of $0.100\,mol\,dm^{-3}$ sodium hydroxide solution. Calculate the partition coefficient for the acid between ethoxyethane and water.

7. A small amount of iodine is shaken in a separating funnel containing $50.0\,cm^3$ of tetrachloromethane and $500\,cm^3$ of water. On titration, $25.0\,cm^3$ of the aqueous layer require $6.7\,cm^3$ of a $0.0550\,mol\,dm^{-3}$ solution of sodium thiosulphate. $25.0\,cm^3$ of the organic solvent require $27.2\,cm^3$ of $1.15\,mol\,dm^{-3}$ sodium thiosulphate solution. Calculate the distribution coefficient for iodine between water and tetrachloromethane.

8. A solid A is three times as soluble in solvent X as in solvent Y. A has the same relative molecular mass in both solvents. Calculate the mass of A that would be extracted from a solution of 4 g of A in $12\,cm^3$ of Y by extracting it with: a) $12\,cm^3$ of X, and b) three successive portions of $4\,cm^3$ of X.

9. a) The partition coefficient of T between ethoxyethane and water at room temperature is 19. A solution of 5.0 g of T in 100 cm³ of water is extracted with 100 cm³ of ethoxyethane at room temperature. What mass of T will be present in the ethoxyethane layer?

   b) What would be the total mass of T extracted if the ethoxyethane were used in two separate 50 cm³ portions, instead of the single 100 cm³ portion?

*10. When an excess of ammonia is added to an aqueous solution of copper(II) sulphate, a complex ion, $Cu(NH_3)_n{}^{2+}$ is formed.

A 100 cm³ portion of a solution of ammonia in 0.75 mol dm⁻³ copper(II) sulphate solution is added to 100 cm³ of trichloromethane. After equilibration, the two layers were titrated against standard alkali. The concentrations of ammonia in the two layers were: aqueous layer, 3.25 mol dm⁻³; trichloromethane layer, 0.0100 mol dm⁻³. Ammonia dissolves in trichloromethane, but the complex ion does not. For every mole of ammonia in the trichloromethane layer, there are 25 moles of ammonia not complexed with copper(II) ions in the aqueous layer.

Calculate the number of moles of ammonia in the trichloromethane layer; the number of moles of ammonia not complexed in the aqueous layer; and the number of moles of ammonia combined with the copper(II) ions. Deduce the formula of the complex ion.

## EXERCISE 33    Questions from A-level Papers

1. Explain the meaning of the following: (i) the mole; (ii) the ideal gas law; (iii) saturated vapour pressure; (iv) the density of a liquid.

   A 1 dm³ flask at 27 °C is half filled with water. The density of water at 27 °C is 1.00 kg dm⁻³ and its saturated vapour pressure is 3570 N m⁻².

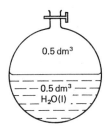

0.5 dm³

0.5 dm³
H₂O(l)

   Assuming ideal behaviour of the vapour, calculate:

   a) the number of molecules of water in the bottom half of the flask

   b) the number of molecules of water in the top half of the flask.

   [$A_r(H) = 1$; $A_r(O) = 16$; $R = 8.31$ J mol⁻¹K⁻¹; 0 °C = 273 K; Avogadro constant = 6.02 × 10²³ molecules mol⁻¹.]    (WJEC80)

2. a) State the meaning of the term *partition coefficient* of a solute between two immiscible solvents.

   b) Outline the principles of an experiment to measure the partition coefficient for a system of your choice.

   c) The partition coefficient of a substance X between methylbenzene (toluene) and water is 4.0, X being more soluble in methylbenzene. Starting with $100 \, cm^3$ of an aqueous solution containing $6.0 \, g$ of X, compare the amounts of X extracted (i) by shaking with one portion of $100 \, cm^3$ of methylbenzene, (ii) by shaking with two successive portions of $50 \, cm^3$.

   d) Explain the principles involved in any one chromatographic technique for the separation of the components of a mixture.

(WJEC80)

*3. Benzoic acid (175 mg) is equilibrated between methylbenzene ($50 \, cm^3$) and water ($100 \, cm^3$) at 285 K. Titration shows that the aqueous layer then contains 50 mg of the acid. Benzoic acid exists as a dimer in the methylbenzene layer. In a second experiment benzoic acid (550 mg) is dissolved in a mixture of methylbenzene ($50 \, cm^3$) and water $100 \, cm^3$) and allowed to equilibrate at 285 K. Calculate the mass of benzoic acid which will be present in the aqueous layer in the second experiment explaining carefully the physico-chemical basis of your calculation.

What intermolecular bonding is responsible for the association of the benzoic acid in the organic solvent, and why is the benzoic acid not associated in aqueous solution?

Describe in outline one chromatographic method which involves the partition of the components of a mixture between two separate phases.

(O80, S)

4. a) What is meant by the term *partition coefficient* for the distribution of a solute between two immiscible liquids?

   b) An acid *A* in aqueous solution is to be extracted with ethoxyethane ('ether'), in which it dissolves without association or dissociation. The concentration of the aqueous solution is $5 \, mol \, dm^{-3}$. The partition coefficient of *A* between ethoxyethene and water is 4.

      i) How many moles of *A* are extracted by shaking $1 \, dm^3$ of the aqueous solution with $500 \, cm^3$ of ethoxyethane in one extraction?

      ii) How many moles of *A* are extracted by shaking $1 \, dm^3$ of the aqueous solution with two successive volumes of $250 \, cm^3$ of ethoxyethane?

   c) What will be the percentage composition by volume of the gas boiled out of water which has been saturated with air at room temperature? Consider air as a mixture of 4 parts of nitrogen to 1 part oxygen only (by volume). At the same partial pressure for each gas, oxygen is twice as soluble as nitrogen in water at room temperature.

(C82)

5. a) Describe a method which could be used to determine experimentally the value of the partition coefficient for iodine distributed between trichloromethane and water at a given temperature.

   b) The value of the partition coefficient for the distribution of iodine between tetrachloromethane and water at 298 K is 90.

   The following results were obtained when a solution of iodine in tetrachloromethane was shaken with aqueous potassium iodide, of concentration $0.250 \, mol \, dm^{-3}$, at 298 K and the system allowed to come to equilibrium.

   The concentration of iodine in the tetrachloromethane layer was $0.180 \, mol \, dm^{-3}$.

   The *total* concentration of iodine in the aqueous layer was $0.150 \, mol \, dm^{-3}$. Use this information to calculate the value of $K_c$, at 298 K, for the equilibrium:

   $$I_2(aq) \; + \; I^-(aq) \; \rightleftharpoons \; I_3^-(aq). \qquad \text{(AEB84S)}$$

6. a) Sketch and label the temperature-composition curves for mixtures of two miscible, volatile liquids which obey Raoult's law. Use your sketch graph to show how, by distillation, the components of the mixture may be separated. (Practical details are not required.)

   Explain the use of a fractionating column in the above separation.

   b) At 20 °C the vapour pressures of pure methanol and pure ethanol are 95.0 and 45.0 mm Hg, respectively. A solution, assumed to be ideal, contains 16.0 g of methanol and 92.0 g of ethanol; calculate
   i) the partial pressures of methanol and ethanol in the mixture;
   ii) the total vapour pressure of the mixture;
   iii) the composition of the vapour.

   c) By reference to intermolecular forces, explain the existence of positive and negative deviations from Raoult's law. Give one piece of experimental evidence in support of your explanation.
   (Relative atomic masses: $C = 12, H = 1, O = 16$.) \qquad (L85)

7. a) Describe how you would determine the relative molecular mass of a volatile liquid by a syringe or other simple laboratory method. State clearly the precautions that must be taken to ensure accuracy.

   b) A liquid $L$ contains 62.1% carbon, 10.3% hydrogen and 27.6% oxygen by mass. When 0.125 g of $L$ was evaporated in a syringe at 100 °C and standard atmospheric pressure, the volume of vapour produced was $66.0 \, cm^3$. Determine
   i) the empirical formula,
   ii) the molecular formula of $L$.

   c) A gaseous hydrocarbon $H$ has the same relative molecular mass as the liquid $L$. Explain the difference in volatility of these two compounds in terms of structure and bonding. \qquad (C84)

8. $25\,g$ of a solid $X$ were dissolved in $100\,cm^3$ water. The solution was shaken with $20\,cm^3$ tetrachloromethane, and, when equilibrium was attained, $10\,g$ of $X$ were found to be dissolved in the tetrachloromethane. The molecular state of $X$ was the same in both solvents.

The partition coefficient for $X$ between tetrachloromethane and water is

a  0.67                    b  1.50                         c  2.00
d  2.50                    e  3.33
                                                                      (O & C82)

9. a) Iodine is present as $I_2$ molecules in its solutions in both tetrachloromethane (carbon tetrachloride) and water.

Solutions of iodine in the two solvents were thoroughly shaken together at room temperature and the layers allowed to separate. The concentrations of iodine in the two layers were then determined in a series of experiments using different concentrations with the following results.

| Experiment | 1 | 2 | 3 |
|---|---|---|---|
| Concentration $I_2$ in $CCl_4$/mol dm$^{-3}$ | $2.50 \times 10^{-2}$ | $2.30 \times 10^{-2}$ | $2.10 \times 10^{-2}$ |
| Concentration $I_2$ in $H_2O$/mol dm$^{-3}$ | $4.40 \times 10^{-4}$ | $4.10 \times 10^{-4}$ | $3.70 \times 10^{-4}$ |

   i) Explain the principle illustrated by these values.
   ii) State any limitations on its use.
   iii) If the concentration of iodine in tetrachloromethane in a fourth experiment were $1.90 \times 10^{-2}\,mol\,dm^{-3}$, what would you expect that in the water to be?

b) Iodine dissolves freely in potassium iodide solution forming a reddish-brown solution containing the $I_3^-$ ion according to the following reaction:

$$I_2 + I^- \rightleftharpoons I_3^-.$$

   i) Give the expression for the equilibrium constant for this reaction, and state its units.
   ii) Three such solutions in equilibrium at room temperature were analysed and the following concentrations were recorded.

| Experiment | 1 | 2 | 3 |
|---|---|---|---|
| Concentration $I_2$/mol dm$^{-3}$ | $1.0 \times 10^{-4}$ | $2.0 \times 10^{-4}$ | $3.0 \times 10^{-4}$ |
| Concentration $I^-$/mol dm$^{-3}$ | $8.0 \times 10^{-2}$ | $1.2 \times 10^{-1}$ | $1.4 \times 10^{-1}$ |
| Concentration $I_3^-$/mol dm$^{-3}$ | $5.7 \times 10^{-3}$ | $1.7 \times 10^{-2}$ | $3.0 \times 10^{-2}$ |

Calculate a value for the equilibrium constant for each solution. Compare the values obtained and comment.

c) i) Devise an experiment by which you would carry out the first series of measurements. Give essential experimental detail and write an equation for the principal reaction involved.

ii) By ensuring that the potassium iodide is in excess, and given that the $I_3^-$ ion is insoluble in tetrachloromethane, what would you expect to be the result of repeating this determination for the reaction mixture in the second series of experiments?

d) After shaking a solution of ammonia in water with trichloro-methane (chloroform) the concentration of ammonia in the water was $5.211 \, g \, dm^{-3}$, and in the trichloromethane, $0.2171 \, g \, dm^{-3}$, the two solvents being immiscible. On repeating the experiment but using ammoniacal $0.0500 \, M$ copper(II) sulphate solution instead of water, the concentration in the aqueous layer was $6.801 \, g \, dm^{-3}$ and in the trichloromethane, $0.1417 \, g \, dm^{-3}$. Write an equation for the reaction between ammonia and copper(II) ions. ($NH_3 = 17$.)

(SUJB82)

10. a) Describe, giving full experimental details, one method for determining the relative molecular mass of a volatile liquid.

b) Volatilisation of $0.1665 \, g$ of a compound ($A$), containing sulphur, oxygen and chlorine, displaced $39.2 \, cm^3$ of air measured over water at $20 \, ^\circ C$ and a total pressure of $88.7 \, kPa$ ($665.5 \, mm \, Hg$). The hydrochloric acid formed by the complete hydrolysis of $0.2505 \, g$ of $A$ gave, when treated with an excess of silver nitrate, $0.6035 \, g$ of AgCl. Deduce
i) the relative molecular mass,
ii) the molecular formula of $A$.

Write an equation for the reaction of $A$ with ethanoic (*acetic*) acid. (The vapour pressure of water at $20 \, ^\circ C$ is $2.35 \, kPa$ ($17.5 \, mm \, Hg$).)

(JMB 83)

11. a) State Raoult's law of vapour pressure lowering, explaining the terms used.

b) Deduce a simpler form of this law when one of the species is involatile or almost involatile at that temperature, and for the latter case sketch a graph to illustrate the law.

c) From the standpoint of the kinetic molecular theory, explain what happens to a liquid maintained at its boiling point during the addition of an involatile solute.

d) Sketch a vapour pressure against temperature graph to illustrate what has been explained.

e) Methanol and ethanol form an almost ideal solution. If equal masses of the two liquids are mixed calculate the mole fraction of ethanol above the mixture, given that the equilibrium vapour pressures at $27 \, ^\circ C$ are for methanol $11.82 \, kPa$ and for ethanol $5.93 \, kPa$. ($H = 1, C = 12, O = 16; Pa = Nm^{-2}$.)

f) Using suitable graphs, explain on the basis of the kinetic molecular theory what happens when a system of two *immiscible* liquids is heated, and compare what happens to what is observed on heating the pure components separately.

g) At 98 °C the vapour pressure of water and phenylamine (aniline) are respectively 94.26 kPa and 7.07 kPa. When a sample of phenylamine was steam distilled at this temperature the distillate contained 27.9% of phenylamine by weight. Calculate the molar mass of phenylamine. ($H_2O = 18.0$.)                    (SUJB82)

**12.**

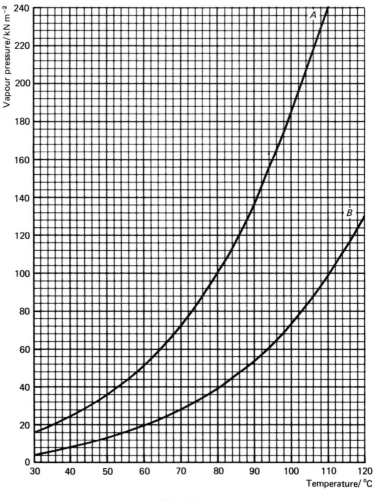

Fig. 6.3

Two miscible organic liquids $A$ and $B$ form an ideal solution.

Figure 6.3 gives the equilibrium vapour pressures of pure $A$ and pure $B$ as a function of temperature (°C). (1 atm = $101\ kN\ m^2$.)

a) i) Give the normal (1 atm) boiling points of pure *A* and pure *B*.
   ii) Give the equilibrium vapour pressures at 90 °C for pure *A* and pure *B*.

b) i) State what is meant by the term *the mole fraction of B in the solution*.
   ii) State Raoult's vapour pressure law.
   iii) Explain the meaning of the term *ideal solution of A and B*.
   iv) Describe how the total vapour pressure of the ideal solution varies with the mole fraction of *B*.

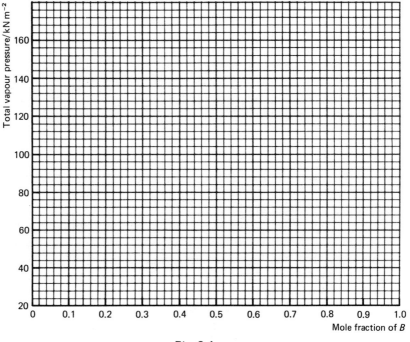

Fig. 6.4

   v) On a copy of Fig. 6.4 sketch, as precisely as you can, the variation of the *total vapour pressure* above the solution with the *mole fraction* of *B* for the ideal solution of *A* and *B* at 90 °C. Mark your line with the letter *L*.
   vi) Calculate the mole fraction of *B* in the vapour when its mole fraction in the liquid mixture (solution) at 90 °C is 0.7.
   vii) On your copy of Fig. 6.4, sketch the variation in the total vapour pressure as a function of the mole fraction of *B* in the *vapour phase*. Mark your line with the letter *V*.

c) A third liquid *C* is completely immiscible with *B*. When a mixture of *B* and *C* was heated (an experiment equivalent to steam distillation) the boiling point was found to be 80 °C.
   i) Give the relationship between the total pressure and the vapour pressures of *B* and *C* at this temperature.
   ii) Estimate the vapour pressure of *C* at 80 °C.          (WJEC84)

**13.** The diagram below shows the boiling point/composition curve for two miscible liquids $X$ and $Y$.

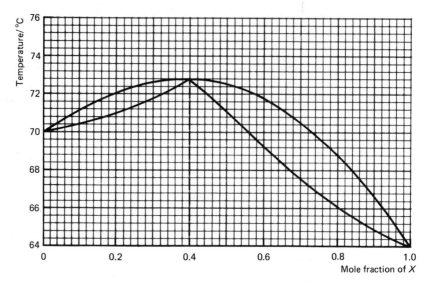

a) Label the liquid and the vapour curves on a copy of the diagram.

b) i) What is the boiling point of a mixture containing 0.6 mole fraction of $X$?

   ii) What is the composition of the vapour produced on boiling this mixture?

c) What will be the composition of the distillate first collected when mixtures of the following compositions are heated under fractional distillation?

   i)   0.3 mole fraction of $X$;

   ii)  0.7 mole fraction of $X$;

   iii) 0.4 mole fraction of $X$.

d) Explain how the interactions between molecules of $X$ and $Y$ give rise to the formation of this boiling point/composition curve.

e) i) Describe the enthalpy and volume changes that occur when $X$ and $Y$ are mixed.

   ii) What mole fraction of $X$ would produce the greatest changes?

(C85)

# 7 Solutions of Solids in Liquids: Colligative Properties

## LOWERING OF THE VAPOUR PRESSURE OF A SOLVENT BY A SOLUTE

For a solution of two liquids, Raoult's law states that the vapour pressure of each component is equal to the product of its mole fraction and the saturated vapour pressure of the pure component. In a solution of a solid in a liquid, the solid is involatile, and the vapour pressure of the solution is equal to the product of the mole fraction of the solvent and the saturated vapour pressure of the pure solvent. It follows that the lowering of the vapour pressure (vapour pressure of solvent minus vapour pressure of solution) is equal to the mole fraction of the solute. This can be expressed as

$$\frac{p^0 - p}{p^0} = \frac{n_1}{n_2 + n_2}$$

where $p^0$ = vapour pressure of pure solvent, $p$ = vapour pressure of solution, $n_1$ = number of moles of solute, and $n_2$ = number of moles of solvent.

In most solutions, $n_2 \gg n_1$, and the expression becomes

$$\frac{p^0 - p}{p^0} = \frac{n_1}{n_2}$$

Substituting $n = m/M$, where $m$ = mass, $M$ = molar mass gives

$$\frac{p^0 - p}{p^0} = \frac{m_1}{M_1} \bigg/ \frac{m_2}{M_2} = \frac{m_1 M_2}{m_2 M_1}$$

If the masses of solute and solvent are known, and the molar mass of the solvent is known, this expression can be used to give $M_1$, the molar mass of the solute.

EXAMPLE 1    The vapour pressure of water at $18\,^\circ\text{C}$ is $2.06 \times 10^3\,\text{N m}^{-2}$. The vapour pressure of a solution of $10.0\,\text{g}$ of a solute in $100\,\text{g}$ of water at $18\,^\circ\text{C}$ is $1.98 \times 10^3\,\text{N m}^{-2}$. Calculate its molar mass.

**METHOD**

$$\frac{p^0 - p}{p^0} = \frac{n_1}{n_2} = \frac{m_1}{m_2} \times \frac{M_2}{M_1}$$

$$\frac{2.06 - 1.98}{2.06} = \frac{10.0}{100} \times \frac{18.0}{M_1}$$

$$M_1 = \frac{18.0 \times 2.06}{0.08 \times 10.0} = 46.35$$

**ANSWER**    The molar mass of the solute is $46 \, g \, mol^{-1}$.

**EXAMPLE 2**  The vapour pressure of water at $18\,^{\circ}C$ is $2.053 \times 10^3 \, Pa$. The vapour pressure of a solution containing $5.132 \, g$ of a solute in $100 \, g$ of water is $2.021 \times 10^3 \, Pa$. Calculate the molar mass of the solute.

**METHOD**

$$\frac{p^0 - p}{p^0} = \frac{n_1}{n_2} = \frac{m_1}{m_2} \times \frac{M_2}{M_1}$$

$$\frac{(2.053 \times 10^3) - (2.021 \times 10^3)}{2.053 \times 10^3} = \frac{5.132}{100} \times \frac{18.0}{M_1}$$

$$\frac{0.032}{2.053} = \frac{5.132 \times 18.0}{100 M_1}$$

$$M_1 = 59.46$$

**ANSWER**    The molar mass of the solute is $59 \, g \, mol^{-1}$.

**EXERCISE 34**    Problems on Vapour Pressure Lowering

1. The vapour pressure of water at $20\,^{\circ}C$ is $2.34 \times 10^3 \, N \, m^{-2}$. Calculate the vapour pressure of the following solutions:
   a)  $34.2 \, g$ of sucrose, $C_{12}H_{22}O_{11}$, in $500 \, g$ of water
   b)  $45.0 \, g$ of glucose, $C_6H_{12}O_6$, in $250 \, g$ of water
   c)  $15.0 \, g$ of urea, $CO(NH_2)_2$, in $200 \, g$ of water
   d)  $50.0 \, g$ of urea in $50.0 \, g$ of water.

2. Which of the following, dissolved in $1 \, dm^3$ of water at $25\,^{\circ}C$, results in the biggest lowering of the vapour pressure?
   a  $1$ mole $C_{12}H_{22}O_{11}$          b  $0.5$ mole $KCl$
   c  $0.5$ mole $CuSO_4$          d  $0.5$ mole $Cu(NH_3)_4SO_4$
   e  $0.5$ mole $KAl(SO_4)_2$

3. The vapour pressure of a solution of a non-volatile compound X $(29.0 \, g)$ in $100 \, g$ of water is $1.12 \times 10^4 \, Pa$ at $50\,^{\circ}C$. At the same temperature, the vapour pressure of water alone was $1.22 \times 10^4 \, Pa$. Deduce the molar mass of X.

4. At $25\,^{\circ}C$, the vapour pressure of water is $3.15 \times 10^3 \, Pa$. Calculate the vapour pressure of a solution of $6.00 \, g$ of urea in $100 \, g$ of water.

# ELEVATION OF THE BOILING POINT OF A SOLVENT BY A SOLUTE

The lowering of the vapour pressure of a solvent depends on the mole fraction of solute in the solution. It does not depend on the nature of the solute. Properties such as vapour pressure lowering, which depend on the concentration of dissolved particles and not on their nature, are called *colligative properties*.

Since the boiling point of a liquid is the temperature at which its vapour pressure is equal to atmospheric pressure, the lowering of the vapour pressure which results from the presence of a solute is accompanied by a rise in boiling point. The boiling points of solutions are simpler to measure than their vapour pressures. A determination of the elevation of the boiling point of a solvent produced by a solute is a useful way of finding the molar mass of the solute.

The boiling point elevation is proportional to the concentration of solute and to a constant which has a certain value for each solvent. The elevation produced by 1 mole of solute in 1 kg of water is 0.52 K. This value is called the ebullioscopic constant for water.

---

1 mole of solute in 1 kg of water gives a boiling point elevation of 0.52 K.

---

For other solutions, the boiling point elevation depends on the concentration of solute and the magnitude of the ebullioscopic constant for the solvent.

---

$$\begin{pmatrix} \text{Boiling point} \\ \text{elevation} \end{pmatrix} = \begin{pmatrix} \text{Boiling point} \\ \text{constant} \end{pmatrix} \times \frac{\text{Moles of solute}}{\text{Mass of solvent in kg}}$$

$$\text{Boiling point elevation} = k \times \frac{m}{M} \times \frac{1}{W}$$

---

where $k$ is the ebullioscopic constant for the solvent (or boiling point constant), $m$ and $M$ are the mass and molar mass of the solute, and $W$ is the mass of the solvent.

If $W$ is expressed in kg, then

$$\text{Boiling point elevation (K)} = k \times \frac{m \text{ (g)}}{M \text{ (g mol}^{-1})} \times \frac{1}{W \text{ (kg)}}$$

and $k$ has the units of $K \, kg \, mol^{-1}$.

**EXAMPLE 1**  3.76 g of solute in 50.0 g of water give a boiling point elevation of 0.95 K. Calculate the molar mass of the solute.

**METHOD 1**    Since 0.95 K elevation is produced by 3.76 g of solute in 50.0 g of water, 1/20 × 0.95 K elevation is produced by 3.76 g of solute in 1000 g of water and 0.52 K elevation is produced by

$$\frac{0.52 \times 3.76}{1/20 \times 0.95} \text{ g of solute} = 41.16 \text{ g of solute}$$

**ANSWER**    The molar mass of the solute is 41 g mol$^{-1}$.

**METHOD 2**    $$\text{Boiling point elevation} = k \times \frac{m}{M} \times \frac{1}{W}$$

$$0.95 = 0.52 \times \frac{3.76}{M} \times \frac{1}{0.0500}$$

**ANSWER**    $M = 41 \text{ g mol}^{-1}$    (as before)

**EXAMPLE 2**    Pure propanone boils at 56.38 °C at 1 atm. A solution of 2.256 g of phenol in 100 g of propanone boils at 56.79 °C, and a solution of 0.635 g of a compound, X, in 50.0 g of propanone boils at 56.52 °C. What is the molar mass of X?

**METHOD**    There are two steps in this calculation: a) find the ebullioscopic constant for propanone, using the data for the compound of known molar mass; b) use the ebullioscopic constant to find the unknown molar mass.

$$\text{Boiling point elevation} = k \times \frac{m}{M} \times \frac{1}{W}$$

a)    $$0.400 = k \times \frac{2.256}{94.1} \times \frac{1}{0.100}$$

$$k = 1.67 \text{ K kg mol}^{-1}$$

b)    $$0.140 = 1.67 \times \frac{0.635}{M} \times \frac{1}{0.0500}$$

**ANSWER**    $$M = 152 \text{ g mol}^{-1}$$

**EXERCISE 35**    Problems on Boiling Point Elevation

1. Calculate the boiling points of the following solutions:
   a) 0.200 mol sucrose, $C_{12}H_{22}O_{11}$, in 500 g water
   b) 0.150 mol sucrose, $C_{12}H_{22}O_{11}$, in 250 g water
   c) 0.850 mol urea, $CO(NH_2)_2$, in 750 g water
   d) 20.0 g glucose, $C_6H_{12}O_6$, in 500 g water
   e) 20.0 g urea, $CO(NH_2)_2$, in 500 g water
   f) 30.0 g propanamide, $C_2H_5CONH_2$, in 400 g water.

2. Calculate the molar masses of the solutes in the following solutions:
   a) 6.00 g A in 100 g water, boiling at 100.52 °C
   b) 518 g B in 750 g water, boiling at 101.05 °C
   c) 60.0 g C in 500 g water, boiling at 100.85 °C
   d) 33.0 g D in 250 g water, boiling at 100.76 °C
   e) 79.0 g E in 200 g water, boiling at 101.14 °C.

## DEPRESSION OF THE FREEZING POINT OF A SOLVENT BY A SOLUTE

A solution has a lower freezing point than the pure solvent. The depression of the freezing point of 1 kg of solvent produced by 1 mole of solute is constant for a particular solvent. Freezing point depression is another colligative property, depending on the concentration of particles in solution and not on their nature.

| 1 mole of solute in 1 kg of water gives a freezing point depression of 1.86 K |
|---|

The molar depression of the freezing point is called the cryoscopic constant or freezing point constant of the solvent.

$$\begin{pmatrix} \text{Freezing point} \\ \text{depression} \end{pmatrix} = \begin{pmatrix} \text{Freezing point} \\ \text{constant} \end{pmatrix} \times \frac{\text{Moles of solute}}{\text{Mass of solvent}}$$

$$= k \times \frac{m}{M} \times \frac{1}{W}$$

where $m$ and $M$ are the mass and molar mass of the solute, $W$ is the mass (kg) of solvent and $k$ is the cryoscopic constant (freezing point constant), which has the units $K \, kg \, mol^{-1}$.

Measurements of freezing point depression can be used to calculate molar masses of solutes in the same way as boiling point elevation.

**EXAMPLE**     What is the molar mass of a solute which, when 1.00 g is dissolved in 15.0 g of water, gives a freezing point of $-0.370$ °C? The cryoscopic constant of water is $1.86 \, K \, kg \, mol^{-1}$.

**METHOD 1**     Since 0.370 °C depression is produced by 1.00 g solute in 15.0 g water, $0.370 \times 15.0/1000$ °C depression is produced by 1.00 g in 1000 g water and 1.86 °C depression is produced by

$$\frac{1.86 \times 1000 \, g}{0.370 \times 15.0} = 335 \, g$$

**ANSWER**     The molar mass of the solute is $335 \, g \, mol^{-1}$.

**METHOD 2**
$$\text{Freezing point depression} = k \times \frac{m}{M} \times \frac{1}{W}$$

$$0.370 = 1.86 \times \frac{1}{M} \times \frac{1}{0.0150}$$

$$M = 335 \, \text{g mol}^{-1} \quad \text{(as before)}$$

## EXERCISE 36    Problems on Freezing Point Depression

1. The freezing point constant for water is $1.86 \, \text{K kg mol}^{-1}$. Calculate the molar masses of the solutes in the following solutions:
   a) $21.0 \, \text{g}$ P in $200 \, \text{g}$ water, freezing at $-3.72 \, ^\circ\text{C}$
   b) $24.0 \, \text{g}$ Q in $200 \, \text{g}$ water, freezing at $-2.79 \, ^\circ\text{C}$
   c) $12.0 \, \text{g}$ R in $250 \, \text{g}$ water, freezing at $-1.49 \, ^\circ\text{C}$
   d) $100.0 \, \text{g}$ S in $500 \, \text{g}$ water, freezing at $-2.07 \, ^\circ\text{C}$
   e) $200.0 \, \text{g}$ T in $333 \, \text{g}$ water, freezing at $-3.26 \, ^\circ\text{C}$.

2. A solution of $3.75 \, \text{g}$ of sucrose, $C_{12}H_{22}O_{11}$, in $100 \, \text{g}$ of water freezes at $-0.204 \, ^\circ\text{C}$. A solution of X containing $27.3 \, \text{g dm}^{-3}$ freezes at $-0.282 \, ^\circ\text{C}$. Calculate the molar mass of X.

## EXERCISE 37    Problems on Freezing Point Depression and Boiling Point Elevation

Use the following data:
$$\begin{aligned}
\text{Freezing point constant for water} &= 1.86 \, \text{K kg mol}^{-1} \\
\text{Boiling point constant for water} &= 0.52 \, \text{K kg mol}^{-1} \\
\text{Freezing point of benzene} &= 5.48 \, ^\circ\text{C} \\
\text{Freezing point constant for benzene} &= 5.12 \, \text{K kg mol}^{-1} \\
\text{Boiling point of benzene} &= 80.1 \, ^\circ\text{C} \\
\text{Boiling point constant for benzene} &= 2.53 \, \text{K kg mol}^{-1}
\end{aligned}$$

1. $12.8 \, \text{g}$ of naphthalene, $C_{10}H_8$, are dissolved in $100 \, \text{g}$ of benzene.
   a) What is the freezing point of the solution?
   b) What is the boiling point of the solution?
   c) If the vapour pressure of pure benzene is $1.33 \times 10^4 \, \text{Nm}^{-2}$ at $25 \, ^\circ\text{C}$, what is the vapour pressure of this solution at $25 \, ^\circ\text{C}$?

2. $8.55 \, \text{g}$ of sucrose, $C_{12}H_{22}O_{11}$, are dissolved in $100 \, \text{g}$ of water.
   a) What is the boiling point of this solution?
   b) What is the freezing point of this solution?
   c) What is the vapour pressure of the solution at $20 \, ^\circ\text{C}$ if the vapour pressure of pure water at this temperature is $2.34 \times 10^3 \, \text{N m}^{-2}$.

3. 0.386 g of A in 50.0 g of benzene depresses the freezing point by 0.270 °C. Calculate the molar mass of A.

4. Calculate the freezing points and the boiling points of the following solutions:

   a) 28.0 g of ethanamide, $CH_3CONH_2$, in 500 g water
   b) 9.0 g of glucose, $C_6H_{12}O_6$, in 250 g water
   c) 17.1 g of sucrose, $C_{12}H_{22}O_{11}$, in 250 g water
   d) 50.0 g of urea, $CO(NH_2)_2$, in 750 g water
   e) 100.0 g of glycerol, $C_3H_8O_3$, in 500 g water.

5. Ethylene glycol (ethane-1, 2-diol, $C_2H_6O_2$) is added to car radiators to prevent freezing. What mass of ethylene glycol must be added to a radiator which holds 8 dm³ of water in order to protect against freezing down to $-15$ °C?

# OSMOTIC PRESSURE

A semipermeable membrane is a film of material which can be penetrated by a solvent but not by a solute. When two solutions are separated by a semipermeable membrane, solvent passes from the more dilute to the more concentrated. This phenomenon is called *osmosis*. The pressure which must be applied to a solution to prevent the solvent from diffusing in is called the osmotic pressure of the solution. There is an analogy with gas pressure. One mole of a solid, A, when vaporised, occupies a volume of 22.4 dm³ at 0 °C and 1.01 $\times$ $10^5$ N m$^{-2}$. One mole of A dissolved in 22.4 dm³ of solvent at 0 °C exerts an osmotic pressure of 1.01 $\times$ $10^5$ N m$^{-2}$.

> 1 mole of solute in 22.4 dm³ of solvent at 0 °C has an osmotic pressure of 1.01 $\times$ $10^5$ N m$^{-2}$ (1 atmosphere).

The expression which relates osmotic pressure to concentration and temperature is similar to the Ideal Gas Equation,

$$\pi V = nRT$$

where $\pi$ is the osmotic pressure, $V$ is the volume, $T$ is the temperature (Kelvin), $n$ is the amount (in mol) of solute, and $R$ is a constant which has the same value as the gas constant, 8.314 J K$^{-1}$ mol$^{-1}$. This equation is obeyed by ideal solutions.

The osmotic pressure of a solution depends on the concentration of solute present: it is a colligative property. Measurements of osmotic pressure can be used to give the molar masses of solutes.

**EXAMPLE**    Calculate the molar mass of a solute, given that 35.0 g of the solute in 1.00 dm$^3$ water have an osmotic pressure of $5.15 \times 10^5 \, \text{N m}^{-2}$ at 20 °C.

**METHOD**

$$\pi V = nRT$$

where $\pi = 5.15 \times 10^5 \, \text{N m}^{-2}$, $V = 1.00 \times 10^{-3} \, \text{m}^3$, $R = 8.314 \, \text{J K}^{-1} \, \text{mol}^{-1}$, and $T = 293 \, \text{K}$.

$$\therefore \qquad 5.15 \times 10^5 \times 1.00 \times 10^3 = n \times 8.314 \times 293$$

$$n = 0.211$$

$$n = \frac{35}{M}$$

$$M = 165 \, \text{g mol}^{-1}.$$

**ANSWER**    The solute has a molar mass of 165 g mol$^{-1}$.

# EXERCISE 38    Problems on Osmotic Pressure

1. Find the osmotic pressure of the following aqueous solutions at 25 °C:
   a) a sucrose solution of concentration 0.213 mol dm$^{-3}$
   b) a solution containing 144 g dm$^{-3}$ of glucose
   c) a solution which freezes at $-0.60$ °C
   d) a solution which boils at 100.30 °C
   e) a solution containing 12.0 g of urea in 200 cm$^3$ of solution.

2. Find the molar masses of the following solutes:
   a) 1.50 g of A in 200 cm$^3$ of aqueous solution, having an osmotic pressure of $2.66 \times 10^5 \, \text{N m}^{-2}$ at 20 °C
   b) 20.0 g of B in 100 cm$^3$ of aqueous solution, having an osmotic pressure of $3.00 \times 10^6 \, \text{N m}^{-2}$ at 27 °C
   c) 5.00 g of C in 200 cm$^3$ of solution, having an osmotic pressure of $2.39 \times 10^5 \, \text{N m}^{-2}$ at 25 °C.

3. The osmotic pressure of blood at 37 °C is $8.03 \times 10^5 \, \text{N m}^{-2}$. What is its freezing point?

4. A polysaccharide has the formula $(C_{12}H_{22}O_{11})_n$. A solution containing 5.00 g dm$^{-3}$ of the sugar has an osmotic pressure of $7.12 \times 10^2 \, \text{N m}^{-2}$ at 20 °C. Find $n$ in the formula.

5. A solution of PVC $(CH_2CHCl)_n$, in dioxan has a concentration of 4.00 g dm$^{-3}$ and an osmotic pressure of $65 \, \text{N m}^{-2}$ at 20 °C. Calculate the value of $n$.

6. A solution of 2.00 g of a polymer in 1 dm³ of water has an osmotic pressure of $300 \, N \, m^{-2}$ at 20 °C. Calculate the molar mass of the polymer.

7. a) Calculate the osmotic pressure of a solution of glucose containing $15.0 \, g \, dm^{-3}$ at 17 °C.
   b) What concentration of urea solution (in $g \, dm^{-3}$) would be isotonic with this solution?

# ANOMALOUS RESULTS FROM MOLAR MASS DETERMINATIONS FROM COLLIGATIVE PROPERTIES

Colligative properties depend on the number of particles in solution. If the number of particles increases through dissociation of the solute, the observed colligative property increases. If association occurs, the number of particles is less than expected, and the colligative property is less than expected.

## Dissociation

Electrolytes give abnormally high values of colligative properties. Van't Hoff discovered this, and the ratio

$$\frac{\text{Observed colligative property}}{\text{Calculated colligative property}}$$

is called the *Van't Hoff factor*, usually represented as *i*.

The reason for the high value of *i* for electrolytes is dissociation into ions. If a solute, such as sodium chloride, NaCl, is completely dissociated, there are two moles of ions for every mole of salt, and the colligative properties are twice the values calculated in the absence of ionisation. If copper(II) nitrate, $Cu(NO_3)_2$ is completely dissociated, there are three moles of ions per mole of solute, and the colligative properties are three times the values calculated in the absence of ionisation.

If the solute is incompletely dissociated, the degree of dissociation can be calculated. If a solute AB dissociates into two ions, and $\alpha$ is the degree of dissociation,

$$AB \;\; \rightleftharpoons \;\; A^+ + B^-$$
$$(1 - \alpha) \qquad\quad \alpha \quad\;\; \alpha$$

1 mole of AB will produce $\alpha$ moles of $A^+$ and $\alpha$ moles of $B^-$, leaving $(1 - \alpha)$ moles of AB. The total number of particles present is $(1 + \alpha)$.

Thus,

$$\frac{\text{Actual no. of particles}}{\text{Expected no. of particles}} = \frac{\text{Observed colligative property}}{\text{Calculated colligative property}} = \frac{1+\alpha}{1}$$

In general, if 1 mole of solute dissociates into $n$ moles of ions,

$$\frac{\text{Observed colligative property}}{\text{Expected colligative property}} = \frac{1+(n-1)\alpha}{1}$$

$$\therefore \qquad \alpha = \frac{i-1}{n-1}$$

It is now believed that strong electrolytes are always completely dissociated in solution. The Debye-Hückel and Onsager theory offers a different explanation for the low values of $\alpha$ obtained in measurements on concentrated solutions of strong electrolytes. The value of $\alpha$ calculated from colligative properties is called the *apparent degree of ionisation.*

## Association

A value for a colligative property lower than the calculated value may indicate association of molecules of solute. If $n$ molecules of solute associate to form $X_n$, and the degree of association is $\alpha$,

$$nX \rightleftharpoons X_n$$
$$1-\alpha \qquad \alpha/n$$

The total amount of particles $= 1 - \alpha + \alpha/n = 1 - \dfrac{(n-1)\alpha}{n}$ mol

and

$$\frac{\text{Observed colligative property}}{\text{Calculated colligative property}} = i = 1 - \frac{(n-1)\alpha}{n}$$

If X dimerises, $n = 2$, and $i = 1 - \alpha/2$.

**EXAMPLE 1** When 2.15 g of calcium nitrate are dissolved in 100 g of water, the solution freezes at $-0.62\,^\circ C$. Find the apparent degree of dissociation of the salt.

**METHOD** The molar mass of $Ca(NO_3)_2 = 164\,g\,mol^{-1}$.

For 1 mole of solute in 1000 g of water, the expected depression $= 1.86\,^\circ C$.

Thus, 164 g of $Ca(NO_3)_2$ in 100 g water should give a depression $= 18.6\,^\circ C$

and 2.15 g of $Ca(NO_3)_2$ in 100 g water should give $18.6 \times \dfrac{2.16}{164} = 0.24\,^\circ C$

$$i = 0.62/0.24 = 2.58$$

Number of ions per formula unit of $Ca(NO_3)_2 = n = 3$

$$\therefore \qquad \alpha = \frac{i-1}{n-1} = \frac{2.58-1}{3-1} = 0.78$$

**ANSWER**     The apparent degree of dissociation of the salt is 0.78 (78%).

**EXAMPLE 2**   A solution of 3.78 g of ethanoic acid in 75.0 g of benzene boiled 1.14 °C above the boiling point of the pure solvent. The ebullioscopic constant of benzene is $2.53 \text{ K kg mol}^{-1}$. What do these results indicate about the molecular condition of ethanoic acid?

**METHOD**     The molar mass of $CH_3CO_2H$ is $60.0 \text{ g mol}^{-1}$

60 g of ethanoic acid in 1 000 g of benzene should give an elevation of 2.53 °C.

3.78 g of ethanoic acid in 75.0 g of solvent should give

$$\frac{3.78}{60.0} \times \frac{1000}{75} \times 2.53 = 2.12 \,^{\circ}\text{C}$$

The observed elevation is only just over half this; therefore the molecules are associated in pairs. The degree of association can be calculated, using the expression:

$$\frac{\text{Observed colligative property}}{\text{Calculated colligative property}} = 1 - \frac{(n-1)\alpha}{n} = 1 - \frac{\alpha}{2}$$

$$\therefore \qquad \frac{1.14}{2.12} = 1 - \frac{\alpha}{2} \qquad \text{and} \qquad \alpha = 0.92$$

**ANSWER**     The degree of association is 0.92 (92%).

# EXERCISE 39     Problems on Electrolytes

Use the physical constants on p. 122.

1. Assume that sodium chloride is completely ionised in aqueous solution. What are the freezing point and the boiling point of a $0.220 \text{ mol dm}^{-3}$ solution of sodium chloride?

2. Assume that sodium nitrate is completely ionised in aqueous solution. Calculate the concentration of a sodium nitrate solution that freezes at $-2.15\,^{\circ}\text{C}$.

3. A solution of calcium nitrate freezes at $-2.45\,^{\circ}\text{C}$. What is the concentration of the solution, assuming calcium nitrate to be completely ionised?

4. What mass of anhydrous calcium chloride must be dissolved in $1.00 \text{ dm}^3$ of water to make a solution with the same freezing point as a solution containing 1.86 g of potassium chloride in $500 \text{ cm}^3$ of water?

5. What is the boiling point of a solution containing 150.0 g of potassium sulphate in 750 cm$^3$ of aqueous solution?

6. A solution of ethanoic acid has a concentration of 0.100 mol dm$^{-3}$ and a freezing point of $-0.190\,^\circ$C. What is the degree of ionisation of ethanoic acid at this temperature?

7. A solution of potassium bromide has a concentration of 0.250 mol dm$^{-3}$ and a freezing point of $-0.83\,^\circ$C. What is the apparent degree of ionisation?

8. A solution of magnesium nitrate has a concentration of 0.500 mol dm$^{-3}$ and a boiling point of 100.665 $^\circ$C. What is the apparent degree of ionisation?

9. Barium hydroxide has an apparent degree of ionisation of 0.92. What is the freezing point of a solution containing 2.50 g of barium hydroxide in 1.00 dm$^3$?

10. An organic compound has the empirical formula $C_3H_2Br$. 10.0 g of the compound dissolved in 1000 g benzene made a solution with a boiling point of 80.21 $^\circ$C. What is the molecular formula of the compound?

11. A 0.200 mol dm$^{-3}$ aqueous solution of the acid HA freezes at $-0.656\,^\circ$C. Calculate the degree of ionisation of the acid.

12. Nitrous acid is 4.25% ionised in a 0.200 mol dm$^{-3}$ solution. What is the freezing point of the solution?

13. Benzoic acid has the empirical formula $C_7H_6O_2$. When 0.2095 g of benzoic acid is dissolved in 100 g of water, the freezing point of the solution is $-0.0320\,^\circ$C. A solution of 2.057 g of benzoic acid in 100 g of benzene froze at 5.05 $^\circ$C. Find the molecular formula for benzoic acid in these two solutions. Explain any differences observed.

14. A solution of car radiator antifreeze ($C_2H_6O_2$, 1,2-dihydroxyethane) freezes at $-10.0\,^\circ$C. What is the concentration of the antifreeze in kg per kg of water?

15. The freezing point of a solution of a salt is $-0.100\,^\circ$C. What is the vapour pressure of the solution at 25 $^\circ$C if the vapour pressure of pure water at 25 $^\circ$C is $3.16 \times 10^3$ Pa?

16. 1.00 g of an acid dissolved in 100 g of water depressed the freezing point by 0.250 $^\circ$C. Dissolved in 100 g of benzene, 1.00 g of the acid depressed the freezing point by 0.350 $^\circ$C. What can you deduce from this information?

17. A solution containing 1.00 g of X in 50.0 g of water boils at 100.135 $^\circ$C at 10$^5$ Pa. Calculate: a) the molar mass of X, and b) the osmotic pressure of the solution.

18. The freezing point of a solution is $-1.150\,^\circ$C. Calculate its osmotic pressure at 20 $^\circ$C.

**EXERCISE 40**    Questions from A-level Papers

1. Explain the term *osmotic pressure*.

   Outline the principles of an osmotic pressure method for the determination of the relative molecular mass of a solute.

   Calculate the osmotic pressure of an aqueous solution containing $25.0\,\mathrm{g\,dm^{-3}}$ of a protein of relative molecular mass $5.00 \times 10^4$ at $27\,^{\circ}\mathrm{C}$.

   Explain carefully why this method is appropriate for solutes of high relative molecular mass.                                    (C80, p)

2. a) Describe the essential features of the phenomenon of osmosis and define the term *osmotic pressure*.

   At $20\,^{\circ}\mathrm{C}$ the osmotic pressure of an aqueous solution containing $3.221 \times 10^{-3}\,\mathrm{g\,cm^{-3}}$ of an enzyme was found to be $5.637 \times 10^2\,\mathrm{N\,m^{-2}}$. What is the relative molecular mass of the enzyme?

   b) State
   i) Henry's law,
   ii) Raoult's law.

   The following table gives partial vapour pressures, $p$, for the two components of mixtures of propanone (acetone) and trichloromethane (chloroform) at $35\,^{\circ}\mathrm{C}$ for a range of mole fractions, $x$, of trichloromethane.

   | $x\,(\mathrm{CHCl_3})$ | 0.00 | 0.20 | 0.40 | 0.60 | 0.80 | 1.00 |
   |---|---|---|---|---|---|---|
   | $p\,(\mathrm{CHCl_3})/\mathrm{mm\,Hg}$ | 0 | 35 | 82 | 142 | 219 | 293 |
   | $p\,((\mathrm{CH_3})_2\mathrm{CO})/\mathrm{mm\,Hg}$ | 347 | 270 | 185 | 102 | 37 | 0 |

   Use a graphical method to show that this system deviates from Raoult's law. How may this deviation be explained on a molecular basis?                                                    (O85)

3. a) An aromatic amine, $A$, and water are immiscible. A mixture of $A$ and water boiled at $98.0\,^{\circ}\mathrm{C}$ when the confining pressure was $101.32\,\mathrm{kN\,m^{-2}}$. The vapour pressure of pure water at this temperature is $94.26\,\mathrm{kN\,m^{-2}}$. The final distillate was found to contain 31.6% by volume of $A$. The densities of $A$ and water are 0.961 and $1.000\,\mathrm{g\,cm^{-3}}$ respectively.

   What is the relative molecular mass of $A$? Give a possible name or structural formula for $A$.

   b) Calculate the freezing point and osmotic pressure (at 298 K) of a solution containing $10.0\,\mathrm{g}$ of sucrose ($\mathrm{C_{12}H_{22}O_{11}}$) in $1000\,\mathrm{g}$ of water.

   Comment briefly on the suitability of measuring freezing point depression and osmotic pressure for the determination of relative molecular masses.

(The freezing point depression constant for water is $1.86 \, \text{K mol}^{-1} \text{kg}$. The density of the sucrose solution should be taken as $1.00 \, \text{g cm}^{-3}$ at 298 K.)                                                                     (O&C85)

4. a) What is meant by the term *colligative property*?

   b) Explain why the vapour pressure of a pure liquid is reduced by the addition of a soluble non-volatile solute.

   c) The vapour pressure of pure water is 23.50 mm Hg at 25 °C. What will be the vapour pressure of a sucrose solution at this temperature which has 1 mol of sucrose dissolved in 39 mol of water?

   d) The osmotic pressure of blood is 7 atm at 37 °C. What is the concentration of the sodium chloride solution which has the same osmotic pressure as that of blood at normal body temperature (37 °C)?                                          (JMB85)

5. a) When one mole of a non-ionised solute is dissolved in 1000 g of water, the freezing point of water falls to −1.86 °C at 101 kPa pressure. 4.84 g of a non-ionised solute dissolved in 250 g of water gives a freezing point of −0.20 °C at 101 kPa pressure.
   i)   Calculate the number of moles of solute used.
   ii)  Calculate the relative molecular mass of the solute.

   b) A solute $AB_2$ forms $A^{2+}$ and $B^-$ ions in solution and the freezing point depression depends on the total number of particles present in solution.
   i)   Write an equation showing the dissociation of $AB_2$ into ions.
   ii)  If the degree of dissociation of $AB_2$ is $\alpha$ at a particular concentration, how many moles of ions will be present if one mole of $AB_2$ is in solution at that concentration?
   iii) How many moles of unionised $AB_2$ will remain?
   iv)  Use your answers to ii) and iii) to write an expression for the ratio

$$\frac{\text{freezing point depression observed}}{\text{freezing point depression calculated if no ionisation}}$$

   v)   Using the information in a), calculate the freezing point depression which would have been caused if 0.01 mol of $AB_2$ were dissolved in 100 g of water and no ionisation occurred.
   vi)  If 0.01 mol of $AB_2$ in 100 g of water is observed to give freezing point depression of 0.487 °C, calculate the degree of dissociation, $\alpha$, of the solute $AB_2$.                               (AEB85)

6. a) The vapour pressure of a solution ($P$) is equal to the product of the vapour pressure of the pure solvent ($P_0$) and its mole fraction.
   i)   Derive the expression:

$$\frac{P_0 - P}{P_0} = \text{mole fraction of solute.}$$

ii) Explain why the vapour pressure of a solution of most solids in a given solvent is less than the vapour pressure of the pure solvent.

b) An organic solid $X$ was made up into two solutions $A$ and $B$, solution $A$ containing 5.0 g of $X$ in 100 g of water and solution $B$ containing 2.3 g of $X$ in 100 g of benzene. Both solutions $A$ and $B$ had the same vapour pressure (100 570 Pa) at the boiling points of the pure solvents at atmospheric pressure (101 300 Pa). Calculate the apparent relative molecular mass of $X$ in each case and suggest a reason for the differing results. You may use a simplified form of the above equation. (H = 1; C = 12; O = 16.)

c) What is a colligative property? Derive and explain the approximate ratios of:
  i) the osmotic pressures of solutions of $X$, of the same concentration (g dm$^{-3}$), in water and benzene;
  ii) the freezing points of two solutions containing the same masses of rhombic and monoclinic sulphur in the same volume of toluene.

d) Sketch vapour pressure-composition diagrams for *one* of the following mixtures showing the contributions made by each component to the total vapour pressure of the mixture:
  i) benzene and methylbenzene;
  ii) phenylamine (aniline) and water;
  iii) propanone and trichloromethane.

e) Draw a boiling point-composition curve for the mixture of nitric acid (boiling point 86 °C) and water which forms a constant boiling mixture (boiling point 121 °C) containing 68% nitric acid by mass. Explain what happens when a mixture containing 30% nitric acid is distilled and state the effect of re-distilling the distillate collected at 121 °C.                                                    (SUJB85)

7. a) The depression in freezing point of a solvent by a solute is known as a *colligative property*. Explain what is meant by this term and give *two* other examples of colligative properties.

b) Describe how you would determine the relative molecular mass of a compound dissolved in a suitable solvent by an experiment involving *one* of these properties.

c) Explain, with reasons, whether your method would be satisfactory for determining the relative molecular mass of
  i) ethanoic (*acetic*) acid in a concentrated solution of the acid in a suitable solvent, and
  ii) a protein molecule.

d) Calculate the freezing point of
  i) a solution of 22.5 g of cane sugar ($C_{12}H_{22}O_{11}$) in 450 g of water, and

ii) a solution of 3.0 g of potassium chloride in 400 g of water. From which solution will ice first separate? (The freezing point (cryoscopic) constant of water is $1.86 \text{ K mol}^{-1} \text{kg}$.)     (JMB84)

8. a) What is meant by the term *colligative properties*? Explain how the measurement of a colligative property can be used to obtain a value of the relative molecular mass of a compound. (Details of apparatus are not required.)

b) Measurement of relative molecular masses by means of colligative properties often gives rise to anomalous results. Discuss the factors, which give rise to such results, other than deficiencies in practical techniques.

c) Explain why salt is spread on the roads in winter to prevent formation of ice. Deduce which salt is the more cost effective for this purpose, sodium chloride or calcium chloride.

Cost/1000 kg     sodium chloride £40     calcium chloride £160

d) The relative molecular mass of a protein was determined by measuring the osmotic pressure of solutions of different concentractions. Why was this method selected in preference to, for example, depression of the freezing point?

Calculate the relative molecular mass of the protein from the following data.

| Concentration/g l$^{-1}$ | 10 | 15 | 25 | 35 | 45 |
|---|---|---|---|---|---|
| Pressure/atm | 0.0040 | 0.0063 | 0.0112 | 0.0167 | 0.0228 |

($R = 0.08\,205 \text{ litre atm mol}^{-1}\text{K}^{-1} = 8.314 \text{ J mol}^{-1}\text{K}^{-1}$, Temperature $= 298 \text{ K}$.)     (JMB81,S)

9. Chlorobenzene and bromobenzene are completely miscible. A certain mixture boils at 410 K when the pressure is $1.013 \times 10^5 \text{ Pa}$ ($\text{Pa} = \text{Nm}^{-2}$). If the vapour pressure of chlorobenzene at this temperature is $1.15 \times 10^5 \text{ Pa}$ and of bromobenzene is $6.04 \times 10^4 \text{ Pa}$, calculate the mole fraction of chlorobenzene

a) in the liquid mixture,

b) in the vapour.     (SUJB84p)

# 8 Electrochemistry

## ELECTROLYSIS

Electrovalent compounds, when molten or in solution, conduct electricity. The conductors which connect the melt or solution with the applied voltage are called the *electrodes*. The positive electrode is called the *anode*; the negative electrode is the cathode. Chemical reactions occur at the electrodes, and elements are deposited as solids or evolved as gases. These reactions are called *electrolysis*.

If the compound is a salt of a metal low in the electrochemical series, the metal ions are discharged, and a layer of metal is deposited on the cathode. Silver, copper and gold are such metals. During electrolysis of their salts, the cathode processes can be represented by

$$Ag^+(aq) \; + \; e^- \; \longrightarrow \; Ag(s)$$

$$Cu^{2+}(aq) \; + \; 2e^- \; \longrightarrow \; Cu(s)$$

$$Au^{3+}(aq) \; + \; 3e^- \; \longrightarrow \; Au(s)$$

One can see from these equations that

1 mole of $e^-$ discharge 1 mole of $Ag^+$ ions to give 1 mole of Ag atoms,

2 moles of $e^-$ discharge 1 mole of $Cu^{2+}$ ions to give 1 mole of Cu atoms,

3 moles of $e^-$ discharge 1 mole of $Au^{3+}$ ions to give 1 mole of Au atoms.

Thus, in general,

$$\left( \begin{array}{l} \text{No. of moles of} \\ \text{element discharged} \end{array} \right) = \frac{\text{No. of moles of electrons}}{\text{No. of charges on one ion of the element}}$$

It is found by experiment that 96 500 coulombs of charge are required to discharge 1 mole of silver (108 g); therefore 96 500 coulombs must be the charge on a mole of electrons. The ratio $96\,500\,C\,mol^{-1}$ is called the *Faraday constant*. Note that

No. of coulombs (C) = No. of amperes (A) × Time in second (s)

**EXAMPLE 1**  A direct current of 10.0 mA flows for 4.00 hours through three cells in series. They contain solutions of silver nitrate, copper(II) sulphate, and gold(III) nitrate. Calculate the mass of metal deposited in each.

**METHOD**     Coulombs = Amperes × Seconds = $0.0100 \times 4.00 \times 60 \times 60$
                                                    $= 144\,C$

Electrical charge passed = 144/96 500 moles of electrons
No. of moles of Ag deposited = 144/96 500 mol
Mass of Ag deposited = $108 \times 144/96\,500 = 0.161\,g$
No. of moles of Cu deposited = $\frac{1}{2} \times 144/96\,500$ mol
Mass of Cu deposited = $63.5 \times \frac{1}{2} \times 144/96\,500 = 0.0474\,g$
No. of moles of Au deposited = $\frac{1}{3} \times 144/96\,500$ mol
Mass of Au deposited = $197 \times \frac{1}{3} \times 144/96\,500 = 0.0980\,g$

**ANSWER**     Deposited are: 0.161g silver; 0.0474g copper; 0.0980g gold.

**EXAMPLE 2**  A metal of relative atomic mass 27 is deposited by electrolysis. If 0.176g of the metal is deposited on the cathode when 0.15 A flows for $3\frac{1}{2}$ hours, what is the charge on the cations of this metal?

**METHOD**     Coulombs = Amperes × seconds = $0.15 \times 3\frac{1}{2} \times 60 \times 60 = 1890\,C$
If 1 890 C deposit 0.176 g of metal,

then 96 500 C deposit $\dfrac{96\,500 \times 0.176}{1\,890}\,g = 8.98\,g$

1 mole of metal = 27 g
Since    8.98 g of metal are discharged by 1 mole of electrons,
              27 g are discharged by 27/8.98 = 3 moles of electrons

**ANSWER**     If 1 mole of metal needs 3 moles of electrons, the charge on a metal ion must be +3.

## EVOLUTION OF GASES

Solutions of salts of metals high in the electrochemical series evolve hydrogen at the cathode on electrolysis. The cathode process is:

$$H^{+}(aq)\ +\ e^{-} \longrightarrow H(g)$$

followed by                  $$2H(g) \longrightarrow H_2(g)$$

Thus, 2 moles of electrons are needed to evolve 1 mole of hydrogen molecules; that is, 2 g of hydrogen (the molar mass) or 22.4 dm³ at s.t.p. (the molar volume).

At the anode, solutions of halides evolve the halogen, and other salts evolve oxygen. When chlorine is evolved, the anode process is

$$Cl^{-}(aq) \longrightarrow Cl(g)\ +\ e^{-}$$

followed by                  $$2Cl(g) \longrightarrow Cl_2(g)$$

Thus, 2 moles of electrons are required for the evolution of 1 mole of chlorine molecules, 71.0 g or 22.4 dm³ at s.t.p.

When oxygen is evolved, the anode process is the discharge of hydroxide ions, derived from the water in the solution:

$$OH^-(aq) \longrightarrow OH(aq) + e^-$$

followed by    $4OH(aq) \longrightarrow O_2(g) + 2H_2O(l)$

Thus, 4 moles of electrons are required for the evolution of 1 mole of oxygen molecules ($22.4 \, dm^3$ oxygen at s.t.p.).

**EXAMPLE 1** State the names and calculate the volumes of gases formed at the cathode and anode at s.t.p. when $0.0250 \, A$ of current are passed for $4.00$ hours through a solution of sulphuric acid.

**METHOD** At the anode, oxygen is evolved, and at the cathode hydrogen.

Coulombs = Amperes $\times$ Seconds = $0.0250 \times 4.00 \times 60 \times 60 = 360 \, C$

No. of moles of electrons = $360/96\,500$

No. of moles of $H_2$ = $\frac{1}{2} \times 360/96\,500$

Vol. of $H_2$ = $22.4 \times \frac{1}{2} \times 360/96\,500 \, dm^3$ = $0.0417 \, dm^3$ = $41.7 \, cm^3$

Vol. of $O_2$ = $\frac{1}{2} \times 41.7$ = $20.9 \, cm^3$

**ANSWER** $41.7 \, cm^3$ hydrogen is formed at the cathode and $20.9 \, cm^3$ of oxygen at the anode.

**EXAMPLE 2** A current of $1.00 \, A$ flowing for 1 hour 50 minutes deposits $2.15 \, g$ of copper from an aqueous solution of copper(II) sulphate. If the Avogadro constant is $6.02 \times 10^{23} \, mol^{-1}$, calculate the charge on an electron.

**METHOD** Coulombs = Amperes $\times$ Seconds = $1.00 \times 110 \times 60 = 6600 \, C$

No. of moles of Cu = $2.15/63.5$

No. of atoms of Cu = $6.02 \times 10^{23} \times 2.15/63.5$

No. of electrons = $2 \times 6.02 \times 10^{23} \times 2.15/63.5$ = $0.406 \times 10^{23}$

$\dfrac{\text{No. of coulombs}}{\text{No. of electrons}}$ = $6600/(0.406 \times 10^{23})$ = $1.63 \times 10^{-19} \, C \, electron^{-1}$

**ANSWER** The charge on an electron is $1.63 \times 10^{-19} \, C$.

## EXERCISE 41    Problems on Electrolysis

In the following problems use Faraday constant = $96\,500 \, C \, mol^{-1}$.

1. Calculate the mass of copper that would be deposited on a copper cathode from an aqueous solution of copper(II) sulphate, if the same current, passed for the same time, liberated $0.900 \, g$ of silver from an aqueous solution of silver nitrate, $AgNO_3$.

2. A current of 100 mA was passed for 1.00 h through an aqueous solution of $1.00 \, \text{mol dm}^{-3}$ silver nitrate. A silver cathode was used. Calculate the increase in mass of the cathode. State how the change in mass would be affected by:
   a) passing a current of 200 mA
   b) passing a current of 100 mA for 2.00 h
   c) using a $2.00 \, \text{mol dm}^{-3}$ solution of silver nitrate.

3. An electric current of 5.00 A was passed through molten anhydrous calcium chloride, $CaCl_2$, for 20.0 minutes between graphite electrodes. Calculate the mass of each product liberated.

4. A current of electricity liberated $6.22 \times 10^{-3}$ mol of silver from a silver nitrate solution. What mass in grams, of aluminium, would be liberated from a suitable aluminium compound, using the same quantity of electricity?

5. A current is passed for 45 minutes through three solutions in series, using platinum electrodes. In one, 0.203 g of silver is deposited from silver nitrate solution. In a second, hydrogen is evolved from a solution of dilute sulphuric acid, and, in a third, lead is deposited from a solution of lead nitrate. Calculate: a) the current passed, b) the volume of hydrogen collected at $0.983 \times 10^5 \, \text{N m}^{-2}$ and 18 °C, c) the mass of lead deposited.

6. A current of 0.750 A passes through 250 $cm^3$ solution of 0.250 mol $dm^{-3}$ copper(II) sulphate solution. How long (in minutes) will it take to deposit all the copper on the cathode?

7. What current is needed to deposit 0.500 g of nickel from a nickel(II) sulphate solution in 1.00 hour?

8. A current of 1.25 A passes for 5.00 h between platinum electrodes in 500 $cm^3$ of copper(II) sulphate solution of concentration 2.00 mol $dm^{-3}$. What will be the concentration of copper(II) sulphate at the end of the time?

9. How long (in hours) will it take a current of 0.100 A to deposit 1.00 kg of silver from an unlimited source of silver ions?

10. A current of 2.05 A passes through a solution of sulphuric acid for 5.00 h. Calculate the volumes of hydrogen and oxygen produced.

11. A steady current was passed through a solution of copper(II) sulphate until 6.05 g of metallic copper were deposited on the cathode. How many coulombs of electricity had been used?

12. In the electrolysis of a solution of potassium chloride, 100 $cm^3$ of chlorine are produced at 20 °C and $9.9 \times 10^4 \, \text{N m}^{-2}$. How many seconds has a current of 0.750 A been flowing through the solution to effect this?

13. A current of 1.75 A passed for 1.00 hour through a solution of copper(II) sulphate. At the cathode, 1.245 g of copper were deposited, and some hydrogen was evolved. Calculate the volume of hydrogen (at s.t.p.) which was evolved.

## ELECTROLYTIC CONDUCTIVITY

Solutions of electrolytes obey Ohm's law, from which it follows that

$$R = V/I$$

where $R$ = Resistance, $V$ = Potential difference, $I$ = Current.

Resistivity is defined by the equation

$$R = \rho \times l/A$$

where $l$ = Length and $A$ = Cross-sectional area of the conductor.

If $R$ is in ohms ($\Omega$), $l$ in metres (m), and $A$ in square metres (m$^2$), then $\rho$ has the dimensions $\Omega$ m. (If $l$ is in cm, and $A$ in cm$^2$, then $\rho$ has the dimensions $\Omega$ cm.)

The reciprocal of resistance is conductance; the reciprocal of resistivity $\rho$ is conductivity $\kappa$.

$$1/\rho = \kappa$$

$\kappa$ has the dimensions $\Omega^{-1}m^{-1}$ or $\Omega^{-1}cm^{-1}$.

---

Molar conductivity $\Lambda$ is defined by the equation

$$\Lambda = \kappa/c$$

---

where $c$ = Molar concentration of solute. If $\kappa$ is in $\Omega^{-1}m^{-1}$ and $c$ is in mol m$^{-3}$, then $\Lambda = \Omega^{-1}m^2\,mol^{-1}$, and $\Lambda$ is numerically equal to the conductivity of 1 mole of the electrolyte.

The molar conductivity, $\Lambda$, of the solute increases as the concentration of the solution decreases, until further dilution has no further effect. The value of molar conductivity at this dilution is called the molar conductivity at infinite dilution, and is represented as $\Lambda_0$ or $\Lambda_\infty$.

**EXAMPLE 1** Calculate the molar conductivity of sodium chloride solution of concentration 0.0100 mol dm$^{-3}$, given that its conductivity is 0.1185 $\Omega^{-1}m^{-1}$.

**METHOD**
$c = 0.0100\,\text{mol dm}^{-3} = 10.0\,\text{mol m}^{-3}$

$\kappa = 0.1185\,\Omega^{-1}m^{-1}$

$\Lambda = \kappa/c = 0.1185/10.0 = 1.185 \times 10^{-2}\,\Omega^{-1}m^2\,mol^{-1}$

**ANSWER** The molar conductivity is $1.185 \times 10^{-2}\,\Omega^{-1}m^2\,mol^{-1}$ or $118.5\,\Omega^{-1}cm^2\,mol^{-1}$.

To find the conductivity of a solution, its resistance is measured, and the equation $R = \rho \times l/A$ is inverted to give

$$1/R = \kappa \times A/l$$

The area of the electrodes, $A$, and the distance between them, $l$, can be measured and inserted in the equation to give $\kappa$. In practice, the method of finding the ratio $l/A$, which is called the cell constant, is to measure the resistance of a solution of known conductivity and insert $\kappa$ and $R$ into the equation to give the value of $l/A$. Once the cell constant is known, this cell can be used to find values of conductivity from measurements of the conductance of solutions.

**EXAMPLE 2**  The resistance of a conductance cell containing $0.100 \, mol \, dm^{-3}$ KCl solution at 25 °C is 47.85 $\Omega$. If the same cell contains sodium nitrate solution of concentration $0.0200 \, mol \, dm^{-3}$, the resistance is 254 $\Omega$. The conductivity of the KCl solution is $0.0129 \, \Omega^{-1} cm^{-1}$. Calculate a) the cell constant, b) the conductivity of the sodium nitrate solution and c) the molar conductivity of sodium nitrate at a concentration of $0.0200 \, mol \, dm^{-3}$.

**METHOD**  a) $R = 47.85 \, \Omega$, and $\kappa = 0.0129 \, \Omega^{-1} cm^{-1} = 1.29 \, \Omega^{-1} m^{-1}$

Cell constant $= l/A$

Since $1/R = \kappa \times A/l$

**ANSWER**  $l/A = 47.85 \times 1.29 = 61.7 \, m^{-1}$

b) $R = 254.0 \, \Omega$

Since $1/R = \kappa \times A/l$

$1/254.0 = \kappa/61.7$

**ANSWER**  $\kappa = 61.7/254.0 = 0.243 \, \Omega^{-1} m^{-1}$

c) Since $\Lambda = \kappa/c$

**ANSWER**  $\Lambda = 0.243/20.0 = 1.215 \times 10^{-2} \, \Omega^{-1} m^2 \, mol^{-1}$.

# EXERCISE 42    Problems on Conductivity

1. A column of electrolyte 3.00 cm long and of cross-section 2.00 cm² has a resistance of 15.0 $\Omega$. Calculate: a) the resistivity and b) the conductivity of the electrolyte.

2. An electrolytic cell is 3.00 cm long and has a cross-section of 1.50 cm². If the resistance is 12.0 $\Omega$, a) what is the resistivity, and b) what is the conductivity of the electrolyte?

3. The resistance of a cell containing $0.0100 \, mol \, dm^{-3}$ KCl solution is 12.5 $\Omega$. The conductivity of KCl at this concentration is $0.1413 \, \Omega^{-1} m^{-1}$. What is the cell constant?

4. An electrolyte solution of concentration $3.00 \times 10^{-3} \, mol \, dm^{-3}$ has a resistance of 45.0 $\Omega$ in a cell with a cell constant of $2.20 \, m^{-1}$. Calculate the molar conductivity of the electrolyte.

5. A solution of an electrolyte of concentration $0.0200\ mol\ dm^{-3}$ has a resistance of $0.357\ \Omega$ in a cell with a cell constant of $1.50\ m^{-1}$. Calculate the molar conductivity of the electrolyte.

## MOLAR CONDUCTIVITY AND CONCENTRATION

The molar conductivity of a solute depends on its concentration. When values of molar conductivity $\Lambda$ are plotted against the volume of solution containing 1 mole of solute (called the dilution), the graphs obtained for different electrolytes fall into two categories. These are shown in Fig. 8.1(a).

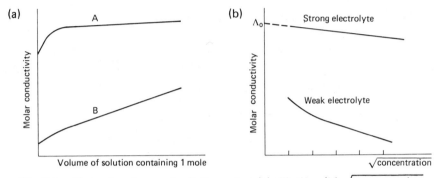

Fig. 8.1   Plots of molar conductivity against (a) dilution, (b) $\sqrt{\text{concentration}}$

Strong electrolytes are those which are completely ionised in solution. They give graphs of shape A, swiftly rising to a maximum. Weak electrolytes, which are incompletely ionised, give graphs of shape B. They do not reach a maximum at dilutions for which conductivities can be measured in practice. Arrhenius explained graphs of shape B as arising from an increase in the ionisation of the solute with dilution. The fraction of the solute which is ionised, $\alpha$, gradually increases with dilution. If measurements could be made in more and more dilute solutions, $\alpha$ would increase up to a value of 1, and $\Lambda$ would reach its limiting value of $\Lambda_0$. This is called the molar conductivity at infinite dilution since further dilution results in no further increase in $\Lambda$. In more concentrated solutions,

$$\alpha = \Lambda/\Lambda_0$$

Strong electrolytes show the steep increase in $\Lambda$ with dilution illustrated in curve A. It seems to indicate that the strong electrolytes are completely ionised except at high concentrations. According to more modern views, strong electrolytes are completely ionised at all concentrations. Another explanation of the low value of $\Lambda$ at high concentration is proposed. It is attributed to the attraction of oppositely charged ions slowing down the ions in a concentrated solution. The value of $\alpha$ given by $\Lambda/\Lambda_0$ is referred to as the *apparent degree of ionisation* or the *conductivity ratio*.

For strong electrolytes, the graph of $\Lambda$ against $\sqrt{c}$ is a straight line, which can be extrapolated to cut the $\Lambda$ axis at $\Lambda_0$. For weak electrolytes, the graph of $\Lambda$ against $\sqrt{c}$ is of the shape shown in Fig. 8.1(b). You can see that it cannot be used to find $\Lambda_0$. Kohlrausch's law helps here. It enables $\Lambda_0$ to be found for weak electrolytes. Kohlrausch's law can be expressed as

$$\Lambda_0 = \Lambda_0(\text{cation}) + \Lambda_0(\text{anion})$$

For every cation and anion, values of $\Lambda_0$ can be looked up in tables and added to give $\Lambda_0$ for the salt. Alternatively, measurements on strong electrolytes can be used to give values of $\Lambda_0$ which can be combined to give $\Lambda_0$ for the required weak electrolyte.

**EXAMPLE**   Calculate the molar conductivity at infinite dilution at $25\,^\circ C$ of ethanoic acid, given that the molar conductivities at infinite dilution of hydrochloric acid, sodium chloride and sodium ethanoate are $4.26 \times 10^{-2}$, $1.26 \times 10^{-2}$ and $9.1 \times 10^{-3}\,\Omega^{-1}\,m^2\,mol^{-1}$ respectively.

**METHOD**   According to Kohlrausch's law,

$$\Lambda_0(CH_3CO_2H) = \Lambda_0(CH_3CO_2^-) + \Lambda_0(H^+)$$

$$\Lambda_0(HCl) = \Lambda_0(H^+) + \Lambda_0(Cl^-)$$

$$\Lambda_0(NaCl) = \Lambda_0(Na^+) + \Lambda_0(Cl^-)$$

$$\Lambda_0(CH_3CO_2Na) = \Lambda_0(CH_3CO_2^-) + \Lambda_0(Na^+)$$

Thus,

$$\Lambda_0(CH_3CO_2H) = \Lambda_0(HCl) + \Lambda_0(CH_3CO_2Na) - \Lambda_0(NaCl)$$

**ANSWER**   $$\Lambda_0(CH_3CO_2H) = (4.26 \times 10^{-2}) + (9.00 \times 10^{-3}) - (1.26 \times 10^{-2})$$

$$= 3.91 \times 10^{-2}\,\Omega^{-1}\,m^2\,mol^{-1}$$

## CALCULATION OF SOLUBILITY FROM CONDUCTIVITY

The solubility of a sparingly soluble salt can be found by measuring the conductance of a saturated solution of the salt. The conductance depends on the concentration of the dissolved ions. From it, the conductivity can be calculated. For a sparingly soluble salt, the concentration is so low, even in a saturated solution, that the salt can be assumed to be completely ionised. The conductivity is compared with the molar conductivity at infinite dilution of the salt. This is calculated as the sum of the molar conductivities at infinite dilution of the cation and anion. Then

$$\Lambda_0 = \kappa/c$$

where $\Lambda_0$ = Molar conductivity at infinite dilution, $\kappa$ = Conductivity, and $c$ = Molar concentration of solute.

Hence $c$ can be calculated. Since the solution is saturated, the concentration is equal to the solubility of the salt.

**EXAMPLE**   After making allowance for the conductivity of water, the conductivity of a saturated solution of silver chloride at $25\,°C = 1.50 \times 10^{-4}\,\Omega^{-1}\,m^{-1}$. If $\Lambda_0(Ag^+) = 6.2 \times 10^{-3}$ and $\Lambda_0(Cl^-) = 7.6 \times 10^{-3}\,\Omega^{-1}\,m^2\,mol^{-1}$, what is the solubility of silver chloride in $mol\,dm^{-3}$?

**METHOD**   The molar conductivity at infinite dilution, $\Lambda_0$, is the sum of those of the ions.

$$\Lambda_0 = \Lambda_0(Ag^+) + \Lambda_0(Cl^-)$$
$$= (6.2 \times 10^{-3}) + (7.7 \times 10^{-3}) = 1.38 \times 10^{-2}$$
$$\Lambda_0 = \kappa/c$$
$$c = \kappa/\Lambda_0$$
$$= (1.50 \times 10^{-4})/(1.38 \times 10^{-2})$$
$$= 1.09 \times 10^{-2}\,mol\,m^{-3} = 1.09 \times 10^{-5}\,mol\,dm^{-3}$$

**ANSWER**   The solubility of silver chloride is $1.09 \times 10^{-5}\,mol\,dm^{-3}$.

## IONIC EQUILIBRIA; WEAK ELECTROLYTES

Electrolytes are ionic compounds. Strong electrolytes consist entirely of ions. Weak electrolytes consist of molecules, some of which dissociate to form ions. The fraction of molecules which dissociate is called the degree of ionisation or degree of dissociation. As the concentration of a solution of a weak electrolyte decreases, the degree of ionisation increases. *Ostwald's dilution law* gives a relationship between the degree of ionisation, $\alpha$, of a weak electrolyte and its concentration $c$. Consider a weak acid, HA, in a solution of concentration $c$. An equilibrium is set up between the undissociated molecules, HA, and the ions $H_3O^+$ and $A^-$. $H_3O^+$ is the oxonium ion, a complex ion formed by the combination of a hydrogen ion and a molecule of water. It can also be written as $H^+(aq)$. Since the degree of ionisation is $\alpha$, the equilibrium concentrations of the species are: $[HA] = (1-\alpha)c$; $[H_3O^+] = [A^-] = \alpha c$ (where the square brackets stand for concentration). The equilibrium can be represented:

$$HA + H_2O \rightleftharpoons H_3O^+ + A^-$$
$$c(1-\alpha) \qquad\qquad c\alpha \quad\; c\alpha$$

The ratio $\dfrac{[H_3O^+]\,[A^-]}{[HA]\,[H_2O]}$ is a constant. (See Chapter 11 on Equilibria.)

The value of $[H_2O]$ is constant as the concentration of water is not significantly altered by the ionisation of HA.

The ratio $\dfrac{[H_3O^+][A^-]}{[HA]}$ is called the dissociation constant, $K_a$, of the acid:

$$K_a = \frac{[H_3O^+][A^-]}{[HA]} = \frac{c\alpha \times c\alpha}{c(1-\alpha)} = \frac{c\alpha^2}{1-\alpha}$$

The Ostwald dilution law is embodied in the equation

$$K_a = \frac{c\alpha^2}{1-\alpha}$$

For some weak electrolytes, $\alpha$ is so small that the error involved in putting $(1-\alpha)$ equal to 1 is negligible. In this case,

$$K_a = c\alpha^2$$
$$\alpha = \sqrt{K_a/c}$$

Ostwald's dilution law can be applied to weak bases. In the case of a weak base B which is partially ionised in solution,

$$B + H_2O \rightleftharpoons BH^+ + OH^-$$

The concentration of water is not significantly altered by the dissociation: $[H_2O]$ = constant.

The base dissociation constant $K_b$ is given by the equation

$$K_b = \frac{[BH^+][OH^-]}{[B]}$$

$$K_b = \frac{c\alpha^2}{1-\alpha}$$

If $\alpha \ll 1$, the expression approximates to $K_b = c\alpha^2$.


## CALCULATION OF THE DEGREE OF DISSOCIATION AND THE DISSOCIATION CONSTANT OF A WEAK ELECTROLYTE FROM CONDUCTANCE MEASUREMENTS

From conductance measurements on a solution of known concentration, the molar conductivity $\Lambda$ is found. The relationship,

$$\alpha = \Lambda/\Lambda_0$$

enables the degree of ionisation $\alpha$ to be found. The value of $\alpha$ can be substituted in the Ostwald equation to give the ionisation constant of the electrolyte.

**EXAMPLE 1**   a) Calculate the degree of dissociation of ethanoic acid of concentration $1.00 \times 10^{-3}\,mol\,dm^{-3}$. The conductivity of the solution is $4.85 \times 10^{-3}\,\Omega^{-1}\,m^{-1}$ and the values of $\Lambda_0$ for ethanoic acid is $3.91 \times 10^{-2}\,\Omega^{-1}\,m^2\,mol^{-1}$.

b) Calculate the dissociation constant of ethanoic acid at the temperature at which the measurements were made.

**METHOD**   a) Conductivity, $\kappa = 4.85 \times 10^{-3}\,\Omega^{-1}\,m^{-1}$

Concentration, $c = 1.00 \times 10^{-3}\,mol\,dm^{-3} = 1.00\,mol\,m^{-3}$

$$\Lambda = \kappa/c = (4.85 \times 10^{-3})/1.00 = 4.85 \times 10^{-3}\,\Omega^{-1}\,m^2\,mol^{-1}$$

$$\Lambda_0(CH_3CO_2H) = 3.91 \times 10^{-2}\,\Omega^{-1}\,m^2\,mol^{-1}$$

$$\alpha = \Lambda/\Lambda_0$$

$$= (4.85 \times 10^{-3})/(3.91 \times 10^{-2}) = 0.124$$

**ANSWER**   The degree of dissociation of ethanoic acid is 0.124.

**METHOD**   b) From the Ostwald equation,

$$K_a = \frac{c\alpha^2}{1-\alpha}$$

With a value of 0.124 for $\alpha$, $(1-\alpha)$ cannot be put $= 1$

$$K_a = \frac{10^{-3} \times (0.124)^2}{0.876} = 1.76 \times 10^{-5}\,mol\,dm^{-3}$$

**ANSWER**   The dissociation constant of ethanoic acid is $1.76 \times 10^{-5}\,mol\,dm^{-3}$.

**EXAMPLE 2**   Calculate the degree of dissociation of hydrogen cyanide in a solution of concentration $0.100\,mol\,dm^{-3}$. $K_a = 7.24 \times 10^{-10}\,mol\,dm^{-3}$.

**METHOD**                        $K_a = \dfrac{c\alpha^2}{(1-\alpha)}$

$$7.24 \times 10^{-10} = \frac{0.100\alpha^2}{(1-\alpha)} \qquad \therefore \ 7.24 \times 10^{-9} = \frac{\alpha^2}{1-\alpha}$$

Inspection shows that $\alpha$ is of the order of $10^{-5}$, and $(1-\alpha)$ can be put equal to 1.

Then             $\alpha = \sqrt{7.24 \times 10^{-9}} = 8.5 \times 10^{-5}$

**ANSWER**   The degree of dissociation is $8.5 \times 10^{-5}$.

# EXERCISE 43     Problems on Molar Conductivity

1. The molar conductivity of a solution of a weak acid of concentration $0.0400\,mol\,dm^{-3}$ at $25\,°C$ is $8.25 \times 10^{-4}\,\Omega^{-1}\,m^2\,mol^{-1}$. If the molar conductivity at infinite dilution is $3.98 \times 10^{-2}\,\Omega^{-1}\,m^2\,mol^{-1}$ at $25\,°C$, calculate the dissociation constant of the acid at $25\,°C$.

2. A solution of a weak monobasic acid has a concentration of 0.0250 $mol\,dm^{-3}$. The conductivity at 25 °C is $2.64 \times 10^{-2}\Omega^{-1}m^{-1}$. The molar conductivity at infinite dilution is $3.91 \times 10^{-2}\Omega^{-1}m^2\,mol^{-1}$. Calculate: a) the degree of dissociation, b) the concentration of hydrogen ions in solution, and c) the dissociation constant of the acid.

3. A solution of a weak monobasic acid of concentration 0.0200 mol $dm^{-3}$ has conductivity $0.184\,\Omega^{-1}m^{-1}$ at 25 °C. The molar conductivity at infinite dilution is $4.00 \times 10^{-2}\Omega^{-1}m^2\,mol^{-1}$. Calculate: a) the degree of dissociation, and b) the dissociation constant of the acid at 25 °C.

4. A solution of methanoic acid of concentration 0.100 mol $dm^{-3}$ has conductivity $0.166\,\Omega^{-1}m^{-1}$ at 25 °C. The molar conductivity at infinite dilution is $4.04 \times 10^{-2}\Omega^{-1}m^2\,mol^{-1}$. Calculate the dissociation constant of methanoic acid at 25 °C.

5. A solution of an organic acid, $RCO_2H$, has a concentration of 0.0300 mol $dm^{-3}$ and a molar conductivity of $1.00 \times 10^{-3}\Omega^{-1}m^2$ $mol^{-1}$ at 25 °C. The molar conductivity at infinite dilution at 25 °C is $39.0 \times 10^{-3}\Omega^{-1}m^2\,mol^{-1}$. Calculate: a) the degree of dissociation, and b) the dissociation constant for the acid at 25 °C.

## CALCULATION OF pH, pOH and pK

Acids react with water to give hydrogen ions. These are not simple $H^+$ ions; they are complex ions of formula $H_3O^+$, formed by combination of a molecule of water and an $H^+$ ion. Although they are properly called *oxonium* ions, in this text, $H_3O^+$ ions will be referred to as hydrogen ions. The hydrogen ion concentration of a solution can be indicated by means of a number on the pH scale.

The pH of a solution is the negative logarithm to the base 10 of the hydrogen ion concentration. pOH, $pK_a$ and $pK_b$ are defined below.

$$pH = -lg[H_3O^+/mol\,dm^{-3}] \qquad pK_a = -lg(K_a/mol\,dm^{-3})$$
$$pOH = -lg[OH^-/mol\,dm^{-3}] \qquad pK_b = -lg(K_b/mol\,dm^{-3})$$

### Strong acids and bases

The pH of a solution of a strong acid or base is simply calculated. If the concentration of hydrochloric acid is 0.1 mol $dm^{-3}$,

$$[H_3O^+] = 0.1\,mol\,dm^{-3} = 10^{-1}\,mol\,dm^{-3}$$
$$lg\,[H_3O^+] = -1$$
$$pH = 1$$

If the concentration of a solution of sodium hydroxide is 0.01 mol dm$^{-3}$,

$$[OH^-] = 0.01 = 10^{-2}\,mol\,dm^{-3}$$

The product of the hydrogen ion concentration and the hydroxide ion concentration in a solution is called the ionic product for water, $K_w$. At 25 °C, the value of $K_w$ is $10^{-14}\,mol^2\,dm^{-6}$.

$$[H_3O^+][OH^-] = K_w = 10^{-14}\,mol^2\,dm^{-6}$$

Therefore, in this solution,

$$[H_3O^+] = 10^{-14}/10^{-2} = 10^{-12}\,mol\,dm^{-2}$$
$$pH = -lg\,[H_3O^+] = 12$$

**EXAMPLE 1** Calculate the pH of a solution of 0.020 mol dm$^{-3}$ hydrochloric acid.

**METHOD** If the concentration of hydrochloric acid is 0.020 mol dm$^{-3}$,

$$[H_3O^+] = 2 \times 10^{-2}\,mol\,dm^{-3}$$

**ANSWER** $pH = -lg\,[H_3O^+] = -(0.301 + \bar{2}) = 1.70$

**EXAMPLE 2** Calculate the pH of a 0.010 mol dm$^{-3}$ solution of calcium hydroxide.

**METHOD** If the concentration of Ca(OH)$_2$ is $10^{-2}\,mol\,dm^{-3}$,

$$[OH^-] = 2 \times 10^{-2}\,mol\,dm^{-3}$$
$$pOH = -lg\,(2 \times 10^{-2}) = 1.7$$

Since $[H_3O^+][OH^-] = K_w = 10^{-14}\,mol^2\,dm^{-6}$

$$lg\,[H_3O^+] + lg\,[OH^-] = lg\,K_w = -14$$
$$\therefore \quad pH + pOH = 14$$

Thus, if pOH = 1.7,

**ANSWER** pH = 14 − 1.7 = 12.3.

(For practice in using a calculator to find logarithms, see pp. 3 and 10.)

## Weak acids and bases

For calculating the pH of a weak acid or base, the concentration of the solution is not sufficient information. The degree of dissociation must be taken into account. The following examples show how the pH can be calculated from the concentration and the degree of dissociation or the dissociation constant of a weak electrolyte.

The converse is also true. If the pH can be measured experimentally, the value of the pH can be used to calculate the dissociation constant of the weak electrolyte.

**EXAMPLE 1**  Calculate the pH of a solution of propanoic acid of concentration $0.0100 \, mol \, dm^{-3}$, given that the degree of dissociation is 0.116.

**METHOD**  Let the concentration of acid be $c$, and the degree of dissociation be $\alpha$. Then in the equilibrium

$$HA + H_2O \rightleftharpoons H_3O^+ + A^-$$

we can put the concentrations

$$[HA] = c(1-\alpha); \quad [H_3O^+] = [A^-] = c\alpha.$$

Then      $[H_3O^+] = 0.0100 \times 0.116 = 1.16 \times 10^{-3}$

$$pH = -\lg [H_3O^+] = 2.9355$$

**ANSWER**  pH = 2.94.


**EXAMPLE 2**  Calculate the pH of a $0.0100 \, mol \, dm^{-3}$ solution of ethanoic acid, which has a dissociation constant of $1.76 \times 10^{-5} \, mol \, dm^{-3}$.

**METHOD**  When a weak acid ionises,

$$HA + H_2O \rightleftharpoons H_3O^+ + A^-$$

the dissociation constant $K_a$ is given by the expression

$$K_a = \frac{[H_3O^+] [A^-]}{[HA]}$$

One $A^-$ is formed for each $H_3O^+$.

$\therefore$      $[H_3O^+] = [A^-]$  and  $[H_3O^+] [A^-] = [H_3O^+]^2$

The degree of dissociation is so small that we can put [HA] equal to the total acid concentration, $0.0100 \, mol \, dm^{-3}$. In fact, it must be slightly less than this. Say the degree of dissociation is 0.001. Then [HA] = $0.0100 - 0.00001 = 0.00999$. The error in assuming [HA] = 0.0100 is 1 part in 1000. This is smaller than the error in the practical measurements from which the values used in the calculation are obtained. For most solutions of weak acids and bases, we can make the approximation safely.

Thus,      $$K_a = \frac{[H_3O^+]^2}{10^{-2}}$$

$$[H_3O^+]^2 = 1.76 \times 10^{-7}$$

$$[H_3O^+] = 4.19 \times 10^{-4} \, mol \, dm^{-3}$$

$$pH = -\lg [H_3O^+]$$

**ANSWER**  pH = 3.34.


**EXAMPLE 3**  Calculate the dissociation constant of phenol, given that a $1.00 \times 10^{-2} \, mol \, dm^{-3}$ solution has a pH of 5.95.

**METHOD**    If pH $= 5.95$,

$$[H_3O^+] = \text{antilg}(-pH) = 1.13 \times 10^{-6}$$

If we write the dissociation of phenol

$$PhOH + H_2O \rightleftharpoons PhO^- + H_3O^+$$

$$K_a = \frac{[H_3O^+][PhO^-]}{[PhOH]} = \frac{[H_3O^+]^2}{[PhOH]}$$

$$K_a = \frac{(1.13 \times 10^{-6})^2}{10^{-2}} = 1.28 \times 10^{-10}\,\text{mol dm}^{-3}$$

**ANSWER**    The dissociation constant of phenol is $1.28 \times 10^{-10}\,\text{mol dm}^{-3}$.

**EXAMPLE 4**    A solution of pyridine of concentration $0.0100\,\text{mol dm}^{-3}$ has a pH 8.63. Calculate the dissociation constant of pyridine.

**METHOD**    We can write the dissociation of pyridine as

$$P + H_2O \rightleftharpoons PH^+ + OH^-$$

Thus,    $$K_b = \frac{[PH^+][OH^-]}{[P]} = \frac{[OH^-]^2}{0.0100}$$

Since    pH $= 8.63$

pOH $= 14.00 - 8.63 = 5.37$

$[OH^-] = 4.26 \times 10^{-6}$

$$K_b = \frac{(4.26 \times 10^{-6})^2}{0.0100} = 1.82 \times 10^{-9}\,\text{mol dm}^{-3}$$

**ANSWER**    The dissociation constant of pyridine is $1.82 \times 10^{-9}\,\text{mol dm}^{-3}$.

**EXAMPLE 5**    Calculate the degree of dissociation of a $0.0100\,\text{mol dm}^{-3}$ solution of ethylamine, given that $K_b = 6.46 \times 10^{-4}\,\text{mol dm}^{-3}$; $K_w = 1.00 \times 10^{-14}\,\text{mol}^2\,\text{dm}^{-6}$.

**METHOD**    Since    $$B + H_2O \rightleftharpoons BH^+ + OH^-$$

then

$$K_b = \frac{c\alpha^2}{(1-\alpha)} \quad (c = \text{Concentration}; \alpha = \text{Degree of dissociation})$$

Making the approximation $(1-\alpha) = 1$ gives

$$K_b = c\alpha^2$$

$$6.46 \times 10^{-4} = 0.0100\,\alpha^2$$

$$\alpha = 0.254$$

This value of $\alpha$ is not small compared with 1: the approximation $(1-\alpha) = 1$ cannot be made. Since $(1-\alpha) = 0.766$,

$$6.46 \times 10^{-4} = \frac{0.0100\alpha^2}{0.766}$$

$$\alpha = 0.222$$

This second value for $\alpha$ makes $(1-\alpha) = 0.778$. Putting this value into the equation gives

$$6.46 \times 10^{-4} = \frac{0.0100\alpha^2}{0.778}$$

$$\alpha = 0.224$$

The value of $(1-\alpha)$ is now 0.776, and

$$6.46 \times 10^{-4} = \frac{0.0100\alpha^2}{0.776}$$

ANSWER    Degree of dissociation        $\alpha = 0.224$

We have finally arrived at a value of $\alpha$ which satisfies the equation

$$6.46 \times 10^{-4} = \frac{0.0100\alpha^2}{(1-\alpha)}$$

This method of calculation is called the *method of successive approximations*.

It is a neater and more convenient method than solving a quadratic equation.

## Conjugate pairs

Some tables list the base dissociation constants, $K_b$, of bases such as amines. Others list the acid dissociation constants, $K_a$.

The value of $K_b$ refers to the equilibrium

$$B + H_2O \rightleftharpoons BH^+ + OH^-$$

where

$$K_b = \frac{[BH^+][OH^-]}{[B]}$$

The species $BH^+$ is referred to as the conjugate acid of the base, B. $BH^+$ dissociates according to the equilibrium

$$BH^+ + H_2O \rightleftharpoons B + H_3O^+$$

One can, therefore, write the acid dissociation constant of $BH^+$ as

$$K_a = \frac{[B][H_3O^+]}{[BH^+]}$$

Multiplying $K_a$ by $K_b$ gives $[H_3O^+][OH^-]$. That is, $K_aK_b = K_w$.

This is a useful relationship between the dissociation constants of conjugate pairs.

**EXAMPLE 1** Calculate the pH of a solution of ammonia ($pK_a = 9.25$) of concentration $0.0100 \, mol \, dm^{-3}$.

**METHOD**

$$pK_a = 9.25 \qquad pK_b = 14.00 - 9.25 = 4.75$$

$$K_b = 1.78 \times 10^{-5}$$

Since

$$K_b = \frac{[NH_4^+][OH^-]}{[NH_3]} = \frac{[OH^-]^2}{[NH_3]}$$

$$[OH^-]^2 = K_b \times c$$

$$= 1.78 \times 10^{-5} \times 0.0100$$

$$[OH^-] = 4.22 \times 10^{-4}$$

$$pOH = 3.37$$

**ANSWER**

$$pH = 10.6$$

**EXAMPLE 2** Calculate the degree of ionisation of phenylmethylamine ($pK_a = 9.37$) in a $0.100 \, mol \, dm^{-3}$ aqueous solution. Quote your answer to two significant figures.

**METHOD**    Since

$$pK_a = 9.37$$

$$pK_b = 14.00 - 9.37 = 4.63$$

$$K_b = 2.34 \times 10^{-5}$$

Since

$$K_b = \frac{[C_6H_5CH_2NH_3^+][OH^-]}{[C_6H_5CH_2NH_2]}$$

$$= \frac{\alpha^2 c}{(1-\alpha)}$$

making the approximation $1 - \alpha = 1$,

$$\alpha^2 = K_b/c$$

$$= 2.34 \times 10^{-5}/0.100 = 2.34 \times 10^{-4}$$

$$\alpha = 1.53 \times 10^{-2}$$

The approximation $(1 - \alpha) = 1$ is not entirely justified, but, since the answer is required to two significant figures only, this is close enough.

**ANSWER**    The degree of ionisation is $1.5 \times 10^{-2}$.

**EXERCISE 44**     Problems on pH

1. Calculate the pH of solutions with following $H_3O^+$ concentrations in mol dm$^{-3}$:

   a) $10^{-8}$          b) $10^{-4}$          c) $10^{-7}$
   d) $6.8 \times 10^{-3}$     e) $3.2 \times 10^{-5}$     f) $0.035$
   g) $0.25$         h) $5.4 \times 10^{-9}$     i) $7.1 \times 10^{-7}$
   j) $9.9 \times 10^{-2}$

   Now calculate the pOH of each of the solutions.

2. Calculate the pH of solutions with the following $OH^-$ concentrations in mol dm$^{-3}$:

   a) $10^{-2}$          b) $10^{-3}$          c) $10^{-8}$
   d) $0.055$        e) $0.0010$      f) $0.083$
   g) $7.6 \times 10^{-3}$     h) $4.9 \times 10^{-5}$     i) $6.4 \times 10^{-8}$
   j) $3.7 \times 10^{-10}$

3. Calculate the $H_3O^+$ concentrations in solutions with the following pH values:

   a) $0.00$         b) $4.30$         c) $2.35$
   d) $1.88$         e) $4.15$         f) $7.84$
   g) $9.21$         h) $13.7$         i) $9.50$
   j) $2.63$

4. Calculate the pH of the solutions made by dissolving the following in distilled water and making up to 500 cm$^3$ of solution:

   a) 3.00 g of hydrogen chloride
   b) 4.50 g of chloric(VII) acid, $HClO_4$
   c) 4.00 g of sodium hydroxide
   d) 1.00 g of calcium hydroxide
   e) 6.30 g of potassium hydroxide

5. An aqueous solution contains the acid, HX, at a concentration 0.100 mol dm$^{-3}$. The degree of ionisation of the acid is 0.0300. Calculate the pH of the solution.

6. If the acid HA is 1% ionised in a solution of concentration 0.0100 mol dm$^{-3}$, calculate: a) $K_a$, b) $pK_a$.

7. The degree of ionisation of phenol in water in a solution of concentration $1.10 \times 10^{-2}$ mol dm$^{-3}$ is $1.1 \times 10^{-4}$. Calculate the value of $pK_a$.

8. Calculate the dissociation constants for each of the weak acids listed below:

| | Electrolyte | Concentration/mol dm$^{-3}$ | $[H_3O^+]$/mol dm$^{-3}$ |
|---|---|---|---|
| *a) | $HCO_2H$ | 0.0100 | $1.33 \times 10^{-3}$ |
| b) | $CH_3CO_2H$ | 0.100 | $1.32 \times 10^{-3}$ |
| c) | HCN | 1.00 | $1.99 \times 10^{-5}$ |
| *d) | HF | 0.200 | $5.97 \times 10^{-3}$ |
| *e) | $HClO_2$ | 0.0200 | $1.41 \times 10^{-2}$ |

9. Calculate the dissociation constants of the weak acids listed below:
   a) a solution of $0.0100 \, mol \, dm^{-3} \, CH_3CO_2H$ has a pH of 3.38
   b) a solution of $0.200 \, mol \, dm^{-3} \, HCN$ has a pH of 5.05
   *c) a solution of $0.500 \, mol \, dm^{-3} \, ClCH_2CO_2H$ has a pH of 3.16
   d) a solution of $0.0100 \, mol \, dm^{-3} \, CH_3CH_2CO_2H$ has a pH of 3.43
   e) a solution of $0.0100 \, mol \, dm^{-3} \, HOBr$ has a pH of 5.35

10. Calculate the values of $K_b$ for the following weak bases from the data:

| Base | Concentration/$mol \, dm^{-3}$ | $[OH^-]$/$mol \, dm^{-3}$ |
|------|------|------|
| a) $CH_3NH_2$ | 0.0100 | $4.78 \times 10^{-7}$ |
| b) $NH_3$ | 1.00 | $2.37 \times 10^{-5}$ |
| c) $C_2H_5NH_2$ | 0.0500 | $9.65 \times 10^{-7}$ |
| d) $C_3H_7NH_2$ | 0.0100 | $3.80 \times 10^{-7}$ |
| e) $C_6H_5CH_2NH_2$ | 0.0100 | $2.06 \times 10^{-6}$ |

11. Calculate the degree of ionisation of each of the following in aqueous solution:
   a) $2.00 \times 10^{-3} \, mol \, dm^{-3} \, HCN$     ($pK_a = 9.40$)
   b) $0.500 \times 10^{-3} \, mol \, dm^{-3} \, HCO_2H$     ($pK_a = 3.75$)
   c) $3.00 \times 10^{-2} \, mol \, dm^{-3} \, CH_3CO_2H$     ($pK_a = 4.76$)
   d) $5.00 \times 10^{-2} \, mol \, dm^{-3} \, NH_3$     ($pK_a = 9.25$)
   e) $1.00 \times 10^{-2} \, mol \, dm^{-3} \, (CH_3)_3N$     ($pK_a = 9.80$)

   (*Note.* You are given the values of $pK_a$ for the conjugate acids of the bases.)

12. Calculate the pH of:
   a) $0.0100 \, mol \, dm^{-3}$ hydrochloric acid,
   b) $0.0100 \, mol \, dm^{-3}$ sodium hydroxide, and
   c) the solution obtained by diluting $5.00 \, cm^3$ of hydrochloric acid of concentration $1.00 \, mol \, dm^{-3}$ with conductivity water to $1.00 \, dm^3$.

13. The dissociation constant of phenol in water at 20 °C is $1.21 \times 10^{-10}$ $mol \, dm^{-3}$. Calculate the percentage of phenol which is ionised in a solution of concentration $0.0100 \, mol \, dm^{-3}$.

14. Given that the $pK_a$ of the ammonium ion, $NH_4^+$, is 9.25 at 25 °C, find the pH of an aqueous solution of ammonia of concentration 0.100 mol $dm^{-3}$. The ionic product for water at 25 °C is $1.00 \times 10^{-14} \, mol^2 \, dm^{-6}$.

15. If the acid HA is 1% ionised in a solution of concentration 0.0100 mol $dm^{-3}$, calculate: a) $K_a$, b) $pK_a$.

16. Calculate the pH of hydrochloric acid of concentration $1.00 \times 10^{-3}$ $mol \, dm^{-3}$. Calculate how many $cm^3$ of a) hydrochloric acid of concentration $1.00 \, mol \, dm^{-3}$, b) sodium hydroxide of concentration $1.00 \, mol \, dm^{-3}$ are needed to change the pH of $1.00 \, dm^3$ of hydrochloric acid of concentration $1.00 \times 10^{-3} \, mol \, dm^{-3}$ by 1 unit. Ignore the small changes in volume.

## BUFFER SOLUTIONS

A buffer solution is one which will resist changes in pH due to the addition of small amounts of acid and alkali. An effective buffer can be made by preparing a solution containing both a weak acid and also one of its salts with a strong base, e.g. ethanoic acid and sodium ethanoate. This will absorb hydrogen ions because they react with ethanoate ions to form molecules of ethanoic acid:

$$CH_3CO_2^- + H_3O^+ \rightleftharpoons CH_3CO_2H + H_2O$$

Hydroxide ions are absorbed by combining with ethanoic acid molecules to form ethanoate ions and water:

$$OH^- + CH_3CO_2H \rightleftharpoons CH_3CO_2^- + H_2O$$

A solution of a weak base and one of its salts formed with a strong acid, e.g. ammonia solution and ammonium chloride, will act as a buffer. If hydrogen ions are added, they combine with ammonia, and, if hydroxide ions are added, they combine with ammonium ions:

$$NH_3 + H_3O^+ \rightleftharpoons NH_4^+ + H_2O$$

$$OH^- + NH_4^+ \rightleftharpoons NH_3 + H_2O$$

The pH of a buffer solution consisting of a weak acid and its salt is calculated from the equation

$$K_a = \frac{[H_3O^+][A^-]}{[HA]}$$

$$[H_3O^+] = K_a \frac{[HA]}{[A^-]}$$

$$pH = pK_a + \lg \frac{[A^-]}{[HA]}$$

Since the salt is completely ionised and the acid only slightly ionised, one can assume that all the anions come from the salt, and put

$$[A^-] = [Salt]$$

$$[HA] = [Acid]$$

$$\therefore \qquad pH = pK_a + \lg \frac{[Salt]}{[Acid]}$$

For a buffer made from a base, B, and its salt with a strong acid, $BH^+X^-$,

$$K_b = \frac{[BH^+][OH^-]}{[B]}$$

$$pOH = pK_b + \lg \frac{[BH^+]}{[B]}$$

and
$$pH = pK_w - pK_b + lg \frac{[B]}{[BH^+]}$$

Since the weak base is only slightly ionised, one can put

$$[B] = [\text{Base added}]$$
$$[BH^+] = [\text{Salt added}]$$

$\therefore$
$$pH = pK_w - pK_b + lg \frac{[\text{Base}]}{[\text{Salt}]}$$

**EXAMPLE 1**  Three solutions contain propanoic acid ($K_a = 1.34 \times 10^{-5} \, mol \, dm^{-3}$) at a concentration of $0.10 \, mol \, dm^{-3}$ and sodium propanoate at concentrations of a) $0.10 \, mol \, dm^{-3}$, b) $0.20 \, mol \, dm^{-3}$, c) $0.50 \, mol \, dm^{-3}$ respectively. Calculate the pH values of the three solutions.

**METHOD**
$$pH = pK_a + lg \frac{[\text{Salt}]}{[\text{Acid}]}$$

In solution a),  $pH = 4.87 + lg \dfrac{0.10}{0.10} = 4.87 + lg \, 1.0$

**ANSWER**    $pH(a) = 4.87$

In solution b),  $pH = 4.87 + lg \dfrac{0.20}{0.10} = 4.87 + lg \, 2.0$

**ANSWER**    $pH(b) = 5.17$

In solution c),  $pH = 4.87 + lg \dfrac{0.50}{0.10} = 4.87 + lg \, 5.0$

**ANSWER**    $pH(c) = 5.57$

**EXAMPLE 2**  Calculate the pH of solutions of $0.25 \, mol \, dm^{-3}$ methylamine ($K_b = 4.54 \times 10^{-4} \, mol \, dm^{-3}$) containing a) $0.25 \, mol \, dm^{-3}$ and b) $0.50 \, mol \, dm^{-3}$ methylamine hydrochloride.

**METHOD**
$$pH = pK_w - pK_b + lg \frac{[\text{Base}]}{[\text{Salt}]}$$

In solution a),    $pH = 14.00 - 3.34 + lg \dfrac{0.25}{0.25}$

**ANSWER**    $pH(a) = 10.66$

In solution b),    $pH = 14.00 - 3.34 + lg \dfrac{0.25}{0.50}$

**ANSWER**    $pH(b) = 10.36$

**\*Calculating the change in the pH of a buffer solution produced by the addition of acid or alkali**

EXAMPLE 3    Calculate the effect of adding a) $10 \, cm^3$ of hydrochloric acid of concentration $1.0 \, mol \, dm^{-3}$ and b) $10 \, cm^3$ of sodium hydroxide of concentration $1.0 \, mol \, dm^{-3}$ to $1.0 \, dm^3$ of a buffer containing $0.10 \, mol \, dm^{-3}$ ethanoic acid and $0.10 \, mol \, dm^{-3}$ sodium ethanoate. ($pK_a$ of ethanoic acid $= 4.75$.)

METHOD    Use the equation    $$pH = pK_a + \lg \frac{[Salt]}{[Acid]}$$

$$pH = 4.75 + \lg \frac{0.10}{0.10} = 4.75$$

a) Amount of hydrochloric acid added $= 10 \, cm^3$ of $1.0 \, mol \, dm^{-3}$ solution $= 0.010 \, mol$.

This amount of hydrogen ion combines with ethanoate ions to form undissociated molecules of ethanoic acid. The amount of salt is therefore decreased by $0.010 \, mol$, and the amount of acid is increased by $0.010 \, mol$. Thus,

$$pH = 4.75 + \lg \frac{(0.10 - 0.010)}{(0.10 + 0.010)} = 4.66$$

ANSWER    The pH has decreased by 0.09.

b) Amount of $OH^-$ ion added $= 10 \times 10^{-3} \times 1.0 = 0.010 \, mol$.

The hydroxide ions react with $0.010 \, mol$ of weak acid. This results in the formation of an additional $0.010 \, mol$ of anions. Thus, [Acid] is decreased by $0.010 \, mol$, and [Salt] is increased by $0.010 \, mol$.

$$pH = 4.75 + \lg \frac{(0.10 + 0.010)}{(0.10 - 0.010)} = 4.84$$

ANSWER    The pH has increased by 0.09.

## EXERCISE 45    Problems on Buffers

1. What fraction of a mole of sodium ethanoate must be added to $1 \, dm^3$ of ethanoic acid of concentration $0.10 \, mol \, dm^{-3}$ and $K_a = 2 \times 10^{-5} \, mol \, dm^{-3}$ in order to produce a buffer solution of pH $= 5$?

   a 0.2          b 0.25          c 0.4          d 0.5          e 0.6

2. What amount of sodium ethanoate must be added to $1.00 \, dm^3$ of ethanoic acid of $pK_a \, 4.73$, concentration $0.100 \, mol \, dm^{-3}$, to produce a buffer of pH $= 5.73$?

3. Calculate the pH values of the following solutions.
   a) $20.0 \, \text{cm}^3$ of $1.00 \, \text{mol dm}^{-3}$ nitrous acid ($pK_a = 3.34$) added to $40.0 \, \text{cm}^3$ of $0.500 \, \text{mol dm}^{-3} \, \text{mol dm}^{-3}$ sodium nitrite solution.
   b) $10.0 \, \text{cm}^3$ of $1.00 \, \text{mol dm}^{-3}$ nitrous acid added to $20.0 \, \text{cm}^3$ of $2.00 \, \text{mol dm}^{-3}$ sodium nitrite solution.

4. a) What is the pH of a solution containing $0.100 \, \text{mol dm}^{-3}$ of ethanoic acid and $0.100 \, \text{mol dm}^{-3}$ of sodium ethanoate? ($K_a(CH_3CO_2H) = 1.86 \times 10^{-5} \, \text{mol dm}^{-3}$.)
   b) How many moles of sodium ethanoate must be added to $1.00 \, \text{dm}^3$ of $0.0100 \, \text{mol dm}^{-3}$ ethanoic acid to produce a buffer solution of pH 5.8?

*5. What is the pH of a buffer which has been made by adding $100 \, \text{cm}^3$ of chloroethanoic acid of concentration $0.500 \, \text{mol dm}^{-3}$ and $100 \, \text{cm}^3$ of sodium hydroxide of concentration $0.250 \, \text{mol dm}^{-3}$? ($pK_a(ClCH_2CO_2H) = 2.86$.)

*6. a) Calculate the effect of adding $10.0 \, \text{cm}^3$ of hydrochloric acid of concentration $1.00 \, \text{mol dm}^{-3}$ to $1.00 \, \text{dm}^3$ of the following buffers ($pK_a$ of ethanoic acid is 4.75):

| Buffer | A | B | C | D |
|---|---|---|---|---|
| Concentration | $\text{mol dm}^{-3}$ | $\text{mol dm}^{-3}$ | $\text{mol dm}^{-3}$ | $\text{mol dm}^{-3}$ |
| Ethanoic acid | 0.10 | 0.10 | 0.10 | 0.20 |
| Sodium ethanoate | 0.10 | 0.050 | 0.020 | 0.20 |

   b) From a comparison of your values, state how the ratio of $\dfrac{[\text{Salt}]}{[\text{Acid}]}$ affects the buffering capacity of the solution and also how the concentration of the buffer affects its buffering capacity.

*7. Methanoic acid has $K_a = 1.60 \times 10^{-4} \, \text{mol dm}^{-3}$. Calculate the pH of:
   a) a solution of concentration $0.100 \, \text{mol dm}^{-3}$
   b) a solution of $3.40 \, \text{g}$ of sodium methanoate in $1.00 \, \text{dm}^3$ of methanoic acid of concentration $0.100 \, \text{mol dm}^{-3}$
   c) the solution obtained when $24.9 \, \text{cm}^3$ of $0.100 \, \text{mol dm}^{-3}$ sodium hydroxide solution has been added to $25.0 \, \text{cm}^3$ of methanoic acid of concentration $0.100 \, \text{mol dm}^{-3}$.

*8. a) Calculate the pH of a solution of $0.200 \, \text{mol}$ ethanoic acid ($K_a = 1.8 \times 10^{-5} \, \text{mol dm}^{-3}$) in $1.00 \, \text{dm}^3$ of aqueous solution.
   b) What does the pH of the solution become after the addition of $0.100 \, \text{mol}$ sodium ethanoate?
   c) If $0.010 \, \text{mol}$ hydrochloric acid is added to the ethanoic acid–sodium ethanoate solution, what change in pH results?

## *SALT HYDROLYSIS

When the salt of a weak acid and a strong base, e.g. sodium ethanoate, dissolves in water, the solution is alkaline. This is because the anions react with water molecules to form undissociated molecules of ethanoic acid. This interaction is called *salt hydrolysis*. An equilibrium is set up:

$$CH_3CO_2^-(aq) + H_2O(l) \rightleftharpoons CH_3CO_2H(aq) + OH^-(aq)$$

Since $CH_3CO_2H$ and $CH_3CO_2^-$ are a conjugate acid–base pair (see p. 136).

$$K_a(CH_3CO_2H) = K_w/K_b(CH_3CO_2^-)$$

$$K_b(CH_3CO_2^-) = \frac{[CH_3CO_2H][OH^-]}{[CH_3CO_2^-]}$$

If the degree of hydrolysis of the salt is $\alpha$ and its concentration is $c$,

$$[CH_3CO_2^-] = c(1-\alpha)$$

$$[CH_3CO_2H] = c\alpha$$

$$[OH^-] = c\alpha$$

If $\alpha$ is small, $(1-\alpha)$ can be put $= 1$, and

$$c\alpha^2 = K_b$$

$$\alpha = \sqrt{K_b/c} = \sqrt{K_w/cK_a}$$

The pH of the solution can now be found:

Since

$$[OH^-] = c\alpha$$

$$[H_3O^+] = K_w/c\alpha = \sqrt{K_w K_a/c}$$

$$\lg[H_3O^+] = \tfrac{1}{2}\lg K_w + \tfrac{1}{2}\lg K_a - \tfrac{1}{2}\lg c$$

$$pH = \tfrac{1}{2}pK_w + \tfrac{1}{2}pK_a + \tfrac{1}{2}\lg c$$

This equation enables one to find the pH of a solution of a salt of a weak acid from the dissociation constant of the acid and the concentration of the salt.

In the case of a salt of a weak base and a strong acid, e.g. ammonium chloride, hydrolysis of the salt will increase the concentration of hydrogen ions:

$$NH_4^+(aq) + H_2O(l) \rightleftharpoons NH_3(aq) + H_3O^+(aq)$$

A treatment similar to that above gives the relationship

$$pH = \tfrac{1}{2}pK_w - \tfrac{1}{2}pK_b - \tfrac{1}{2}\lg c$$

Thus, the pH of a solution of a salt of a weak base and a strong acid can be found from the dissociation constant of the base and the concentration of the salt.

**EXAMPLE 1**   Calculate the pH at 25 °C of a solution of 0.100 mol dm$^{-3}$ solution of sodium propanoate. ($K_a = 1.34 \times 10^{-5}$ mol dm$^{-3}$, $K_w = 1.00 \times 10^{-14}$ mol$^2$ dm$^{-6}$ at 25 °C.)

**METHOD**    Since          $K_a = 1.34 \times 10^{-5}$,   $pK_a = 4.87$

$$pH = \tfrac{1}{2}pK_w + \tfrac{1}{2}pK_a + \tfrac{1}{2}\lg c$$

$$= 7.00 + 2.44 - 0.50$$

**ANSWER**              $pH = 8.94$

**EXAMPLE 2**   Calculate the pH of a solution of methylamine hydrochloride of concentration 0.100 mol dm$^{-3}$ at 25 °C. ($K_b = 4.54 \times 10^{-4}$ mol dm$^{-3}$; $K_w = 1.00 \times 10^{-14}$ mol$^2$ dm$^{-6}$ at 25 °C.)

**METHOD**    Since          $K_b = 4.54 \times 10^{-4}$,   $pK_b = 3.34$

$$pH = \tfrac{1}{2}pK_w - \tfrac{1}{2}pK_b - \tfrac{1}{2}\lg c$$

$$= 7.00 - 1.67 + 0.50$$

**ANSWER**              $pH = 5.83$

## EXERCISE 46     Problems on Salt Hydrolysis

**\*1.** Calculate the pH of the following solutions:

a) a 0.100 mol dm$^{-3}$ solution of ammonium chloride
b) a 0.0100 mol dm$^{-3}$ solution of methylamine hydrochloride
c) a 0.100 mol dm$^{-3}$ solution of potassium cyanide
d) a 0.100 mol dm$^{-3}$ solution of sodium methanoate
e) a 0.0100 mol dm$^{-3}$ solution of sodium ethanoate

(p$K_a$ values are: $NH_3$, 9.25; $CH_3NH_2$, 10.64; HCN, 9.40; $HCO_2H$, 3.75; $CH_3CO_2H$, 4.76. p$K_w = 14.00$.)

**\*2.** Use the following dissociation constants (mol dm$^{-3}$):

| | |
|---|---|
| $HNO_2$ | $K_a = 4.6 \times 10^{-4}$ |
| HCN | $K_a = 4.9 \times 10^{-10}$ |
| $NH_3$ | $K_b = 1.8 \times 10^{-5}$ |
| $HCO_2H$ | $K_a = 1.8 \times 10^{-4}$ |

a) Find the pH of a 0.50 mol dm$^{-3}$ solution of sodium methanoate.
b) What is the pH of a solution of ammonium nitrate of concentration 0.25 mol dm$^{-3}$?
c) Calculate the pH of a 0.25 mol dm$^{-3}$ solution of potassium cyanide.
d) Calculate the pH at the equivalence point of a titration of 0.40 mol dm$^{-3}$ nitrous acid and 0.40 mol dm$^{-3}$ sodium hydroxide solution.
e) Find the pH at the equivalence point of a titration of 0.25 mol dm$^{-3}$ ammonia solution with 0.25 mol dm$^{-3}$ hydrochloric acid.
f) Calculate the concentration of a solution of ammonium chloride which has a pH of 4.75.

**\*3.** Use the following dissociation constants $(mol\,dm^{-3})$:

| | | | |
|---|---|---|---|
| Ethanoic acid | $K_a = 1.76 \times 10^{-5}$ | Ethylamine | $K_b = 6.46 \times 10^{-4}$ |
| Methanoic acid | $K_a = 1.77 \times 10^{-4}$ | Pyridine | $K_b = 5.62 \times 10^{-9}$ |
| Hydrogen cyanide | $K_a = 4.9 \times 10^{-10}$ | Phenylamine | $K_b = 4.17 \times 10^{-10}$ |

Calculate the values of pH in the following solutions:
a) $0.100\,mol\,dm^{-3}$ sodium ethanoate
b) $0.0200\,mol\,dm^{-3}$ sodium methanoate
c) $0.0200\,mol\,dm^{-3}$ sodium cyanide
d) $0.100\,mol\,dm^{-3}$ ethylamine hydrochloride
e) $0.0200\,mol\,dm^{-3}$ pyridine hydrochloride
f) $0.0500\,mol\,dm^{-3}$ phenylamine hydrochloride

## COMPLEX ION FORMATION

Another type of equilibrium between ions is due to complex ion formation. A complex ion is formed by the combination of an ion with an oppositely charged ion or ions or with a neutral molecule or molecules. For example, when copper(II) ions combine with ammonia to form tetraamminecopper ions, there is set up an equilibrium:

$$Cu(NH_3)_4^{2+}(aq) \rightleftharpoons Cu^{2+}(aq) + 4NH_3(aq)$$

The equilibrium constant or dissociation constant for the reaction is

$$K_d = \frac{[Cu^{2+}][NH_3]^4}{[Cu(NH_3)_4^{2+}]} = 5.0 \times 10^{-14}\,mol^4\,dm^{-12}$$

The inverse of the dissociation constant is called the *stability constant* of the complex ion.

**EXAMPLE 1**  The complex $Ag(CN)_2^-$ has a dissociation constant of $1.4 \times 10^{-20}$ $mol^2\,dm^{-6}$. Find the concentration of silver ions in a $2.0 \times 10^{-2}\,mol$ $dm^{-3}$ solution of $KAg(CN)_2$.

**METHOD**
$$Ag(CN)_2^- \rightleftharpoons Ag^+ + 2CN^-$$

$$K_d = \frac{[Ag^+][CN^-]^2}{[Ag(CN)_2^-]}\,mol^2\,dm^{-6}$$

There are two $CN^-$ ions for every $Ag^+$ ion

$$\therefore \qquad 1.4 \times 10^{-20} = \frac{[Ag^+](2[Ag^+])^2}{2.0 \times 10^{-2}}$$

**ANSWER**
$$[Ag^+] = 4.1 \times 10^{-8}\,mol\,dm^{-3}$$

**EXAMPLE 2**   In a $2.0 \times 10^{-2} \, \text{mol dm}^{-3}$ solution of diammine silver nitrate, the concentration of free silver ions is $6.8 \times 10^{-4} \, \text{mol dm}^{-3}$. What is the dissociation constant of the complex ion?

**METHOD**

$$Ag(NH_3)_2^+(aq) \rightleftharpoons Ag^+(aq) + 2NH_3(aq)$$

$$K_d = \frac{[Ag^+][NH_3]^2}{[Ag(NH_3)_2^+]} = \frac{[Ag^+](2[Ag^+])^2}{[Ag(NH_3)_2^+]} = \frac{4[Ag^+]^3}{[Ag(NH_3)_2^+]}$$

**ANSWER**

$$K_d = \frac{4(6.8 \times 10^{-4})^3}{2.0 \times 10^{-2}} = 6.3 \times 10^{-8} \, \text{mol}^2 \, \text{dm}^{-6}$$

## EXERCISE 47     Problems on Complex Ions

1. The dissociation constant for the hexaaquo aluminium ion is $1.0 \times 10^{-5} \, \text{mol dm}^{-3}$.

$$Al(H_2O)_6^{3+}(aq) \rightleftharpoons Al(H_2O)_5(OH)^{2+}(aq) + H^+(aq)$$

Calculate the pH of a $0.10 \, \text{mol dm}^{-3}$ solution of aluminium nitrate.

2. A solution was made by dissolving $0.025 \, \text{mol} \, KAu(CN)_2$ in $1 \, \text{dm}^3$ of solution. If the dissociation constant of the complex ion, $Au(CN)_2^-$ is $1.1 \times 10^{-47} \, \text{mol}^2 \, \text{dm}^{-6}$, calculate the concentration of free $Au^+$ ions in the solution.

3. Ammonia is added to a solution of $0.50 \, \text{mol dm}^{-3}$ silver nitrate solution. What is the concentration of free ammonia when 95% of the $Ag^+$ ions have been converted to the complex $Ag(NH_3)_2^+$ ion? The dissociation constant of the complex ion is $6.0 \times 10^{-8} \, \text{mol}^2 \, \text{dm}^{-6}$.

4. A solution is made by adding equal volumes of a solution of cadmium ions at a concentration of $0.20 \, \text{mol dm}^{-3}$ and ammonia at a concentration of $0.80 \, \text{mol dm}^{-3}$. At equilibrium, the concentration of free $Cd^{2+}$ ions is $1.0 \times 10^{-2} \, \text{mol dm}^{-3}$. Calculate the dissociation constant for the equilibrium

$$Cd(NH_3)_4^{2+} \rightleftharpoons Cd^{2+} + 4NH_3$$

5. The salt $Cu(NH_3)_4(NO_3)_2$ was dissolved at a concentration of $0.100 \, \text{mol dm}^{-3}$. The equilibrium concentration of ammonia was found to be $2.85 \times 10^{-3} \, \text{mol dm}^{-3}$. Find the dissociation constant for the equilibrium

$$Cu(NH_3)_4^{2+} \rightleftharpoons Cu^{2+} + 4NH_3$$

## SOLUBILITY AND SOLUBILITY PRODUCT

Many salts which we refer to as insoluble do in fact dissolve to a small extent. In a saturated solution, an equilibrium exists between the dissolved ions and the undissolved salt. For example, in a saturated solution of silver chloride in contact with undissolved silver chloride,

$$AgCl(s) \rightleftharpoons Ag^+(aq) + Cl^-(aq)$$

The product of the concentrations of silver ions and chloride ions is called the solubility product of silver chloride.

$$K_{sp} = [Ag^+][Cl^-]$$

*The solubility product of a salt is the product of the concentrations of all the ions in a saturated solution of the salt.*

It is different from solubility. *The solubility of a salt is expressed as either the amount of solute (in mol) or the mass of solute (in g) dissolved in 1 dm³ of solution at a stated temperature.*

Another example of a sparingly soluble salt is lead(II) chloride.

$$PbCl_2(s) \rightleftharpoons Pb^{2+}(aq) + 2Cl^-(aq)$$

$$K_{sp} = [Pb^{2+}][Cl^-]^2$$

If the solubility of $PbCl_2$ is $a$ mol dm⁻³, then $[Pb^{2+}] = a$ and $[Cl^-] = 2a$

$$K_{sp} = a \times (2a)^2 = 4a^3$$

## Calculation of solubility product

**EXAMPLE 1**   The solubility of lead(II) hydroxide at 25 °C is $6.64 \times 10^{-4}$ mol dm⁻³. Calculate its solubility product.

**METHOD**   $$Pb(OH)_2(s) \rightleftharpoons Pb^{2+}(aq) + 2OH^-(aq)$$

Since the solubility of $Pb(OH)_2$ is $6.64 \times 10^{-4}$ mol dm⁻³

$$[Pb^{2+}] = 6.64 \times 10^{-4} \text{ mol dm}^{-3}$$

$$[OH^-] = 2 \times 6.64 \times 10^{-4} \text{ mol dm}^{-3}$$

$$K_{sp} = [Pb^{2+}][OH^-]^2 = 6.64 \times 10^{-4} \times (1.33 \times 10^{-3})^2$$

**ANSWER**   $$K_{sp} = 1.17 \times 10^{-9} \text{ mol}^3 \text{ dm}^{-9}$$

*Note* that the units in which the solubility product is expressed are the result of multiplying three concentrations together: $(mol\,dm^{-3})^3 = mol^3\,dm^{-9}$.

**EXAMPLE 2**   Given that the solubility product of lead(II) sulphate at 25 °C is $1.60 \times 10^{-8}$ mol² dm⁻⁶, calculate the solubility at this temperature.

**METHOD**   Solubility of $PbSO_4$ = Concn of $PbSO_4$ in solution.

All the dissolved $PbSO_4$ is in the form of $Pb^{2+}$ ions and $SO_4^{2-}$ ions.

$$\therefore \text{ Solubility of } PbSO_4 = [Pb^{2+}] = [SO_4^{2-}]$$

$$\text{Solubility product} = K_{sp} = [Pb^{2+}][SO_4^{2-}] = [PbSO_4]^2$$

$$[PbSO_4] = \sqrt{K_{sp}} = \sqrt{1.60 \times 10^{-8}} \text{ mol dm}^{-3}$$

**ANSWER**   $$[PbSO_4] = 1.26 \times 10^{-4} \text{ mol dm}^{-3}$$

**EXAMPLE 3**  $1.00 \, dm^3$ of a solution of calcium chloride of concentration $0.100 \, mol$ $dm^{-3}$ is added to $1.00 \, dm^3$ of sodium hydroxide of concentration $0.100 \, mol \, dm^{-3}$. If the solubility product of calcium hydroxide is $5.50 \times 10^{-6} \, mol^3 \, dm^{-9}$, calculate the mass of calcium hydroxide that will be precipitated.

**METHOD**  The maximum concentration of $Ca^{2+}$ which can remain in solution is given by

$$[Ca^{2+}][OH^-]^2 = 5.5 \times 10^{-6} \, mol^3 \, dm^{-9}$$

Let $[Ca^{2+}] = a$;  then  $4a^3 = 5.5 \times 10^{-6} \, mol^3 \, dm^{-9}$

$$a = 1.11 \times 10^{-2} \, mol \, dm^{-3}$$

Amount of $Ca^{2+}$ in $2.00 \, dm^3 = 2.22 \times 10^{-2} \, mol$

Amount of $Ca^{2+}$ added as $CaCl_2 = 0.100 \, mol$

Amount of $Ca^{2+}$ precipitated as $Ca(OH)_2 = 0.100 - 0.0222$
$$= 0.0778 \, mol$$

**ANSWER**  Mass of $Ca(OH)_2$ precipitated $= 0.0778 \times 74.0 = 5.76 \, g$

## THE COMMON ION EFFECT

In a saturated solution of a salt, MA, in equilibrium with solid MA,

$$MA(s) \rightleftharpoons M^{2+}(aq) + A^{2-}(aq)$$
$$K_{sp} = [M^{2+}][A^{2-}]$$

If a solution containing $M^{2+}$ ions is added, $[M^{2+}]$ is increased. The solubility product, $K_{sp}$, remains the same, even when the ions are not present in equimolar concentrations. So that the product $[M^{2+}][A^{2-}]$ shall not exceed $K_{sp}$, $M^{2+}$ ions will be removed from solution as solid MA. Solute will be precipitated from solution. The addition of a solution containing $A^{2-}$ ions will have the same effect. The separation of a solute from a solution on addition of an electrolyte solution which has an ion in common with the solute is an example of the *common ion effect.*

Another example of the common ion effect is the change in the concentration of ions produced by the dissociation of a weak acid in the presence of a solution of one of its ions.

**EXAMPLE 1**  Calculate the solubility at $25 \, ^\circ C$ of silver chloride: a) in water, and b) in $0.10 \, mol \, dm^{-3}$ hydrochloric acid. The solubility product of silver chloride at $25 \, ^\circ C$ is $1.8 \times 10^{-10} \, mol^2 \, dm^{-6}$.

**METHOD**   a)  Since $K_{sp} = [Ag^+][Cl^-]$

$$[Ag^+]^2 = 1.8 \times 10^{-10} \, mol^2 \, dm^{-6}$$
$$[Ag^+] = 1.3 \times 10^{-5} \, mol \, dm^{-3}$$

Concn of AgCl $= 1.3 \times 10^{-5} \, mol \, dm^{-3}$
Molar mass of AgCl $= 143.5 \, g \, mol^{-1}$

**ANSWER**   Solubility of AgCl $= 1.3 \times 10^{-5} \times 143.5 = 1.9 \times 10^{-3} \, g \, dm^{-3}$.

b) The value of $[Cl^-]$ is the sum of that from the $0.10 \, mol \, dm^{-3}$ HCl, and that from the dissolved AgCl. The latter is of the order of $10^{-5} \, mol \, dm^{-3}$, and can be neglected in comparison with $0.10 \, mol \, dm^{-3}$ from the acid

$$K_{sp} = [Ag^+][Cl^-] = [Ag^+](0.10) \, mol^2 \, dm^{-6}$$
$$[Ag^+] = 1.8 \times 10^{-10}/0.10 = 1.8 \times 10^{-9} \, mol \, dm^{-3}$$

Concn of AgCl $= 1.8 \times 10^{-9} \, mol \, dm^{-3}$

**ANSWER**   Solubility of AgCl $= 1.8 \times 10^{-9} \times 143.5 = 2.6 \times 10^{-7} \, g \, dm^{-3}$.

The qualitative analysis scheme for the identification of metal cations is an application of solubility products. Many metals are precipitated as sulphides when a solution of the metal cations is treated with hydrogen sulphide. This large group of sparingly soluble sulphides can be divided into two groups. Hydrogen sulphide in acid solution brings down the least soluble sulphides in Group II of the qualitative analysis scheme. The rest are precipitated in Group IV from an alkaline solution of hydrogen sulphide. The effect of hydrogen ion concentration on the ionisation of hydrogen sulphide can be calculated.

**EXAMPLE 2**  In a saturated solution of hydrogen sulphide,

$$[H_3O^+]^2[S^{2-}] = 1.1 \times 10^{-23} \, mol^3 \, dm^{-9}$$

Calculate the sulphide ion concentration: a) at pH 7, b) at pH 8, and c) at pH 2.

**METHOD**   a)  At pH 7,

$$[H_3O^+] = 10^{-7}$$

**ANSWER**   $[S^{2-}] = 1.1 \times 10^{-23}/(10^{-7})^2 = 1.1 \times 10^{-9} \, mol \, dm^{-3}$

b) At pH 8,

**ANSWER**   $[S^{2-}] = 1.1 \times 10^{-23}/(10^{-8})^2 = 1.1 \times 10^{-7} \, mol \, dm^{-3}$

c) At pH 2,

**ANSWER**   $[S^{2-}] = 1.1 \times 10^{-23}/(10^{-2})^2 = 1.1 \times 10^{-19} \, mol \, dm^{-3}$

The very low sulphide ion concentration at pH 2 will precipitate only the most insoluble metal sulphides.

**EXERCISE 48**   Problems on Solubility Products

1. Given the following solubilities in $mol\,dm^{-3}$ of solution, calculate the solubility products of the solids listed:

   a) CaS        $1.3 \times 10^{-14}$         b) CoS       $6.3 \times 10^{-10}$
   c) $Ag_2S$      $1.1 \times 10^{-17}$         d) $Sb_2S_3$   $1.0 \times 10^{-19}$
   e) $Pb(OH)_2$    $5.0 \times 10^{-6}$

2. Given the following solubilities in g per $dm^3$ of solution, calculate the solubility products of the solids listed:

   a) PbS        $1.20 \times 10^{-11}$         b) AgI       $2.14 \times 10^{-6}$
   c) $BaSO_4$     $2.41 \times 10^{-3}$          d) $CaF_2$     $1.47 \times 10^{-2}$
   e) AgCN        $1.50 \times 10^{-6}$

   The following questions require a knowledge of the solubility products listed below:

   | | | | |
   |---|---|---|---|
   | CuS | $6.3 \times 10^{-36}\,mol^2\,dm^{-6}$ | $Bi_2S_3$ | $1.0 \times 10^{-97}\,mol^5\,dm^{-15}$ |
   | $Ag_2S$ | $6.3 \times 10^{-51}\,mol^3\,dm^{-9}$ | HgS | $1.6 \times 10^{-52}\,mol^2\,dm^{-6}$ |
   | NiS | $3.2 \times 10^{-19}\,mol^2\,dm^{-6}$ | FeS | $6.3 \times 10^{-18}\,mol^2\,dm^{-6}$ |
   | $BaSO_4$ | $1.1 \times 10^{-10}\,mol^2\,dm^{-6}$ | $Al(OH)_3$ | $6.3 \times 10^{-32}\,mol^4\,dm^{-12}$ |
   | $CaSO_4$ | $2.4 \times 10^{-5}\,mol^2\,dm^{-6}$ | $SrF_2$ | $2.4 \times 10^{-9}\,mol^3\,dm^{-9}$ |
   | $Ag_2SO_4$ | $1.7 \times 10^{-5}\,mol^3\,dm^{-9}$ | $PbCl_2$ | $1.6 \times 10^{-5}\,mol^3\,dm^{-9}$ |

3. What concentration of sulphide ion is needed to precipitate the metal as its sulphide from each of the following solutions?

   a) $CuSO_4(aq)$       $1.0 \times 10^{-2}\,mol\,dm^{-3}$
   b) $AgNO_3(aq)$       $1.0 \times 10^{-4}\,mol\,dm^{-3}$
   c) $NiSO_4(aq)$       $1.0 \times 10^{-5}\,mol\,dm^{-3}$
   d) $Bi(NO_3)_3(aq)$    $1.0 \times 10^{-4}\,mol\,dm^{-3}$
   e) $Hg(NO_3)_2(aq)$    $1.0 \times 10^{-6}\,mol\,dm^{-3}$
   f) $FeSO_4(aq)$       $1.0 \times 10^{-6}\,mol\,dm^{-3}$

4. Will a precipitate appear when the following solutions are added?

   a) $10\,cm^3\,BaCl_2$ $(0.01\,mol\,dm^{-3})$ and
      $10\,cm^3\,Na_2SO_4$ $(0.1\,mol\,dm^{-3})$

   b) $25\,cm^3\,Ca(OH)_2$ $(8 \times 10^{-3}\,mol\,dm^{-3})$ and
      $25\,cm^3\,Na_2SO_4$ $(0.01\,mol\,dm^{-3})$

   c) $50\,cm^3\,AlCl_3$ $(10^{-3}\,mol\,dm^{-3})$ and
      $50\,cm^3\,NaOH$ $(10^{-2}\,mol\,dm^{-3})$

   d) $10\,cm^3\,AgNO_3$ $(10^{-3}\,mol\,dm^{-3})$ and
      $40\,cm^3\,Na_2SO_4$ $(0.1\,mol\,dm^{-3})$

   e) $100\,cm^3\,Sr(NO_3)_2$ $(10^{-2}\,mol\,dm^{-3})$ and
      $100\,cm^3\,KF$ $(2 \times 10^{-2}\,mol\,dm^{-3})$

   f) $250\,cm^3\,Pb(NO_3)_2$ $(2 \times 10^{-2}\,mol\,dm^{-3})$ and
      $150\,cm^3\,NaCl$ $(0.01\,mol\,dm^{-3})$

   Show how you arrive at your conclusions.

5. In a saturated aqueous solution of hydrogen sulphide, the product
$$[H_3O^+]^2[S^{2-}] = 1.1 \times 10^{-23} \, mol^3 \, dm^{-9}$$
The solubility products of four sulphides are:

$$CdS, \quad 3.6 \times 10^{-29} \, mol^2 \, dm^{-6}$$
$$FeS, \quad 3.7 \times 10^{-19} \, mol^2 \, dm^{-6}$$
$$MnS, \quad 1.4 \times 10^{-15} \, mol^2 \, dm^{-6}$$
$$NiS, \quad 1.4 \times 10^{-24} \, mol^2 \, dm^{-6}$$

A solution contains each of the metal ions at a concentration of $0.10 \, mol \, dm^{-3}$ and $0.25 \, mol \, dm^{-3}$ hydrochloric acid. The solution is saturated with hydrogen sulphide. Calculate which of the sulphides will be precipitated.

6. In the estimation of chlorides by titration with a standard silver nitrate solution, using a chromate indicator, the precipitation of silver chloride is complete before the precipitation of silver chromate begins. Explain why this is so, using the solubility of silver chloride ($2.009 \times 10^{-3} \, g \, dm^{-3}$) and silver chromate ($3.207 \times 10^{-2} \, g \, dm^{-3}$) at $25 \, °C$ to calculate the solubility products and then: a) the concentration of silver ions needed to precipitate silver chloride from a neutral solution of chloride ions of concentration $0.100 \, mol \, dm^{-3}$, and b) the concentration of silver ions required to precipitate silver chromate, $Ag_2CrO_4$, from a neutral solution of chromate ions containing $5.00 \times 10^{-3} \, mol \, dm^{-3}$.

7. A solution contains $0.10 \, mol \, dm^{-3}$ of sodium carbonate and $0.10 \, mol \, dm^{-3}$ of sodium sulphate. To $1 \, dm^3$ of the solution is added $0.10 \, mol$ calcium chloride. The solubility products are: $CaCO_3$, $1.7 \times 10^{-8}$; $CaSO_4$, $2.3 \times 10^{-4} \, mol^2 \, dm^{-6}$. Find out which salt will be precipitated and calculate the mass of the precipitate.

8. The solubility product of mercury(II) sulphide is quoted in one reference book as $2 \times 10^{-49} \, mol^2 \, dm^{-6}$. If this value is correct, how many mercury(II) ions will be present in $1 \, dm^3$ of a saturated solution of this salt? (The Avogadro constant is $6 \times 10^{23} \, mol^{-1}$.)

9. The solubility product of $PbBr_2$ is $7.9 \times 10^{-5}$; that of $PbI_2$ is $1.0 \times 10^{-9} \, mol^3 \, dm^{-9}$; $1.0 \, dm^3$ of a solution containing $0.20 \, mol \, dm^{-3}$ of sodium bromide and $0.20 \, mol \, dm^{-3}$ of sodium iodide is added to $1.0 \, dm^3$ of lead(II) nitrate solution of concentration $0.10 \, mol \, dm^{-3}$. Which salt is precipitated? What is the mass of the precipitate?

10. Calculate the solubility of magnesium hydroxide: a) in water, b) in $0.10 \, mol \, dm^{-3}$ sodium hydroxide solution, c) in $0.010 \, mol \, dm^{-3}$ magnesium chloride solution, all at $25 \, °C$. The solubility product of magnesium hydroxide is $1.1 \times 10^{-11} \, mol^3 \, dm^{-9}$ at $25 \, °C$.

11. $K_{sp}(SrSO_4) = 4.0 \times 10^{-7} \, mol^2 \, dm^{-6}$. Calculate the solubility in mol $dm^{-3}$ of $SrSO_4$ a) in water, b) in $0.10 \, mol \, dm^{-3}$ aqueous sodium sulphate.

12. $K_{sp}(MgF_2) = 7.2 \times 10^{-9} \, mol^3 \, dm^{-9}$. Calculate the solubility in mol $dm^{-3}$ of $MgF_2$ a) in water, b) in a $0.20 \, mol \, dm^{-3}$ solution of NaF.

13. Calculate the solubility of calcium fluoride in mol $dm^{-3}$ a) in water, b) in a $0.010 \, mol \, dm^{-3}$ solution of sodium fluoride, c) in a $1.0 \, mol$ $dm^{-3}$ solution of hydrogen fluoride. ($K_{sp}(CaF_2) = 4.0 \times 10^{-11} \, mol^3$ $dm^{-9}$, $K_a(HF) = 5.6 \times 10^{-4} \, mol \, dm^{-3}$.)

## ELECTRODE POTENTIALS

If a strip of metal is placed in a solution of its ions, atoms of the metal may dissolve as positive ions, leaving a build-up of electrons on the metal:

$$M(s) \longrightarrow M^{2+}(aq) + 2e^-$$

The metal will become negatively charged. Alternatively, metal ions may take electrons from the strip of metal and be discharged as metal atoms:

$$M^{2+}(aq) + 2e^- \longrightarrow M(s)$$

In this case, the metal will become positively charged. The potential difference between the strip of metal and the solution depends on the nature of the metal and on the concentration of the ions involved in the equilibrium at the metal surface. Zinc acquires a more negative potential than copper, since it has a greater tendency to dissolve as ions and a smaller tendency to be deposited as metal. In order to compare electrode potentials for different metals, *standard electrode potentials* are quoted at 25 °C with an ionic concentration of 1 mol $dm^{-3}$. The zero on the standard electrode potential scale is the potential of a strip of platinum in contact with hydrogen gas at 1 atm pressure and hydrogen ions at a concentration of 1 mol $dm^{-3}$.

Metals are reducing agents. Other oxidation–reduction systems also have electrode potentials, the value of which depend on the standard electrode potential for the system and the concentrations of the ions in the equilibrium. The standard electrode potential of a redox system is the potential acquired by a piece of platinum immersed in a solution of the redox system in which the concentration of each dissolved component is 1 mol $dm^{-3}$. A powerful oxidising agent removes electrons and gives the platinum a high positive potential. When all the redox systems are arranged in order of their standard electrode potentials, the *electrochemical series* is obtained. Table 8.1 shows some of the redox systems in the series.

Table 8.1    Values of standard electrode potential $E^\circ$ at 298 K

| *Reaction* | $E^\circ/V$ |
|---|---|
| $K^+(aq) + e^- \longrightarrow K(s)$ | $-2.92$ |
| $Ca^{2+}(aq) + 2e^- \longrightarrow Ca(s)$ | $-2.87$ |
| $Na^+(aq) + e^- \longrightarrow Na(s)$ | $-2.71$ |
| $Mg^{2+}(aq) + 2e^- \longrightarrow Mg(s)$ | $-2.36$ |
| $Al^{3+}(aq) + 3e^- \longrightarrow Al(s)$ | $-1.66$ |
| $Zn^{2+}(aq) + 2e^- \longrightarrow Zn(s)$ | $-0.76$ |
| $Fe^{2+}(aq) + 2e^- \longrightarrow Fe(s)$ | $-0.44$ |
| $Cr^{3+}(aq) + e^- \longrightarrow Cr^{2+}(aq)$ | $-0.41$ |
| $Ni^{2+}(aq) + 2e^- \longrightarrow Ni(s)$ | $-0.25$ |
| $Sn^{2+}(aq) + 2e^- \longrightarrow Sn(s)$ | $-0.14$ |
| $Pb^{2+}(aq) + 2e^- \longrightarrow Pb(s)$ | $-0.13$ |
| $2H_3O^+(aq) + 2e^- \longrightarrow H_2(g) + 2H_2O(l)$ | $0.00$ |
| $Sn^{4+}(aq) + 2e^- \longrightarrow Sn^{2+}(aq)$ | $0.15$ |
| $Cu^{2+}(aq) + 2e^- \longrightarrow Cu(s)$ | $0.34$ |
| $I_2(s) + 2e^- \longrightarrow 2I^-(aq)$ | $0.54$ |
| $Fe^{3+}(aq) + e^- \longrightarrow Fe^{2+}(aq)$ | $0.77$ |
| $Ag^+(aq) + e^- \longrightarrow Ag(s)$ | $0.80$ |
| $Br_2(l) + 2e^- \longrightarrow 2Br^-(aq)$ | $1.09$ |
| $MnO_2(s) + 4H^+(aq) + 2e^- \longrightarrow Mn^{2+}(aq) + 2H_2O(l)$ | $1.23$ |
| $Cr_2O_7^{2-}(aq) + 14H^+(aq) + 6e^- \longrightarrow 2Cr^{3+}(aq) + 7H_2O(l)$ | $1.33$ |
| $Cl_2(g) + 2e^- \longrightarrow 2Cl^-(aq)$ | $1.36$ |
| $Ce^{4+}(aq) + e^- \longrightarrow Ce^{3+}(aq)$ (in $H_2SO_4(aq)$) | $1.44$ |
| $PbO_2(s) + 4H^+(aq) + 2e^- \longrightarrow Pb^{2+}(aq) + 2H_2O(l)$ | $1.46$ |
| $MnO_4^-(aq) + 8H^+(aq) + 5e^- \longrightarrow Mn^{2+}(aq) + 4H_2O(l)$ | $1.51$ |
| $Ce^{4+}(aq) + e^- \longrightarrow Ce^{3+}(aq)$ (in $HNO_3(aq)$) | $1.61$ |
| $H_2O_2(aq) + 2H^+(aq) + 2e^- \longrightarrow 2H_2O(l)$ | $1.78$ |
| $F_2(g) + 2e^- \longrightarrow 2F^-(aq)$ | $2.85$ |

## GALVANIC CELLS

When two electrodes are combined to form a cell, their standard electrode potentials will tell you which will be the positive and which the negative electrode. An easy way to work out which of the possible reactions will happen is to use the *anticlockwise rule*. Write down the two redox systems, with the more negative standard electrode potential at the top. Then draw a circle anticlockwise. For example, when copper and silver are in contact with solutions of their ions,

$$Cu^{2+}(aq) + 2e^- \rightleftharpoons Cu(s) \quad E^{\ominus} = +0.34\,V$$

$$Ag^+(aq) + e^- \rightleftharpoons Ag(s) \quad E^{\ominus} = +0.80\,V$$

The circle tells you that the reaction which takes place is

$$Cu(s) + 2Ag^+(aq) \longrightarrow Cu^{2+}(aq) + 2Ag(s)$$

The silver electrode is positive; the copper electrode is negative.

Reaction will take place between two redox systems which differ by 0.3 V or more.

Fig. 8.2 shows two metals inserted into solutions of their ions. The two solutions are joined by a salt bridge, and the two metal electrodes are connected by an external circuit. The cell has an e.m.f. which is equal to the difference between the standard electrode potentials of the two metals, and a current flows through the external circuit.

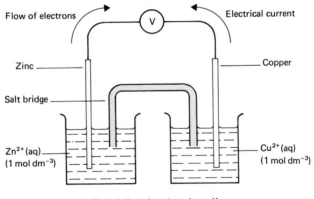

Fig. 8.2   A galvanic cell

The cell shown in Fig. 8.2 can be represented by

$$Zn(s) \mid Zn^{2+}(aq) \,(1\,mol\,dm^{-3}) \mathrel{\vdots} Cu^{2+}(aq) \,(1\,mol\,dm^{-3}) \mid Cu(s)$$

By convention, the e.m.f. of the cell is taken as

$$E = E(\text{RHS electrode}) - E(\text{LHS electrode})$$

$$= E^{\ominus}_{Cu} - E^{\ominus}_{Zn} \quad \text{where } E^{\ominus} \text{ is the standard electrode potential}$$

$$= 0.34 - (-0.76) = +1.10\,V$$

The flow of electrons is clockwise through the external circuit (from zinc to copper). Conventional electricity flows anticlockwise through the external circuit (from copper to zinc).

If the cell is written as

$$Cu(s) \mid Cu^{2+}(aq) \ (1 \ mol \ dm^{-3}) \mid Zn^{2+}(aq) \ (1 \ mol \ dm^{-3}) \mid Zn(s)$$

then the e.m.f. is given by

$$E = E_{Zn}^{\ominus} - E_{Cu}^{\ominus} = -0.76 - 0.34 = -1.10 \ V$$

In general, in a cell A | B | C | D if the reactions which occur are A ⟶ B and C ⟶ D, then the e.m.f. is positive; if B ⟶ A and D ⟶ C, then the e.m.f. is negative.

## EXERCISE 49     Problems on Standard Electrode Potentials

Refer to the table of values on p.154.

1. Which of the following species are oxidised by manganese(IV) oxide?
   $Br^-$,     Ag,     $I^-$,     $Cl^-$

2. Which of the following species are reduced by $Sn^{2+}$?
   $I_2$,     $Ni^{2+}$,     $Cu^{2+}$,     $Fe^{3+}$

3. Calculate the standard e.m.f.'s of the following cells at 298 K:
   a) $Ni(s) \mid Ni^{2+}(aq) \mid Sn^{2+}(aq), Sn^{4+}(aq) \mid Pt$
   b) $Pt \mid I_2(s), I^-(aq) \mid Ag^+(aq) \mid Ag(s)$
   c) $Pt \mid Cl_2(g), Cl^-(aq) \mid Br_2(l), Br^-(aq) \mid Pt$

4. Calculate the standard e.m.f. of each of the cells:
   a) $Sn(s) \mid Sn^{2+}(aq) \mid Ag^+(aq) \mid Ag(s)$
   b) $Ag(s) \mid Ag^+(aq) \mid Cu^{2+}(aq) \mid Cu(s)$
   c) $Ce^{3+}(aq) \mid Ce^{4+}(aq) \mid Fe^{3+}(aq) \mid Fe^{2+}(aq)$
   d) $Fe(s) \mid Fe^{2+}(aq) \mid Cu^{2+}(aq) \mid Cu(s)$
   e) $Zn(s) \mid Zn^{2+}(aq) \mid Pb^{2+}(aq) \mid Pb(s)$

5. Iron filings are added to a solution containg the ions $Cu^{2+}$, $Fe^{2+}$, $Fe^{3+}$, $H_3O^+$ and $Zn^{2+}$, all at a concentration of $1 \ mol \ dm^{-3}$. From the standard electrode potentials of the redox systems, deduce what reaction occurs, and write the equation.

6. A solution contains $Fe^{2+}$, $Fe^{3+}$. $Cr^{3+}$ and $Cr_2O_7^{2-}$ in their standard states and dilute sulphuric acid. Deduce what happens, and write the equation for the reaction.

7. Predict the reactions between:
   a) $Fe^{3+}(aq)$ and $I^-(aq)$          b) $Ag^+(aq)$ and $Cu(s)$
   c) $Fe^{3+}(aq)$ and $Br^-(aq)$          d) $Ag(s)$ and $Fe^{3+}(aq)$
   e) $Br_2(aq)$ and $Fe^{2+}(aq)$

   From the standard electrode potentials, predict which of the halogens, $Cl_2$, $Br_2$, $I_2$, will oxidise i) $Fe^{2+}$ to $Fe^{3+}$ ii) $Sn^{2+}$ to $Sn^{4+}$.

**EXERCISE 50**    Questions from A-level Papers

*1. The following data were obtained from a sample of an aqueous solution of aminoethane (ethylamine) of concentration $0.1 \, mol \, dm^{-3}$ at 25 °C:

   i) Electrolytic conductivity $1.5 \times 10^{-1} \, S \, m^{-1}$ ($1 \, S \, m^{-1} \equiv 1 \, \Omega^{-1} m^{-1}$).

   ii) Molar conductivity at infinite dilution $2.04 \times 10^{-2} \, S \, m^2 \, mol^{-1}$ ($1 \, S \, m^2 \, mol^{-1} \equiv 1 \, \Omega^{-1} m^2 \, mol^{-1}$).

   iii) Addition of an equal volume of a solution of compound X of concentration $0.01 \, mol \, dm^{-3}$ produced a solution of pH 11.73. This pH value remained constant in spite of small additions of $H_3O^+$ and $OH^-$ ions.

   a) Calculate the dissociation constant and pH value for the aqueous solution of aminoethane. Comment on the pH value you obtain.

   b) Suggest a named compound which could be represented by X. What is the name given to such a mixed solution?

   c) Explain the action of such a solution.

   d) Describe concisely how the values i) and ii) in the data above could be obtained.                                          (AEB80, S)

2. a) A buffer solution can be prepared by mixing solutions of a weak acid and one of its salts. Derive the relationship between the hydrogen ion concentration, the acid dissociation constant ($K_a$) and the concentrations of the weak acid and its salt.

   b) Calculate $K_a$ for a weak monobasic acid given the following information. A buffer solution made up of $10 \, cm^3$ of acid of concentration $0.09 \, mol \, dm^{-3}$ mixed with $20 \, cm^3$ of the potassium salt of the acid of concentration $0.15 \, mol \, dm^{-3}$ has a pH of 5.85. (O & C80, p)

3. a) Write an expression for the solubility product of lead(II) chloride.

   b) The solubility product of lead(II) chloride is $1.6 \times 10^{-5} \, mol^3 \, l^{-3}$ at a given temperature.
     i) What is the solubility in $mol \, l^{-1}$ of lead(II) chloride in water at the same temperature?

     ii) How many moles of chloride ion must be added to a 1.0 M solution of lead(II) nitrate at the same temperature in order just to cause a precipitate of lead(II) chloride? Assume that no change in volume occurs on adding the chloride ion.

   c) When a 1.0 M solution of hydrochloric acid is added to a saturated solution of lead(II) chloride, a permanent precipitate is obtained. However, if concentrated hydrochloric acid is used, the precipitate initially formed dissolves. How do you account for this?

d) When an excess of aqueous ammonia is added to a solution containing copper(II) ions, a complex ion is formed.

  i) Describe what you would see when the addition is carried out slowly and write down the equations for the reactions occurring.

  ii) Write an expression for the stability constant of the complex ion.

  iii) If the stability constant for the complex ion is $1 \times 10^{13}\,\text{mol}^{-4}\text{l}^4$, what will be the concentration of $Cu^{2+}$ ions in a solution where the concentration of the complex ion is 1.0 M and that of ammonia is 3.0 M?                    (L80)

**4.** a) Balance the following redox equations using the principles of *either* electron transfer *or* change in oxidation state (number):

  i) $Ag(s) + NO_3^-(aq) + H^+(aq) \longrightarrow Ag^+(aq) + NO(g)\ H_2O(l)$

  ii) $Fe(CN)_6^{4-}(aq) + Cl_2(g) \longrightarrow Fe(CN)_6^{3-}(aq) + 2Cl^-(aq)$

b) Discuss *briefly* the electrolysis of each of the following solutions:

| Electrolyte | Cathode | Anode |
|---|---|---|
| Sodium chloride | Carbon | Carbon |
| Sodium hydroxide | Carbon | Carbon |
| Sulphuric acid | Platinum | Silver |
| Copper sulphate | Copper | Copper |

  i) A current of 3.21 A was passed through fused aluminium oxide for 10 minutes. The volume of oxygen collected at the anode was 112 cm³ measured at s.t.p. Calculate the mass of aluminium obtained at the cathode and the charge on 1 mole of electrons (the Faraday). (O = 16; Al = 27; molar volume = 22.4 dm³ at s.t.p.)

  ii) When the same quantity of electricity was passed through the fused chloride of a metal M (relative atomic mass = 137.3), the mass of M obtained was 1.373 g. Calculate the charge on the cation $M^{x+}$.

  iii) The charge on the electron is $1.602 \times 10^{-19}$ coulombs. Calculate a value for the Avogadro constant ($L$).

  iv) The standard electrode potentials of three metals X, Y, and Fe are $-0.14\,\text{V}$, $-0.76\,\text{V}$ and $-0.44\,\text{V}$ respectively. Explain which one of X or Y would be a more effective protection against the corrosion of iron.                    (SUJB85)

5. Define the term *pH*.

a) Write an expression for the acidity constant (acid ionisation constant, dissociation constant), $K_a$, of ethanoic acid (acetic acid), $CH_3COOH$, given that

$$K_a = 1.8 \times 10^{-5} \, mol \, dm^{-3} \text{ at } 291 \text{ K.}$$

b) Explain what is understood by the term $pK_a$ and calculate its value for ethanoic acid.

c) Calculate the percentage ionisation (mol %) of ethanoic acid in a 0.1 M solution of the acid at 291 K.

d) Sketch a titration curve to show the pH change (vertical axis) which occurs on adding 30 cm³ 0.1 M sodium hydroxide solution to 25 cm³ 0.1 M ethanoic acid at 291 K. The pH of the acid should be calculated initially and after addition of 5, 12.5, 20 and 30 cm³ of alkali.

e) What is a buffer solution? Give at least one example, with an indication of the molar proportions of the mixture. Relate what you have described to the titration curve in d) and indicate where you consider buffer capacity to be a maximum. Explain your opinion.

(SUJB82)

6. a)  i) What is a buffer solution?
       ii) What is meant by a weak acid?

b) Ethanoic acid is a weak acid with a $pK_a$ value of 4.76 at 298 K.
   i) Write an expression for $K_a$ for aqueous ethanoic acid.
   ii) Calculate the value of $K_a$.
   iii) Calculate the pH of an aqueous solution of ethanoic acid, $CH_3COOH$, containing 0.25 mol dm⁻³.

c) Describe how an aqueous solution of ethanoic acid and sodium ethanoate behaves as a buffer solution. (AEB85)

7. a) Explain the term solubility product and write expressions for, and the units of the solubility products of, calcium sulphate, aluminium hydroxide and lead bromide.

b) Discuss *each* of the following:
   i) the solubility of silver chloride in water decreases when dilute hydrochloric acid is added but increases when concentrated hydrochloric acid or aqueous ammonia is added;
   ii) aqueous ammonia can precipitate certain metals as their hydroxides but the presence of ammonium chloride often prevents the precipitation.

c) The solubility of strontium hydroxide ($Sr(OH)_2$) is 0.524 g in 100 cm³ water. Calculate:
   i) the solubility of strontium hydroxide in water, mol dm⁻³;
   ii) the hydroxide ion concentration (mol dm⁻³) in a saturated solution of strontium hydroxide;
   iii) the solubility product of strontium hydroxide;

iv) the approximate solubility of strontium hydroxide ($g\,dm^{-3}$) in $1\,dm^3$ of $2 \times 10^{-1}\,mol\,dm^{-3}$ strontium chloride solution;

v) the volume of $1 \times 10^{-2}\,mol\,dm^{-3}$ potassium chromate solution which must be added to $1\,dm^3$ saturated strontium hydroxide solution to precipitate strontium chromate ($SrCrO_4$).

(H = 1; O = 16; Sr = 87.6. Solubility product of strontium chromate is $3.6 \times 10^{-5}\,mol^2\,dm^{-6}$.)    (SUJB85)

**\*8.** In working out each of the following calculations, state clearly the physico-chemical principles upon which they depend, and show all working clearly so that marks can be allocated even if errors occur.

a) Distinguish carefully between the terms 'solubility' and 'solubility product' as applied to a sparingly soluble electrolyte. If the *numerical* values of the solubility products of silver chloride, AgCl, and silver chromate(VI), $Ag_2CrO_4$, are respectively $1.6 \times 10^{-10}$ and $1.0 \times 10^{-12}$, calculate their respective molar solubilities in water at this temperature. (AgCl = 143.5, $Ag_2CrO_4$ = 332.)

b) The numerical value for the solubility product of silver ethanoate (acetate), $CH_3COOAg$, is $2.0 \times 10^{-3}$.

If $4.1\,g$ of sodium ethanoate is added to $100\,cm^3$ of a saturated solution of silver ethanoate in water, calculate the concentration of silver ions remaining in solution and the mass of silver ethanoate precipitated. ($CH_3COONa$ = 82, $CH_3COOAg$ = 167.)

c) If the numerical value of the solubility product of silver iodide, AgI, is $1.0 \times 10^{-16}$ and the instability constant for the ion

$$Ag(NH_3)_2^+(aq) \rightleftharpoons 2NH_3(aq) + Ag^+(aq)$$

has the numerical value $6.0 \times 10^{-8}$, calculate the mass of silver iodide which will dissolve in $1\,dm^3$ (1 litre) of $1.0\,M$ ammonia solution.

$$AgI(s) + 2NH_3(aq) \rightleftharpoons Ag(NH_3)_2^+(aq) + I^-(aq)$$
(AgI = 235.)    (SUJB83)

**9.** a) Define pH.

b) At $25\,°C$, $K_a$ for the dissociation (acetic) acid in water is $1.8 \times 10^{-5}$ $mol\,l^{-1}$ and the ionic product ($K_w$) of water is $1 \times 10^{-14}\,mol^2\,l^{-2}$.

i) Calculate the basicity constant ($K_b$) of the ethanoate (acetate) ion in water at $25\,°C$.

ii) Calculate the pH of a $0.05\,M$ aqueous solution of ethanoic acid at $25\,°C$.

c) The following data refer to aqueous solutions at $25\,°C$.

| Substance | $pK_b$ |
|---|---|
| Ammonia | 4.75 |
| Methylamine | 3.38 |
| Phenylamine (*aniline*) | 9.40 |

Account for the base strengths of methylamine and phenylamine relative to that of ammonia.    (JMB84)

**\*10.** a) Describe and explain the effect of dilution on the molar conductivities of weak electrolytes and strong electrolytes. Why cannot the
molar conductivity of a weak electrolyte at infinite dilution be
determined by direct experimental measurement?

b) The conductivity of a $1.25 \times 10^{-1} \, \text{mol dm}^{-3}$ solution of a monobasic acid is $2.39 \times 10^{-2} \, \text{ohm}^{-1} \, \text{cm}^{-1}$. Its molar conductivity at
infinite dilution is $400 \, \text{ohm}^{-1} \, \text{cm}^2 \, \text{mol}^{-1}$. Calculate the degree of
dissociation of the acid and its dissociation constant.

c) The conductivity of a saturated solution of sparingly soluble silver
bromide at 298 K is $1.79 \times 10^{-7} \, \text{ohm}^{-1} \, \text{cm}^{-1}$, the solvent water
having a conductivity of $5.5 \times 10^{-8} \, \text{ohm}^{-1} \, \text{cm}^{-1}$ at this temperature. The molar conductivities at infinite dilution of silver nitrate,
potassium nitrate, and potassium bromide are 133.4, 145.0, and
$151.9 \, \text{ohm}^{-1} \, \text{cm}^2 \, \text{mol}^{-1}$ respectively. Calculate:
   i)   the molar conductivity of silver bromide at infinite dilution;
   ii)  the solubility of silver bromide $(\text{mol dm}^{-3})$;
   iii) the solubility product of silver bromide.          (SUJB85)

**11.** a) Fill in a copy of the table below to show how each of the species
listed can be estimated volumetrically. State the volumetric reagent
used and any indicator which must be added (if no indicator is
required, state 'none').

| Species to be estimated | Volumetric reagent | Indicator |
|---|---|---|
| $Cl^-$ | | |
| $I_2$ | | |
| $C_2O_4{}^{2-}$ | | |

b) Use the data below to answer the questions which follow.

$$MnO_4{}^-(aq) \; + \; 8H^+(aq) \; + \; 5e^- \longrightarrow Mn^{2+}(aq) \; + \; 4H_2O(l)$$
$$E^\ominus = +1.52 \, V$$

$$Ce^{4+}(aq) \; + \; e^- \longrightarrow Ce^{3+}(aq)$$
$$E^\ominus = +1.45 \, V$$

$$\tfrac{1}{2}Cl_2(g) \; + \; e^- \longrightarrow Cl^-(aq)$$
$$E^\ominus = +1.36 \, V$$

$$\tfrac{1}{2}Cr_2O_7{}^{2-}(aq) \; + \; 7H^+(aq) \; + \; 3e^- \longrightarrow Cr^{3+}(aq) \; + \; 3\tfrac{1}{2}H_2O(l)$$
$$E^\ominus = +1.33 \, V$$

$$VO_2{}^+(aq) \; + \; 2H^+(aq) \; + \; e^- \longrightarrow VO^{2+}(aq) \; + \; H_2O(l)$$
$$E^\ominus = +1.00 \, V$$

$$Fe^{3+}(aq) \; + \; e^- \longrightarrow Fe^{2+}(aq)$$
$$E^\ominus = +0.77 \, V$$

    i) State which species in the above list is/are capable of liberating chlorine from an acidified ($1\,M\,H^+$) aqueous solution of sodium chloride.

    ii) Write a balanced equation for the reaction which you would expect to occur between manganate(VII) ions and oxovanadium(IV), $VO^{2+}$, ions in acidic aqueous solution.

    iii) Explain why dilute sulphuric acid, rather than dilute hydrochloric acid, is used as a source of $H^+(aq)$ in the volumetric estimation of iron(II) ions and with potassium manganate(VII).

                                      (JMB83)

12. a) From a consideration of the potentials given below predict what reactions occur, if any, between
    i) $Fe_2(SO_4)_3(aq)$ and $KI(aq)$,
    ii) $FeSO_4(aq)$ and $I_2$ (dissolved in an aqueous solution of KI), and
    iii) the reagents in ii) with the addition of $KCN(aq)$.

    Comment on your predictions.

$$Fe^{3+}(aq) + e^- \rightleftharpoons Fe^{2+}(aq) \qquad E^\ominus = +0.77\,V$$

$$[Fe(CN)_6]^{3-} + e^- \rightleftharpoons [Fe(CN)_6]^{4-} \qquad E^\ominus = +0.36\,V$$

$$\tfrac{1}{2}I_2(s) + e^- \rightleftharpoons I^-(aq) \qquad E^\ominus = +0.54\,V$$

  b) i) Describe the structure of, and the bonding in, $HClO_4$ and state the oxidation state of Cl in this compound.

    ii) Careful dehydration of $HClO_4$ gives an oily liquid A having the composition Cl 38.8%, O 61.2% and an $M_r$ of 183. What is the molecular formula of A? Suggest a structure for this compound.
                                (JMB85,Sp)

13. In working out each of the following calculations state clearly the physico-chemical principles upon which they depend and show all working clearly so that marks can be allocated even if errors occur.

  a) State what you understand by the term 'Faraday constant',

$$F = 9.649 \times 10^4\,C\,mol^{-1}.$$

An electric current of 4 amperes was passed through molten anhydrous strontium bromide, $SrBr_2$, for 25 minutes between graphite electrodes. State, as equations, the reactions which occur at the surfaces of the electrodes, and calculate the mass of each product liberated. (Br = 80, Sr = 88.)

  b) Define the term 'acidity constant (acid ionisation constant, dissociation constant) of an acid'. The acidity constant of ethanoic acid (acetic acid) at room temperature is $1.85 \times 10^{-5}\,mol\,dm^{-3}$. Assuming that the extent of the ionisation is extremely small, calculate the pH of a 0.010 M solution of ethanoic acid at that temperature.

If an amount of sodium ethanoate, sufficient to make the solution 0.100 M with respect to that salt, is now dissolved in the acid, calculate the change in pH which occurs. Comment.

c) Explain what you understand by the term, 'ideal gas constant, $R' = 8.314\,J\,mol^{-1}\,K^{-1}$, and relate it to the 'kinetic energy' of a mole of ideal gas and to the 'root mean square velocity' of the molecules in such a gas, explaining these terms.

Assuming ideal behaviour, calculate the kinetic energy (kJ) of the molecules in 1.50 mol of
i) sulphur dioxide,
ii) hydrogen iodide at 27 °C,
and the ratio of the *root* mean square velocity of the molecules of sulphur dioxide to those of hydrogen iodide at this temperature. (H = 1, O = 16, S = 32, I = 127; 0 K = −273 °C.)

(SUJB82)

14. a) The table below gives standard electrode potentials in acid solution. Use the data, when appropriate, to answer the questions which follow.

$$E^{\ominus}/V$$

$$Ce^{4+}(aq) + e^- \longrightarrow Ce^{3+}(aq) \qquad +1.61$$

$$\tfrac{1}{2}Cl_2(g) + e^- \longrightarrow Cl^-(aq) \qquad +1.36$$

$$IO_3^-(aq) + 6H^+ + 6e^- \longrightarrow I^-(aq) + 3H_2O \qquad +1.09$$

$$\tfrac{1}{2}Br_2(l) + e^- \longrightarrow Br^-(aq) \qquad +1.06$$

$$Fe^{3+}(aq) + e^- \longrightarrow Fe^{2+}(aq) \qquad +0.77$$

$$\tfrac{1}{2}I_2(s) + e^- \longrightarrow I^-(aq) \qquad +0.54$$

$$Cr^{3+}(aq) + e^- \longrightarrow Cr^{2+}(aq) \qquad -0.41$$

i) Write the chemical equation for which $E^{\ominus}$ is arbitrarily taken to be zero.
ii) Which of the halogens listed in the table is the least electronegative?
iii) Which of the halogens, listed in the table, is (are) capable of oxidising $Fe^{2+}(aq)$ to $Fe^{3+}(aq)$ under standard conditions?
iv) Construct a balanced equation for the reaction between iodate(V) and iodide ions in acid solution.
v) State and explain briefly what change, if any, may occur if a solution of $Cr^{2+}(aq)$ in $1\,M\,H^+(aq)$ is stored in the absence of air.
vi) Which of the species, if any, listed in the table, is (are) capable of reducing $Cr^{3+}(aq)$ to $Cr^{2+}(aq)$ (if none, write 'none')?

b) Indicate how, by means of a simple chemical test or tests, you could distinguish between aqueous solutions of hydrogen bromide and hydrogen iodide.

(JMB85)

**15. a)** Use the following example

$$Ag^+(aq) + e^- \rightleftharpoons Ag(s); \qquad E^\ominus = +0.8 \text{ V}$$

to define the terms *standard electrode potential, oxidation, reduction, oxidising agent* and *reducing agent*.

**b)**

|  | $E^\ominus/V$ |
|---|---|
| $Co^{2+}(aq) + 2e^- \rightleftharpoons Co(s)$ | $-0.28$ |
| $Co^{3+}(aq) + e^- \rightleftharpoons Co^{2+}(aq)$ | $+1.80$ |
| $O_2(g) + 4H^+(aq) + 4e^- \rightleftharpoons 2H_2O(l)$ | $+1.20$ |
| $Fe^{3+}(aq) + e^- \rightleftharpoons Fe^{2+}(aq)$ | $+0.77$ |
| $[Co(NH_3)_6]^{3+}(aq) + e^- \rightleftharpoons [Co(NH_3)_6]^{2+}(aq)$ | $+0.10$ |
| $2H^+(aq) + 2e^- \rightleftharpoons H_2(g)$ | $0.00$ |

Use the values of *standard* redox potentials to explain the following.

i) Both cobalt metal and the cobalt(III) ion are unstable in aqueous acid of concentration $1 \text{ mol dm}^{-3}$, but the cobalt(II) ion is stable.

ii) Complexing cobalt(III) with ammonia gives an ion which is stable in aqueous acid of concentration $1 \text{ mol dm}^{-3}$.

iii) A solution of an iron(II) salt is stable in aqueous acid of concentration $1 \text{ mol dm}^{-3}$ *in the absence of oxygen.*

iv) The electrode potential of the reaction

$$O_2(g) + 4H^+(aq) + 4e^- \rightleftharpoons 2H_2O,$$

decreases as the pH is increased.

**c) i)** For the half reaction,

$$4H^+(aq) + MnO_4^-(aq) + 3e^- \longrightarrow MnO_2(s) + 2H_2O(l);$$
$$E^\ominus = +1.695 \text{ V},$$

calculate the electrode potential, $E$, in a neutral solution, i.e. at pH $= 7$. (You may assume that $E = E^\ominus + \dfrac{0.059}{3} \log[H^+(aq)]^4$)

ii) Using your answer to i), and that at pH $= 7$, $E = +0.81$ V, for the reaction

$$O_2(g) + 4H^+(aq) + 4e^- \rightleftharpoons 2H_2O(l)$$

state whether or not manganate(VII) should oxidise water in a neutral solution.                                    (AEB82)

# 9 Thermochemistry

## INTERNAL ENERGY AND ENTHALPY

Matter contains energy. It is the kinetic energy of molecular motion and the potential energy associated with chemical bonds. These together make up the *internal energy* of matter. Frequently during the course of a chemical reaction heat is either given out or taken in from the surroundings. The heat absorbed during a reaction is equal to the internal energy of the products minus the internal energy of the reactants plus any work done by the system on the surroundings. Since most laboratory work is done at constant pressure, any gases formed are allowed to escape into the atmosphere and work is done in expansion:

$$\begin{pmatrix} \text{Heat absorbed at} \\ \text{constant pressure} \end{pmatrix} = \begin{pmatrix} \text{Change in} \\ \text{internal energy} \end{pmatrix} + \begin{pmatrix} \text{Work done on} \\ \text{surroundings} \end{pmatrix}$$

The heat absorbed at constant pressure is given the name *change in enthalpy* and the symbol $\Delta H$. Enthalpy is defined by the equation

$$H = U + PV$$

where $H$ = Enthalpy, $U$ = Internal energy, $P$ = Pressure, and $V$ = Volume.

Then,
$$\Delta H = \Delta U + P\Delta V$$

When expansion occurs, $\Delta V$ is positive and $\Delta H > \Delta U$.

When contraction occurs, $\Delta V$ is negative and $\Delta H < \Delta U$.

If reaction takes place at constant volume, $\Delta V = 0$, and $\Delta H = \Delta U$.

Reactions of solids and liquids do not involve large changes in volume, and $\Delta H$ is close to $\Delta U$. Reactions in which $\Delta V$ is large are those involving gases, and the value of $\Delta V$ can be calculated from the ideal gas equation. Since

$$PV = nRT$$

$$P\Delta V = \Delta nRT$$

$\Delta n$, the increase in the number of molecules of gas, is indicated by the equation for the reaction. For example, in the reaction

$$CaCO_3(s) \longrightarrow CaO(s) + CO_2(g)$$

$\Delta n = 1$.

The enthalpy of a substance is quoted for the substance in its standard state. The *standard state* of a substance is 1 mole of the substance in a specified state (solid, liquid or gas) at 1 atmosphere pressure. The value of an enthalpy change is quoted for standard conditions: gases at 1 atmosphere, solutions at unit concentration, and substances in their normal states at a specified temperature. $\Delta H_T^\ominus$ means the standard enthalpy change at a temperature $T$. $\Delta H_{298}^\ominus$ is sometimes written as $\Delta H^\ominus$.

Definitions of some standard enthalpy changes follow:

*Standard enthalpy of formation* is the heat absorbed when 1 mole of a substance is formed from its elements *in their standard states* at constant pressure. (If the reaction is exothermic, the heat absorbed is negative, and $\Delta H_F^\ominus$ has a negative value. All elements in their standard states are assigned a value of zero for their standard enthalpies of formation.)

*Standard enthalpy of combustion* is the heat absorbed when 1 mole of a substance is completely burned in oxygen at constant pressure.

*Standard enthalpy of hydrogenation* is the heat absorbed when 1 mole of an unsaturated compound is converted into a saturated compound by reaction with gaseous hydrogen at constant pressure.

*Standard enthalpy of neutralisation* is the heat absorbed when an acid and a base react at constant pressure to form 1 mole of water.

*Standard enthalpy of reaction* is the heat absorbed in a reaction at constant pressure between the number of moles of reactants shown in the equation for the reaction. In the reaction

$$4H_2O(g) \ + \ 3Fe(s) \ \longrightarrow \ Fe_3O_4(s) \ + \ 4H_2(g)$$

the standard enthalpy change refers to the reaction between 4 moles of steam and 3 moles of iron.

*Standard enthalpy of solution* is the heat absorbed when 1 mole of a substance is dissolved at constant pressure in a stated amount of solvent. This may be 100 g or 1 000 g of solvent or it may be an 'infinite' amount of solvent, i.e. a volume so large that on further dilution there is no further heat change.

## STANDARD ENTHALPY CHANGE FOR A CHEMICAL REACTION

The standard enthalpy change for a chemical reaction can be calculated from the standard enthalpies of formation of all the products and reactants involved. For example, in the addition of hydrogen chloride to ethene,

$$CH_2 \!=\! CH_2(g) \ + \ HCl(g) \ \longrightarrow \ C_2H_5Cl(g)$$
$$(+52.3) \qquad\qquad (-92.3) \qquad\qquad (-105)$$

The standard enthalpies of formation in $kJ \ mol^{-1}$ are shown under each species.

The standard enthalpy of reaction $\Delta H^\ominus$ is given by

$$\Delta H^\ominus = (-105) - (52.3 + (-92.3)) = -65 \, \text{kJ mol}^{-1}$$

The negative sign means that the products contain less energy than the reactants and the difference is the heat energy given out in the reaction: the reaction is exothermic. A positive value for $\Delta H^\ominus$ indicates an endothermic reaction.

The standard enthalpy of reaction depends only on the difference between the standard enthalpy of the reactants and the standard enthalpy of the products and not on the route by which the reaction occurs.

*This idea is embodied in Hess's law, which states that, if a reaction can take place by more than one route, the overall change in enthalpy is the same, whichever route is followed.*

## STANDARD ENTHALPY OF NEUTRALISATION

**EXAMPLE**  $250 \, \text{cm}^3$ of sodium hydroxide of concentration $0.400 \, \text{mol dm}^{-3}$ were added to $250 \, \text{cm}^3$ of hydrochloric acid of concentration $0.400 \, \text{mol dm}^{-3}$ in a calorimeter. The temperature of the two solutions and the calorimeter was $17.05\,^\circ\text{C}$. The mass of the calorimeter was $500 \, \text{g}$, and its specific heat capacity was $400 \, \text{J kg}^{-1}\text{K}^{-1}$. The temperature rose to $19.55\,^\circ\text{C}$. Assuming that the specific heat capacity† of all the solutions is $4200 \, \text{J kg}^{-1}\text{K}^{-1}$ calculate the standard enthalpy of neutralisation.

**METHOD**  Mass of solutions $= 500 \, \text{g}$
Heat capacity of solutions $= 0.500 \times 4200 = 2100 \, \text{J}$
Mass of calorimeter $= 500 \, \text{g}$
Heat capacity of calorimeter $= 0.500 \times 400 = 200 \, \text{J}$
Rise in temperature $= 2.50\,^\circ\text{C}$
Heat evolved $= (2100 + 200) \times 2.50 = 5750 \, \text{J}$
Amount of water formed $= 250 \times 10^{-3} \times 0.400 = 0.100 \, \text{mol}$
Heat evolved per mole $= 5750/0.100 = 57500 \, \text{J}$

**ANSWER**  The standard enthalpy of neutralisation $= 57.5 \, \text{kJ mol}^{-1}$.

---

†The *heat capacity* of a mass of substance is the quantity of heat needed to raise its temperature by 1 K or 1 °C.

The *specific heat capacity* of a substance is the quantity of heat required to raise the temperature of 1 kg of the substance by 1 K or 1 °C.

Heat capacity = Mass × Specific heat capacity.

## EXERCISE 51    Problems on Standard Enthalpy of Neutralisation

1. $50.0 \, cm^3$ of sodium hydroxide solution of concentration $0.400 \, mol$ $dm^{-3}$ required $20.0 \, cm^3$ of sulphuric acid of concentration $0.500 \, mol$ $dm^{-3}$ for neutralisation. A temperature rise of $3.6 \,°C$ was observed if both solutions and the calorimeter were initially at the same temperature. Calculate the standard enthalpy of neutralisation of sodium hydroxide with sulphuric acid. The heat capacity of the calorimeter is $39.0 \, J \, K^{-1}$. (The specific heat capacity of all the solutions is $4.2 \, J \, K^{-1} g^{-1}$.)

2. $100 \, cm^3$ of potassium hydroxide solution of concentration $1.00 \, mol$ $dm^{-3}$ and $100 \, cm^3$ of hydrochloric acid of concentration $1.00 \, mol$ $dm^{-3}$ were mixed in a calorimeter. All three were at the same temperature. The heat capacity of the calorimeter was $95 \, J \, K^{-1}$, and the rise in temperature was $6.25 \, K$. Calculate the standard enthalpy of neutralisation. (Specific heat capacity of water $= 4.2 \, J \, K^{-1} g^{-1}$.)

3. $100 \, cm^3$ of $1.00 \, mol \, dm^{-3}$ sodium hydroxide solution and $100 \, cm^3$ of $1.00 \, mol \, dm^{-3}$ ethanoic acid were mixed in a calorimeter. All three were at the same temperature. The heat capacity of the calorimeter was $90 \, J \, K^{-1}$, and the rise in temperature was $5.3 \, K$. Calculate the standard enthalpy of neutralisation.

4. A calorimeter has a mass of $200 \, g$ and a specific heat capacity of $0.42 \, J \, g^{-1}$. Into it are put $50 \, cm^3$ of $1.25 \, mol \, dm^{-3}$ hydrochloric acid and $50 \, cm^3$ of $1.25 \, mol \, dm^{-3}$ potassium hydroxide solution at the same temperature. The temperature of the calorimeter and contents rises by $7.0 \,°C$. Calculate the standard enthalpy of neutralisation.

5. Fig. 9.1 shows the results of a thermometric titration to find a value of the standard enthalpy of neutralisation. $50.0 \, cm^3$ of a solution of

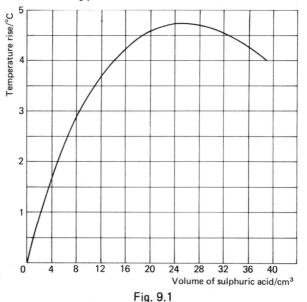

Fig. 9.1

sodium hydroxide of concentration $0.500 \, mol \, dm^{-3}$ were titrated against a $0.568 \, mol \, dm^{-3}$ solution of sulphuric acid. Calculate the standard enthalpy of neutralisation. Assume that the specific heat capacity of the solutions is $4.2 \, J \, K^{-1} g^{-1}$, and assume that no heat passes to the container.

## STANDARD ENTHALPY OF COMBUSTION

**EXAMPLE** The standard enthalpy of combustion of liquid ethanol is $-1370 \, kJ \, mol^{-1}$. Calculate the standard heat of combustion at constant volume.

**METHOD** $$C_2H_5OH(l) \; + \; 3O_2(g) \longrightarrow 2CO_2(g) \; + \; 3H_2O(l)$$

The increase in the number of gaseous molecules $= \Delta n = 2 - 3 = -1$

$$\Delta H^{\ominus} = \Delta U^{\ominus} + \Delta n RT$$

Putting $\quad \Delta H^{\ominus} = -1370 \quad$ and $\quad R = 8.31 \times 10^{-3} \, kJ \, mol^{-1} K^{-1}$

$$-1370 = \Delta U^{\ominus} + (-1 \times 8.31 \times 10^{-3} \times 298)$$

$$\Delta U^{\ominus} = -1367.5 \, kJ \, mol^{-1}$$

**ANSWER** The standard heat of combustion at constant volume is $-1368 \, kJ \, mol^{-1}$.

**EXERCISE 52** Problems on Standard Enthalpy of Combustion

1. In the combustion of $1.00 \, g$ of benzoic acid in a bomb calorimeter the temperature rises by $3.39 \, K$. The equation is
$$C_6H_5CO_2H(s) \; + \; 7\tfrac{1}{2}O_2(g) \longrightarrow 7CO_2(g) \; + \; 3H_2O(l)$$
   a) If the heat capacity of the calorimeter and contents is $7800 \, J \, K^{-1}$, what is the molar heat of combustion of benzoic acid?
   b) What is the value of the standard enthalpy of combustion?

2. a) Calculate the value of $RT\Delta n$ in the combustion of pentane at $298 \, K$. The equation is
$$C_5H_{12}(g) \; + \; 8O_2(g) \longrightarrow 5CO_2(g) \; + \; 6H_2O(l)$$
   b) If the standard enthalpy of combustion is $-3530 \, kJ \, mol^{-1}$, what would be the value of the molar heat of combustion obtained in a bomb calorimeter (at constant volume)?

3. a) Calculate the value of $\Delta n RT$ for the combustion of butan-1,4-dioic acid at $298 \, K$. The equation is
$$C_4H_6O_4(s) \; + \; 3\tfrac{1}{2}O_2(g) \longrightarrow 4CO_2(g) \; + \; 3H_2O(l)$$
   b) Hence find the difference between $\Delta H^{\ominus}$ and the heat absorbed in the combustion of 1 mole of butan-1,4-dioic acid at constant volume in a bomb calorimeter.

4. Calculate the standard enthalpies of combustion of hexane and propane from the following data obtained from measurements in a bomb calorimeter:

   a) The combustion of 1.720 g of hexane released 84.06 kJ of heat.

   b) The combustion of 1.100 g of propane raised the temperature of the calorimeter and contents by 6.4 K. The heat capacity of the calorimeter and contents is 8.575 kJ $K^{-1}$.

## FINDING THE STANDARD ENTHALPY OF A COMPOUND INDIRECTLY

Sometimes, the standard enthalpy of formation of a compound can be measured directly by allowing known amounts of elements to combine and measuring the amount of heat evolved. Other reactions are difficult to study, and the standard enthalpy of reaction must be found indirectly.

To find the standard enthalpy of formation of ethyne from practical measurements is impossible as attempts to make ethyne from carbon and hydrogen,

$$2C(s) \ + \ H_2(g) \longrightarrow C_2H_2(g)$$

will result in the formation of a mixture of hydrocarbons. The standard enthalpy of combustion of ethyne can, however, be measured experimentally, and from it can be calculated the standard enthalpy of formation. The standard enthalpies of combustion of carbon and hydrogen are also required.

**EXAMPLE 1**  Find the standard enthalpy of formation of ethyne, given the standard enthalpies of combustion (in kJ $mol^{-1}$): $C_2H_2(g) = -1300$; $C(s) = -394$; $H_2(g) = -286$.

**METHOD 1**  The method of calculation is based on the three equations for the combustion of ethyne, carbon and hydrogen:

$$C(s) + O_2(g) \longrightarrow CO_2(g); \qquad \Delta H_1^{\ominus} = -394 \text{ kJ mol}^{-1} \quad [1]$$

$$H_2(g) + \tfrac{1}{2}O_2(g) \longrightarrow H_2O(l) \qquad \Delta H_2^{\ominus} = -286 \text{ kJ mol}^{-1} \quad [2]$$

$$C_2H_2(g) + 2\tfrac{1}{2}O_2(g) \longrightarrow 2CO_2(g) + H_2O(l) \quad \Delta H_3^{\ominus} = -1300 \text{ kJ mol}^{-1} \quad [3]$$

Looking at equation [1] one can see that the standard enthalpy of combustion of carbon is the same as the standard enthalpy of formation of carbon dioxide.

Likewise, equation [2] shows that the standard enthalpy of combustion of hydrogen is the same as the standard enthalpy of formation of water.

The standard enthalpy content of a substance is equal to the standard enthalpy of formation of the substance from its elements in their standard states.

Putting the standard enthalpy content of each substance into equation [3] gives

$$C_2H_2(g) + 2\tfrac{1}{2}O_2(g) \longrightarrow 2CO_2(g) + H_2O(l); \quad \Delta H_3^{\ominus} = -1300 \text{ kJ mol}^{-1}$$
$$\Delta H_F^{\ominus}(C_2H_2) \quad 0 \qquad\qquad\qquad 2(-394) \ (-286)$$

Since

$$\begin{pmatrix} \text{Standard} \\ \text{enthalpy change} \end{pmatrix} = \begin{pmatrix} \text{Standard} \\ \text{enthalpy content} \\ \text{of products} \end{pmatrix} - \begin{pmatrix} \text{Standard} \\ \text{enthalpy content} \\ \text{of reactants} \end{pmatrix}$$

$$\Delta H_3^{\ominus} = -1300 = 2(-394) + (-286) - \Delta H_F^{\ominus}(C_2H_2)$$

$$\Delta H_F^{\ominus}(C_2H_2) = 226 \text{ kJ mol}^{-1}$$

**ANSWER** The standard enthalpy of formation of ethyne is $226 \text{ kJ mol}^{-1}$. Since $\Delta H_F^{\ominus}$ is positive, ethyne is referred to as an endothermic compound.

**METHOD 2** Another method of tackling the problem is to construct an enthalpy diagram:

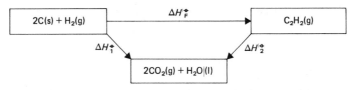

According to Hess's law, the change in standard enthalpy when carbon and hydrogen burn to form carbon dioxide and water is the same as the sum of the standard enthalpy changes when carbon and hydrogen combine to form ethyne and then ethyne burns to form carbon dioxide and water. Thus, in the above diagram,

$$\Delta H_1^{\ominus} = \Delta H_F^{\ominus} + \Delta H_2^{\ominus}$$

Putting

$$\Delta H_1^{\ominus} = 2(\Delta H^{\ominus} \text{ for combustion of C}) + (\Delta H^{\ominus} \text{ for combustion of } H_2)$$

gives

$$\Delta H_1^{\ominus} = 2(-394) + (-286) = -1074$$

$$\Delta H_F^{\ominus} = \Delta H_1^{\ominus} - \Delta H_2^{\ominus} = -1074 - (-1300)$$

**ANSWER** $\Delta H_F^{\ominus} = +226 \text{ kJ mol}^{-1}$ (as before)

**EXAMPLE 2** Calculate the standard enthalpy of formation of propan-1-ol, given the standard enthalpies of combustion, in $\text{kJ mol}^{-1}$: $C_3H_7OH(l)$, $-2010$; $C(s)$, $-394$; $H_2(g)$, $-286$.

**METHOD 1** Again, as the equation for combustion is the basis for the calculation, it must be carefully balanced:

$$C_3H_7OH(l) + 4\tfrac{1}{2}O_2(g) \longrightarrow 3CO_2(g) + 4H_2O(l); \quad \Delta H^{\ominus} = -2010 \text{ kJ mol}^{-1}$$

Putting the standard enthalpies of formation of $CO_2(g)$ and $H_2O(l)$ into the equation, as in Example 1, gives

$$C_3H_7OH(l) + 4\tfrac{1}{2}O_2(g) \longrightarrow 3CO_2(g) + 4H_2O(l); \quad \Delta H^\ominus = -2010 \text{ kJ mol}^{-1}$$
$$\Delta H_F^\ominus(C_3H_7OH) \quad 0 \qquad\qquad 3(-394)\ 4(-286)$$

Since

$$\begin{pmatrix}\text{Standard} \\ \text{enthalpy change} \\ \text{for reaction}\end{pmatrix} = \begin{pmatrix}\text{Standard} \\ \text{enthalpy content} \\ \text{of products}\end{pmatrix} - \begin{pmatrix}\text{Standard} \\ \text{enthalpy content} \\ \text{of reactants}\end{pmatrix}$$

$$-2010 = 3(-394) + 4(-286) - \Delta H_F^\ominus(C_3H_7OH(l))$$

$$\Delta H_F^\ominus(C_3H_7OH(l)) = -316 \text{ kJ mol}^{-1}$$

**ANSWER**    The standard enthalpy of formation of liquid propan-1-ol is $-316$ kJ mol$^{-1}$.

**METHOD 2**    The enthalpy diagram for the formation of propanol is

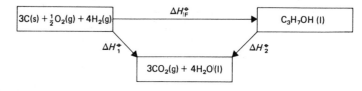

$$\Delta H_1^\ominus = 3(\Delta H^\ominus \text{ for combustion of C}) + 4(\Delta H^\ominus \text{ for combustion of } H_2)$$

$$= 3(-394) + 4(-286) = -2326$$

According to Hess's law,

$$\Delta H_1^\ominus = \Delta H_F^\ominus + \Delta H_2^\ominus$$

$$\Delta H_F^\ominus = \Delta H_1^\ominus - \Delta H_2^\ominus$$

$$\Delta H_F^\ominus = -2326 - (-2010)$$

**ANSWER**    $\Delta H_F^\ominus = -316 \text{ kJ mol}^{-1}$    (as before)

You will have noticed in both Examples 1 and 2 that

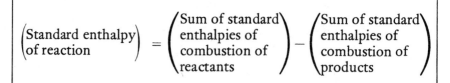

## STANDARD ENTHALPY OF REACTION FROM STANDARD ENTHALPIES OF FORMATION

The standard enthalpies of formation of the reactants and products can be used to give the standard enthalpy of a reaction.

**EXAMPLE 1**   Calculate the standard enthalpy of the reaction

$$CH_2 = CH_2(g) \ + \ H_2(g) \ \longrightarrow \ CH_3CH_3(g)$$

given that the standard enthalpies of formation are: ethene, $+52$, ethane, $-85$ kJ mol$^{-1}$.

**METHOD**   Put the standard enthalpy content of each species into the equation (units kJ mol$^{-1}$):

$$CH_2 = CH_2(g) \ + \ H_2(g) \ \longrightarrow \ CH_3CH_3(g)$$
$$+52 \qquad\qquad 0 \qquad\qquad\qquad -85$$

$$\begin{pmatrix} \text{Standard enthalpy} \\ \text{of reaction} \end{pmatrix} = \begin{pmatrix} \text{Standard enthalpy} \\ \text{of product} \end{pmatrix} - \begin{pmatrix} \text{Standard enthalpy} \\ \text{of reactants} \end{pmatrix}$$

$$= -85 - (52 + 0) = -137$$

**ANSWER**   Standard enthalpy $= -137$ kJ mol$^{-1}$

The method of calculation is simply:

$$\begin{pmatrix} \text{Standard enthalpy} \\ \text{of reaction} \end{pmatrix} = \begin{pmatrix} \text{Sum of standard} \\ \text{enthalpies of} \\ \text{formation of} \\ \text{products} \end{pmatrix} - \begin{pmatrix} \text{Sum of standard} \\ \text{enthalpies of} \\ \text{formation of} \\ \text{reactants} \end{pmatrix}$$

**EXAMPLE 2**   Calculate the standard enthalpy change in the reaction

$$SO_2(g) \ + \ 2H_2S(g) \ \longrightarrow \ 3S(s) \ + \ 2H_2O(l)$$

The standard enthalpy of combustion of sulphur is $-297$ kJ mol$^{-1}$, and the standard enthalpies of formation of hydrogen sulphide and water are $-20.2$ kJ mol$^{-1}$ and $-286$ kJ mol$^{-1}$.

**METHOD**   This problem is tackled by putting the standard enthalpies of formation of each species into the equation (units kJ mol$^{-1}$):

$$SO_2(g) \ + \ 2H_2S(g) \ \longrightarrow \ 3S(s) \ + \ 2H_2O(l)$$
$$(-297) \ + \ 2(-20.2) \qquad\qquad 3 \times 0 \qquad 2(-286)$$

$$\begin{pmatrix} \text{Standard enthalpy} \\ \text{of reaction} \end{pmatrix} = \begin{pmatrix} \text{Standard enthalpy} \\ \text{of products} \end{pmatrix} - \begin{pmatrix} \text{Standard enthalpy} \\ \text{of reactants} \end{pmatrix}$$

$$= -572 + 297 + 40.4$$

**ANSWER**   Standard enthalpy change $= -235$ kJ (mol of the equation)$^{-1}$

## STANDARD BOND DISSOCIATION ENTHALPIES

The standard bond dissociation enthalpy is the energy that must be absorbed to separate the two atoms in a bond. When hydrogen chloride dissociates,

$$HCl(g) \longrightarrow H(g) + Cl(g); \qquad \Delta H^\ominus = 429.7 \, kJ \, mol^{-1}$$

The standard bond dissociation enthalpy of the H—Cl bond in HCl is 429.7 kJ mol$^{-1}$.

## AVERAGE STANDARD BOND ENTHALPIES

When you want to assign a value to the standard enthalpy of dissociation of the C—H bond in methane, the problem is different. The energy required to break the first C—H bond in methane is not the same as that required to remove a hydrogen atom from a methyl radical. In the dissociation,

$$CH_4(g) \longrightarrow C(g) + 4H(g); \quad \Delta H^\ominus = +1662 \, kJ \, mol^{-1}$$

Dividing the standard enthalpy change between the four bonds gives an average value for the C—H bond of 416 kJ mol$^{-1}$. This value is called the average standard bond enthalpy for the C—H bond.

Tables of average standard bond enthalpies make the assumption that the standard enthalpy of a bond is independent of the molecule in which it exists. This is only roughly true. Since standard bond enthalpies vary from one compound to another, the use of average standard bond enthalpies gives only approximate values for standard enthalpies of reaction calculated from them. Experimental methods are used to obtain standard enthalpies of reaction whenever possible. Calculations based on average standard bond enthalpies are used only for reactions which cannot be studied experimentally – for example, the reactions of a substance which has not been isolated in a pure state.

Average standard bond enthalpy is often called the *bond energy term*. One can say that the bond energy term for the C—H bond is 416 kJ mol$^{-1}$. The sum of all the bond energy terms for a compound is the standard enthalpy change absorbed in atomising that compound *in the gaseous state*. The standard enthalpy of formation of a compound includes the sum of the bond energy terms and also the standard enthalpy of atomisation of the carbon atoms and the standard enthalpy of atomisation of the hydrogen atoms.

EXAMPLE    Calculate the standard enthalpy of formation of methane. C—H bond energy term = 416 kJ mol$^{-1}$; standard enthalpies of atomisation are C(s) = 716 kJ mol$^{-1}$; $\frac{1}{2}H_2(g)$ = 217.5 kJ (mol H atoms)$^{-1}$.

**METHOD 1**  The sum of the bond energy terms in methane $= 1662\,kJ\,mol^{-1}$. Putting this information into the form of an equation, and writing the standard enthalpy content of each species underneath its formula, we get

$$C(g) \;+\; 4H(g) \longrightarrow CH_4(g); \quad \Delta H^{\ominus} = -1662\,kJ\,mol^{-1}$$
$$(716) \quad 4(217.5) \qquad\qquad \Delta H_F^{\ominus}$$

The values 716 and 217.5 are the standard enthalpies of formation of gaseous carbon and hydrogen atoms from the elements in their standard states.

Since

$$\begin{pmatrix}\text{Standard enthalpy}\\\text{change}\end{pmatrix} = \begin{pmatrix}\text{Sum of standard}\\\text{enthalpies of}\\\text{products}\end{pmatrix} - \begin{pmatrix}\text{Sum of standard}\\\text{enthalpies of}\\\text{reactants}\end{pmatrix}$$

$$-1662 = \Delta H_F^{\ominus} - 716 - 4(217.5)$$

**ANSWER**  $$\Delta H_F^{\ominus} = -78\,kJ\,mol^{-1}$$

**METHOD 2**  The information can also be represented in the form of an enthalpy diagram:

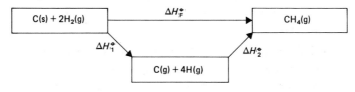

$$\Delta H_1^{\ominus} = \Delta H^{\ominus} \text{ of atomisation of } C + 4\Delta H^{\ominus} \text{ of atomisation of } H$$
$$\Delta H_2^{\ominus} = -(\text{Sum of bond energy terms for } CH_4)$$

According to Hess's law,

$$\Delta H_F^{\ominus} = \Delta H_1^{\ominus} + \Delta H_2^{\ominus}$$
$$= 716 + 4(217.5) - (4 \times 416)$$

**ANSWER**  $$\Delta H_F^{\ominus} = -78\,kJ\,mol^{-1} \quad \text{(as before)}$$

## STANDARD ENTHALPY OF REACTION FROM AVERAGE STANDARD BOND ENTHALPIES

Mean standard bond enthalpies can be used to give an approximate estimate of the standard enthalpy change which occurs in a reaction. During a reaction, energy is supplied to break the bonds in the reactants, and energy is given out when the bonds in the products form. The difference between the sum of the standard bond enthalpies of the products and the standard bond enthalpies of the reactants is the standard enthalpy of the reaction. The value obtained is less reliable than an experimental measurement.

**EXAMPLE 1**   Calculate the standard enthalpy of the reaction

$$CH_2\!=\!CH_2(g) \;+\; H_2(g) \longrightarrow CH_3CH_3(g)$$

Mean standard bond enthalpies are (in $kJ\,mol^{-1}$): C—H, 416; C$=$C, 612; C—C, 348; H—H, 436.

**METHOD**   Bonds broken are:

| | |
|---|---|
| one  C$=$C bond, of standard enthalpy | $= 612\,kJ\,mol^{-1}$ |
| one  H—H bond, of standard enthalpy | $= 436$ |
| Total enthalpy absorbed | $= 1048\,kJ\,mol^{-1}$ |

Bonds created are:

| | |
|---|---|
| one  C—C bond, of standard enthalpy | $= 348\,kJ\,mol^{-1}$ |
| two  C—H bonds, of standard enthalpy | $= 832$ |
| Total enthalpy released | $= -1180\,kJ\,mol^{-1}$ |

**ANSWER**   Standard enthalpy of reaction $= -1180 + 1048 = -132\,kJ\,mol^{-1}$

**EXAMPLE 2**   Benzene has a standard enthalpy of formation of $83\,kJ\,mol^{-1}$. Calculate the standard enthalpy of formation from the following data:

Mean standard bond enthalpies are: (C—C) $= 348$; (C$=$C) $= 615$; (C—H) $= 412\,kJ\,mol^{-1}$.

$\Delta H^{\ominus}$ for vaporisation of carbon $= 715\,kJ\,mol^{-1}$

$\Delta H^{\ominus}$ for atomisation of hydrogen (per mole of H atoms) $= 217.5\,kJ\,mol^{-1}$

Compare the experimental value and the theoretical value for $\Delta H_F^{\ominus}$.

**METHOD**   Enthalpy is absorbed in atomising carbon and hydrogen.

Standard enthalpy absorbed $= (6 \times 715) + (6 \times 217.5)$
$$= 5595\,kJ\,mol^{-1}$$

Enthalpy is released when bonds are formed.

| Standard enthalpy released $=$ 6(C—H) | $= -2472\,kJ\,mol^{-1}$ |
|---|---|
| $+3$(C—C) | $= -1044\,kJ\,mol^{-1}$ |
| $+3$(C$=$C) | $= -1845\,kJ\,mol^{-1}$ |
| Total | $= -5361\,kJ\,mol^{-1}$ |

**ANSWER**   $\Delta H_F^{\ominus} = +5595 - 5361 = 234\,kJ\,mol^{-1}$.

The calculated value for the standard enthalpy of formation is higher than the experimental value: the benzene molecule is more stable than it is calculated to be. The difference is the value of the energy of electron delocalisation or 'resonance', $151\,kJ\,mol^{-1}$.

# THE BORN-HABER CYCLE

The Born–Haber cycle is a technique for applying Hess's law to the standard enthalpy changes which occur when an ionic compound is formed. Consider the reaction between sodium and chlorine to form sodium chloride. The steps which are involved in this reaction are:

a) Vaporisation of sodium

$$Na(s) \longrightarrow Na(g); \quad \Delta H_s^{\ominus} = \text{standard enthalpy of sublimation}$$

b) Ionisation of sodium

$$Na(g) \longrightarrow Na^+(g) + e^-; \quad \Delta H_I^{\ominus} = \text{ionisation energy of sodium}$$

c) Dissociation of chlorine molecules

$$\tfrac{1}{2}Cl_2(g) \longrightarrow Cl(g); \quad \Delta H_D^{\ominus} = \tfrac{1}{2} \text{ standard bond dissociation enthalpy of chlorine}$$

d) Ionisation of chlorine atoms

$$Cl(g) + e^- \longrightarrow Cl^-(g); \quad \Delta H_E^{\ominus} = \text{electron affinity of chlorine}$$

e) Reaction between ions

$$Na^+(g) + Cl^-(g) \longrightarrow NaCl(s); \quad \Delta H_L^{\ominus} = \text{standard lattice enthalpy}$$

Definitions of the standard enthalpies used above are:

The *standard enthalpy of sublimation* is the heat absorbed when one mole of sodium atoms are vaporised.

The *ionisation energy* of sodium is the energy required to remove a mole of electrons from a mole of sodium atoms in the gas phase.

The *standard enthalpy of bond dissociation* of chlorine is the enthalpy required to dissociate one mole of chlorine molecules into atoms.

The *electron affinity* of chlorine is the energy absorbed when a mole of chlorine atoms form chloride ions. It has a negative value, showing that this reaction is exothermic.

The *standard lattice enthalpy* is the energy absorbed when one mole of gaseous sodium ions and one mole of gaseous chloride ions form one mole of crystalline sodium chloride. It has a negative value.

The steps in the Born–Haber cycle are represented as going upwards if they absorb energy and downwards if they give out energy (see Fig. 9.2).

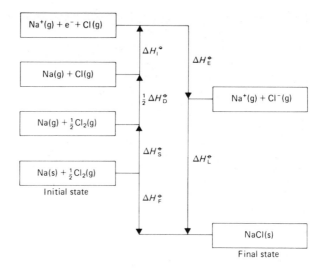

Fig. 9.2

According to Hess's law, the standard enthalpy of formation of sodium chloride is equal to the sum of the enthalpy changes in the various steps:

$$\Delta H_F^\ominus = \Delta H_S^\ominus + \tfrac{1}{2}\Delta H_D^\ominus + \Delta H_I^\ominus + \Delta H_E^\ominus + \Delta H_L^\ominus$$

$$= +109 + 121 + 494 - 380 - 755 = -411 \text{ kJ mol}^{-1}$$

In practice, it is easier to measure standard enthalpies of formation than to measure some of the other steps. The electron affinity is the hardest term to measure experimentally, and the Born–Haber cycle is often used to calculate electron affinities.

**EXERCISE 53**    Problems on Standard Enthalpy of Reaction and Average Standard Bond Enthalpies

1. The following are standard enthalpies of combustion at 298 K, in kJ mol$^{-1}$:

| | | | | | |
|---|---|---|---|---|---|
| C(graphite) | −394 | $C_2H_6(g)$ | −1561 | $C_4H_{10}(l)$ | −3510 |
| $H_2(g)$ | −286 | $CH_2{=}CH_2(g)$ | −1393 | $CH{\equiv}CH(g)$ | −1299 |
| $CH_3CO_2H(l)$ | −876 | $C_2H_5OH(l)$ | −1400 | $CH_3OH(l)$ | −715 |
| $C_4H_6(g)$ | −2542 | $CH_3OCH_3(g)$ | −1455 | $C_2H_5OH(g)$ | −1444 |
| $CH_4(g)$ | −891 | $C_3H_8(g)$ | −2220 | | |
| $CH_3CO_2C_2H_5(l)$ | −2246 | $C_6H_{12}(l)$ | −3924 | | |

a) Calculate the standard enthalpy change for the reaction:

$$2C(\text{graphite}) + 2H_2(g) + O_2(g) \longrightarrow CH_3CO_2H(l)$$

b) Calculate the standard enthalpy change of formation of buta-1, 3-diene, $C_4H_6(g)$.

c) Calculate the standard enthalpy of formation of methane, $CH_4(g)$ and of ethene, $CH_2{=}CH_2(g)$.

d) Calculate the standard enthalpy change in the hydrogenation of ethene(g) to ethane(g).

e) Calculate the standard enthalpy change for the theoretical reaction:

$$CH_3OCH_3(g) \longrightarrow C_2H_5OH(g)$$

f) Calculate the standard enthalpy of formation of propane(g) and of butane(l).

g) Calculate the standard enthalpy of formation of methanol(l), ethanol(l), ethylethanoate(l) and cyclohexane(l).

2. Calculate the standard enthalpy change of the reaction

Anhydrous copper(II) sulphate $+$ Water $\longrightarrow$ Copper(II) sulphate-5-water

Use the values for the standard enthalpy of solution:

a) anhydrous copper(II) sulphate, $-66.5 \text{ kJ mol}^{-1}$

b) copper(II) sulphate-5-water, $11.7 \text{ kJ mol}^{-1}$.

3. Calculate the standard enthalpies of formation of: a) sulphur dioxide, b) carbon dioxide, and c) steam. On burning in excess oxygen under standard conditions (1 atm, 298 K): 1.00 g of sulphur evolves 9.28 kJ; 1.00 g of carbon evolves 32.8 kJ; and $1.00 \text{ dm}^3$ (at 1 atm, 298 K) of hydrogen evolves 12.76 kJ of heat.

4. Calculate the standard enthalpy change in the reaction

$$PbO(s) + CO(g) \longrightarrow Pb(s) + CO_2(g)$$

The standard enthalpies of formation of lead(II) oxide, carbon monoxide and carbon dioxide are $-219$, $-111$, and $-394 \text{ kJ mol}^{-1}$, respectively.

5. Calculate the standard enthalpy change for the reaction

$$Fe_2O_3(s) + 2Al(s) \longrightarrow Al_2O_3(s) + 2Fe(s)$$

The standard enthalpies of formation of iron(III) oxide and aluminium oxide are $-822$ and $-1669 \text{ kJ mol}^{-1}$. State whether the reaction is exothermic or endothermic.

6. The standard enthalpy of combustion of rhombic sulphur is $-296.9 \text{ kJ mol}^{-1}$ and the standard enthalpy of combustion of monoclinic sulphur is $-297.2 \text{ kJ mol}^{-1}$. Calculate the standard enthalpy of conversion of monoclinic sulphur to rhombic sulphur.

7. The standard enthalpies of formation of $CO_2(g)$ and $H_2O(g)$ are $-394$ and $-242 \text{ kJ mol}^{-1}$. The standard enthalpy of combustion of ethane is $-1560 \text{ kJ mol}^{-1}$. The standard enthalpy of reduction of ethene to ethane by gaseous hydrogen is $-138 \text{ kJ mol}^{-1}$. Calculate the standard enthalpy of formation of ethene.

8. Given the standard enthalpy change of formation of MgO $= -602$ kJ mol$^{-1}$ and of Al$_2$O$_3 = -1700$ kJ mol$^{-1}$, calculate the standard enthalpy change for the reaction

$$Al_2O_3 + 3Mg \longrightarrow 2Al + 3MgO$$

Does your answer tell you whether magnesium will reduce aluminium oxide?

9. The following are standard enthalpies of formation, $\Delta H_F^{\ominus}$, in kJ mol$^{-1}$ at 298 K:

CH$_4$(g); $-76$; CO$_2$(g), $-394$; H$_2$O(l), $-286$; H$_2$O(g), $-242$; NH$_3$(g), $-46.2$; HNO$_3$(l), $-176$; C$_2$H$_5$OH(l), $-278$; C$_8$H$_{18}$(l), $-210$.

a) Calculate the standard enthalpy change at 298 K for the reaction

$$CH_4(g) + 2O_2(g) \longrightarrow CO_2(g) + 2H_2O(l)$$

b) Calculate the standard enthalpy change for the reaction

$$\tfrac{1}{2}N_2(g) + \tfrac{3}{2}H_2O(g) \longrightarrow NH_3(g) + \tfrac{3}{4}O_2(g)$$

c) Calculate the standard enthalpy change for the reaction

$$\tfrac{1}{2}N_2(g) + \tfrac{1}{2}H_2O(g) + \tfrac{5}{4}O_2(g) \longrightarrow HNO_3(l)$$

d) Calculate the enthalpy change which occurs when each of the following is burned completely under standard conditions:
i) 1.00 kg hydrogen, ii) 1.00 kg ethanol(l), iii) 1.00 kg octane(l).

10. What is meant by the terms *standard bond dissociation enthalpy* and *bond energy term?*

The standard bond dissociation enthalpies for the first, second, third and fourth C—H bonds in methane are 423, 480, 425 and 335 kJ mol$^{-1}$ respectively. Calculate the C—H bond energy term for methane.

11. Consult the average standard bond enthalpies and standard enthalpies of atomisation (in kJ mol$^{-1}$) listed below:

| C—C | 348 | C=O | 743 | C(graphite) | 718 |
|-----|-----|-----|-----|-------------|-----|
| C=C | 612 | H—Cl | 432 | $\tfrac{1}{2}$H$_2$(g) | 218 |
| C≡C | 837 | C—Cl | 338 | $\tfrac{1}{2}$O$_2$(g) | 248 |
| C—H | 412 | C—Br | 276 | $\tfrac{1}{2}$Br$_2$(g) | 96.5 |
| C—O | 360 | H—Br | 366 | $\tfrac{1}{2}$Cl$_2$(g) | 121 |
| H—O | 463 | | | | |

a) Calculate the standard enthalpy of formation of ethane and of ethene.

b) Find the standard enthalpy change for the reaction,

$$CH_2{=}CH{-}CH_3(g) + Br_2(g) \longrightarrow CH_2BrCHBrCH_3(g)$$

c) Find the standard enthalpy of formation of methoxymethane, CH$_3$OCH$_3$(g).

d) Calculate the standard enthalpy of formation of gaseous ethyl ethanoate, CH$_3$CO$_2$C$_2$H$_5$(g).

e) Calculate the standard enthalpy of formation of benzene, assuming its structure is

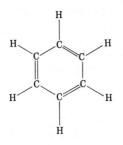

Explain the difference between the value you have calculated and the value of 83 kJ mol⁻¹ obtained from measurements of the standard enthalpy of combustion.

f) Find the standard enthalpy of formation of gaseous buta-1,3-diene, $CH_2{=}CH{-}CH{=}CH_2(g)$. How does this value compare with the value you obtained in Question 1(b) from the standard enthalpy of combustion? How do you explain the difference?

g) Estimate the standard enthalpy changes for the reactions:
   i) $Cl\cdot\ +\ CH_4 \longrightarrow CH_3Cl\ +\ H\cdot$
   ii) $Cl\cdot\ +\ CH_4 \longrightarrow CH_3\cdot\ +\ HCl$
   Which of the two reactions will occur more readily?

12. Use the data below to draw an energy diagram for the formation of potassium chloride. Calculate the electron affinity of chlorine.
    Standard enthalpy of sublimation of potassium       =    90 kJ mol⁻¹
    Standard enthalpy of ionisation of potassium        =   420 kJ mol⁻¹
    Standard enthalpy of dissociation of chlorine       =   244 kJ mol⁻¹
    Standard lattice enthalpy of potassium chloride     = −706 kJ mol⁻¹
    Standard enthalpy of formation of potassium chloride = −436 kJ mol⁻¹

13. Using the following data, which is a set of standard enthalpy changes, calculate the standard enthalpy of formation of potassium chloride, KCl(s):

|  |  | $\Delta H^{\ominus}$/kJ mol⁻¹ |
|---|---|---|
| $KOH(aq)\ +\ HCl(aq) \longrightarrow KCl(aq)\ +\ H_2O(l)$ | | −57.3 |
| $H_2(g)\ +\ \frac{1}{2}O_2(g) \longrightarrow H_2O(l)$ | | −286 |
| $\frac{1}{2}H_2(g)\ +\ \frac{1}{2}Cl_2(g)\ +\ aq \longrightarrow HCl(aq)$ | | −164 |
| $K(s)\ +\ \frac{1}{2}O_2(g)\ +\ \frac{1}{2}H_2(g)\ +\ aq \longrightarrow KOH(aq)$ | | −487 |
| $KCl(s)\ +\ aq \longrightarrow KCl(aq)$ | | +18 |

14. Use the data below to calculate the electron affinity of chlorine:

| | |
|---|---|
| Standard enthalpy of formation of rubidium chloride | $-431\,\text{kJ mol}^{-1}$ |
| Lattice energy of rubidium chloride | $-675\,\text{kJ mol}^{-1}$ |
| First ionisation energy of rubidium | $+408\,\text{kJ mol}^{-1}$ |
| Standard enthalpy of atomisation of rubidium | $+86\,\text{kJ mol}^{-1}$ |
| Bond dissociation enthalpy of molecular chlorine | $+242\,\text{kJ mol}^{-1}$ |

15. From a Born–Haber cycle calculation, it can be estimated that the standard enthalpy of formation of magnesium(I) chloride, MgCl, would be $-130\,\text{kJ mol}^{-1}$. The standard enthalpy of formation of magnesium(II) chloride $MgCl_2$, is $-640\,\text{kJ mol}^{-1}$.

   a) Why do you think that $MgCl_2$ is formed, and not MgCl, when magnesium reacts with chlorine?

   b) Calculate the standard enthalpy change in the theoretical reaction

$$2MgCl(s) \longrightarrow Mg(s) + MgCl_2(s)$$

16. Calculate the lattice energy of sodium chloride from the following data:

| | $\Delta H^{\ominus}/\text{kJ mol}^{-1}$ |
|---|---|
| $Na(s) \longrightarrow Na(g)$ | $+109$ |
| $Na(g) \longrightarrow Na^+(g) + e^-$ | $+494$ |
| $Cl_2(g) \longrightarrow 2Cl(g)$ | $+242$ |
| $Cl(g) + e^- \longrightarrow Cl^-(g)$ | $-360$ |
| $Na(s) + \frac{1}{2}Cl_2(s) \longrightarrow NaCl(s)$ | $-411$ |

17. a) Data for the Born–Haber cycle for the formation of calcium chloride are

   $Ca(s) \longrightarrow Ca(g)$ $\qquad\qquad \Delta H^{\ominus} = +190\,\text{kJ mol}^{-1}$

   $Ca(g) \longrightarrow Ca^{2+}(g) + 2e^-$ $\qquad \Delta H^{\ominus} = +1730\,\text{kJ mol}^{-1}$

   $\frac{1}{2}Cl_2(g) \longrightarrow Cl(g)$ $\qquad\qquad \Delta H^{\ominus} = +121\,\text{kJ mol}^{-1}$

   $Ca^{2+}(g) + 2Cl^-(g) \longrightarrow CaCl_2(s)$ $\quad \Delta H^{\ominus} = -2184\,\text{kJ mol}^{-1}$

   $Ca(s) + Cl_2(g) \longrightarrow CaCl_2(s)$ $\qquad \Delta H^{\ominus} = -795\,\text{kJ mol}^{-1}$

   Calculate the electron affinity of chlorine.

   b) For the reactions

   $Ca(g) \longrightarrow Ca^+(g) + e^-$ $\qquad\qquad \Delta H^{\ominus} = +590\,\text{kJ mol}^{-1}$

   $Ca^+(g) + Cl^-(g) \longrightarrow CaCl(s)$ $\qquad \Delta H^{\ominus} = -760\,\text{kJ mol}^{-1}$

   use these standard enthalpy changes and those given in a) to calculate the standard enthalpy of formation of CaCl(s). Why do you think $CaCl_2$ is formed in preference to CaCl?

18. When an ionic compound dissolves, an amount of energy equal to the lattice energy must be supplied to separate the ions. When the ions dissolve they are hydrated by water molecules, and energy is released. If the enthalpy of hydration is greater than the lattice enthalpy, there is a net release of energy and a decrease in the enthalpy content of the system, and this favours solution.

The values below (in kJ mol$^{-1}$) relate to the solubility of lithium chloride, sodium chloride and sodium fluoride:

|                                                   | LiCl | NaCl | NaF  |
|---------------------------------------------------|------|------|------|
| Standard lattice enthalpy                         | −843 | −775 | −968 |
| Sum of standard hydration enthalpies of separate ions | −883 | −778 | −965 |

What can you predict from these values for the standard enthalpy changes about the relative solubilities of: a) LiCl and NaCl, b) NaF and NaCl? Explain your answer.

19. Given the standard enthalpy changes for the reactions

$$H_2(g) \longrightarrow 2H(g); \qquad \Delta H^{\ominus} = 436 \text{ kJ mol}^{-1}$$
$$Br_2(g) \longrightarrow 2Br(g); \qquad \Delta H^{\ominus} = 193 \text{ kJ mol}^{-1}$$
$$H_2(g) + Br_2(g) \longrightarrow 2HBr(g); \qquad \Delta H^{\ominus} = -104 \text{ kJ mol}^{-1}$$

calculate the standard enthalpy change for the reaction

$$H(g) + Br(g) \longrightarrow HBr(g)$$

20. The following values for standard enthalpy change relate to the hydrogenation of cyclohexene and benzene. Comment on the values of $\Delta H^{\ominus}$.

$$C_6H_{10}(l) + H_2(g) \longrightarrow C_6H_{12}(l) \qquad \Delta H^{\ominus} = -120 \text{ kJ mol}^{-1}$$
$$C_6H_6(l) + H_2(g) \longrightarrow C_6H_8(l) \qquad \Delta H^{\ominus} = +31 \text{ kJ mol}^{-1}$$
$$C_6H_6(l) + 3H_2(g) \longrightarrow C_6H_{12}(l) \qquad \Delta H_c^{\ominus} = -208 \text{ kJ mol}^{-1}$$

## FREE ENERGY AND ENTROPY

Some reactions which happen spontaneously are endothermic. The difference in enthalpy between the products and the reactants cannot be the only factor which decides whether a chemical reaction takes place. There must be an additional factor involved. It is often observed that reactions which occur spontaneously increase the randomness or disorder of the system. For example, when an ionic solid dissolves, it passes from the regular arrangement of a crystalline lattice to a random solution of ions. This is termed an increase in *entropy* of the system. The two factors combine to give the change in the *free energy* of the system:

$$\text{Free energy } G = \text{Enthalpy } H - \text{Temperature/K} \times \text{Entropy } S$$
$$G = H - TS$$

It follows that $\Delta G = \Delta H - T\Delta S$

For a physical or a chemical change to occur, $\Delta G$ for that change must be negative. The change is therefore assisted by a decrease in enthalpy ($\Delta H$ negative) and by an increase in entropy ($\Delta S$ positive).

If the change takes place under standard conditions, i.e. with each reactant and product at unit concentration (or pressure), then the free energy change is equal to the standard free energy change, $\Delta G^{\ominus}$. When reaction takes place under non-standard conditions, $\Delta G$, the free energy change differs from $\Delta G^{\ominus}$ as $\Delta G$ depends on the concentrations (or pressures) of the reactants and products. It is easy to obtain $\Delta G^{\ominus}$ from tables of standard enthalpies and standard entropies, but one really wants to know the value of $\Delta G$ for the real conditions, and this is not easy to compute. However, if $\Delta G^{\ominus}$ has a sufficiently large positive or negative value, $\Delta G^{\ominus}$ may determine the feasibility of reaction over a large range of concentrations (or pressures).

## CALCULATION OF CHANGE IN STANDARD ENTROPY

One method of calculating the standard entropy change of a process is to use the expression

$$\begin{pmatrix} \text{Standard} \\ \text{entropy change} \end{pmatrix} = \begin{pmatrix} \text{Sum of standard} \\ \text{entropies of products} \end{pmatrix} - \begin{pmatrix} \text{Sum of standard} \\ \text{entropies of reactants} \end{pmatrix}$$

**EXAMPLE 1**    Calculate the standard entropy change for the reaction of chlorine and ethene, given the values (in $J\,K^{-1}\,mol^{-1}$):

$S^{\ominus}(Cl_2(g)) = 223; S^{\ominus}(CH_2{=}CH_2(g)) = 219; S^{\ominus}(CH_2ClCH_2Cl(l)) = 208.$

**METHOD**    The equation for the reaction is

$$CH_2{=}CH_2(g) \ + \ Cl_2(g) \longrightarrow CH_2ClCH_2Cl(l)$$

$$S^{\ominus}(\text{product}) \ = \ 208\,J\,K^{-1}\,mol^{-1}$$

$$S^{\ominus}(\text{reactants}) \ = \ 219 + 223 \ = \ 442\,J\,K^{-1}\,mol^{-1}$$

$$\Delta S^{\ominus} \ = \ 208 - 442 \ = \ -234\,J\,K^{-1}\,mol^{-1}$$

**ANSWER**    The standard entropy change for the reaction is $-234\,J\,K^{-1}\,mol^{-1}$. The negative sign means a decrease in disorder. Since two moles of gas have formed one mole of liquid, this is what one would expect.

The other method of calculating the change in standard entropy for a process is to derive it from the changes in standard enthalpy and standard free energy. The equation relating these quantities is

$$\Delta G^{\ominus} \ = \ \Delta H^{\ominus} - T\Delta S^{\ominus}$$

If a system is at equilibrium, $\Delta G^{\ominus} = 0$, and the standard entropy change is simply the standard enthalpy change divided by the temperature:

$$\Delta S^{\ominus} = \Delta H^{\ominus}/T$$

One process during which equilibrium obtains is the vaporisation of a liquid at its boiling point, since this process takes place under reversible conditions at a constant temperature. Another process for which $\Delta G^{\ominus} = 0$ is the melting of a solid at its melting point. For chemical reactions, $\Delta G^{\ominus} \neq 0$, and the change in standard entropy must be found from the equation

$$\Delta S^{\ominus} = \frac{\Delta H^{\ominus} - \Delta G^{\ominus}}{T}$$

**\*EXAMPLE 2** Calculate the standard entropy of melting (or fusion) of ice. The standard enthalpy of melting of ice is $6.00 \, \text{kJ mol}^{-1}$.

**METHOD** $\qquad\qquad\qquad \Delta S^{\ominus} = \Delta H_m^{\ominus}/T = 6.00/273$

**ANSWER** $\qquad\qquad\qquad \Delta S^{\ominus} = 2.20 \times 10^{-2} \, \text{kJ mol}^{-1} \text{K}^{-1}$

**\*EXAMPLE 3** Calculate the standard entropy of vaporisation of water. The standard enthalpy of vaporisation is $41.0 \, \text{kJ mol}^{-1}$.

**METHOD** $\qquad\qquad\qquad \Delta S^{\ominus} = \Delta H_v^{\ominus}/T = 41.0/373$

**ANSWER** $\qquad\qquad\qquad \Delta S^{\ominus} = 0.110 \, \text{kJ mol}^{-1} \text{K}^{-1}$

*Note.* Standard enthalpy of melting was formerly called molar latent heat of fusion. Standard enthalpy of vaporisation (or evaporation) was formerly called molar latent heat of vaporisation.

## CALCULATION OF CHANGE IN STANDARD FREE ENERGY

The change in standard enthalpy, the change in standard entropy and the temperature must be known and inserted into the equation

$$\Delta G^{\ominus} = \Delta H^{\ominus} - T\Delta S^{\ominus}$$

**EXAMPLE** Calculate the change in standard free energy and determine whether the reaction

$$\text{Fe}_2\text{O}_3(s) + 3\text{H}_2(g) \longrightarrow 2\text{Fe}(s) + 3\text{H}_2\text{O}(g)$$

will take place at a) $20 \, ^\circ\text{C}$, b) $500 \, ^\circ\text{C}$. Use the values (in $\text{kJ mol}^{-1}$):

| | $\text{Fe}_2\text{O}_3$ | $\text{H}_2$ | $\text{Fe}$ | $\text{H}_2\text{O}$ |
|---|---|---|---|---|
| Standard enthalpy: | −822 | 0 | 0 | −242 |
| Standard entropy: | 0.090 | 0.131 | 0.027 | 0.189 |

**METHOD**    $\Delta G^{\ominus} = \Delta H^{\ominus} - T\Delta S^{\ominus}$

$\Delta H^{\ominus} = (0 + 3(-242)) - (-822 + 0) = +96 \text{ kJ mol}^{-1}$

$\Delta S^{\ominus} = (2 \times 0.027) + (3 \times 0.189) - 0.090 - (3 \times 0.131)$

$= 0.054 + 0.567 - 0.090 - 0.393 = +0.138 \text{ kJ mol}^{-1}$

a) At 20 °C,

$\Delta G^{\ominus} = \Delta H^{\ominus} - T\Delta S^{\ominus}$

$= +96 - (293 \times 0.138) = 96 - 52.74 = +43.26 \text{ kJ mol}^{-1}$

**ANSWER**    $\Delta G^{\ominus}$ is 42.3 kJ mol$^{-1}$ which is positive, and the reaction will therefore not occur at 20 °C.

b) At 500 °C,

$\Delta G^{\ominus} = +96 - (773 \times 0.138) = -10.7 \text{ kJ mol}^{-1}$

**ANSWER**    $\Delta G^{\ominus}$ is $-10.7$ kJ mol$^{-1}$ which is negative, the reaction will occur at 500 °C.

(*Note* the assumption that $\Delta H^{\ominus}$ does not vary with temperature.)

**EXERCISE 54**    Problems on Standard Entropy Change and Standard Free Energy Change

1. Find the standard entropy of vaporisation of lead at its boiling point, 2017 K, given that its standard enthalpy of vaporisation is 177 kJ mol$^{-1}$.

2. Find the standard entropy of melting of lead at its melting point, 600 K, given that its standard enthalpy of melting is 5.10 kJ mol$^{-1}$.

3. Calculate the standard entropies of vaporisation of the hydrogen halides at their respective boiling points from the data given.

| Hydrogen halide | Boiling point/K | $\Delta H^{\ominus}_{\text{vaporisation}}$/kJ mol$^{-1}$ |
|---|---|---|
| HF | 293 | 7.5 |
| HCl | 188 | 16.2 |
| HBr | 206 | 17.6 |
| HI | 238 | 19.8 |

Comment on the values you obtain.

4. Calculate the standard entropy of sublimation of arsenic at its sublimation temperature of 886 K, given that $\Delta H^{\ominus}_{\text{sublimation}} = 32.4$ kJ mol$^{-1}$.

5. Find the standard entropy of vaporisation of ethanol at its boiling point of 352 K, given $\Delta H^{\ominus}_{\text{vaporisation}} = 43.5$ kJ mol$^{-1}$.

6. Refer to the following values of standard entropy $(J\,mol^{-1}\,K^{-1})$ at 298 K:

| | | | | | |
|---|---|---|---|---|---|
| $H_2(g)$ | 131 | $H_2O(l)$ | 70 | $NH_4Cl(s)$ | 94.6 |
| $Cl_2(g)$ | 223 | $H_2O(g)$ | 189 | $N_2O_4(g)$ | 304 |
| $N_2(g)$ | 192 | $HCl(g)$ | 187 | $C_2H_4(g)$ | 220 |
| $O_2(g)$ | 205 | $NH_3(g)$ | 193 | $C_2H_6(g)$ | 230 |
| $Na(s)$ | 51 | $NO_2(g)$ | 240 | $HNO_3(l)$ | 156 |
| | | $NaCl(s)$ | 72.4 | | |

Calculate the standard entropy changes for the following reactions:

a) $H_2(g) + Cl_2(g) \longrightarrow 2HCl(g)$

b) $N_2(g) + 3H_2(g) \longrightarrow 2NH_3(g)$

c) $H_2(g) + \frac{1}{2}O_2(g) \longrightarrow H_2O(l)$

d) $H_2(g) + C_2H_4(g) \longrightarrow C_2H_6(g)$

e) $N_2O_4(g) \longrightarrow 2NO_2(g)$

f) $Na(s) + \frac{1}{2}Cl_2(g) \longrightarrow NaCl(s)$

g) $NH_4Cl(s) \longrightarrow NH_3(g) + HCl(g)$

h) $4HNO_3(l) \longrightarrow 4NO_2(g) + O_2(g) + 2H_2O(l)$

7. Predict whether the following reactions will have a positive or negative value of $\Delta S^{\ominus}$:

a) $NH_4NO_3(s) \longrightarrow N_2O(g) + 2H_2O(g)$

b) $2H_2O_2(aq) \longrightarrow 2H_2O(l) + O_2(g)$

c) $PH_3(g) + HI(g) \longrightarrow PH_4I(s)$

d) $3O_2(g) \longrightarrow 2O_3(g)$

e) $CO_2(g) + C(s) \longrightarrow 2CO(g)$

f) $Ni(s) + 4CO(g) \longrightarrow Ni(CO)_4(g)$

8. Refer to this list of values:

| Substance | $\Delta H_F^{\ominus}/kJ\,mol^{-1}$ | $\Delta G_F^{\ominus}/kJ\,mol^{-1}$ |
|---|---|---|
| $C_6H_6(g)$ | 82.9 | 130 |
| $C_6H_6(l)$ | 49.0 | 125 |
| $I_2(s)$ | 0 | 0 |
| $I_2(g)$ | 62.6 | 19.4 |
| $Hg(l)$ | 0 | 0 |
| $Hg(g)$ | 60.8 | 31.8 |

a) Find the standard entropy change of vaporisation of benzene at 298 K.

b) Find the standard entropy of vaporisation of iodine at 298 K.

c) Calculate the standard entropy of vaporisation of mercury at 298 K.

9. Use the following values of standard entropy content and standard enthalpy of formation to calculate standard free energy changes:

| Substance | $\Delta H_F^{\ominus}/kJ\,mol^{-1}$ | $S^{\ominus}/J\,K^{-1}\,mol^{-1}$ |
|---|---|---|
| HgO(s) (red) | −90.7 | 72.0 |
| HgO(s) (yellow) | −90.2 | 73.0 |
| HgS(s) (red) | −58.2 | 77.8 |
| HgS(s) (black) | −54.0 | 83.3 |

a) Calculate the value of $\Delta G^{\ominus}$ for the change

$$HgO(s) \text{ (red)} \longrightarrow HgO(s) \text{ (yellow)}$$

at 25 °C and at 100 °C. At what temperature will the change take place?

b) Calculate the value of $\Delta G^{\ominus}$ for the change

$$HgS(s) \text{ (red)} \longrightarrow HgS(s) \text{ (black)}$$

at 25 °C. At what temperature will the change occur?

10. Cis-but-2-ene has $\Delta H_F^{\ominus} = -5.7\,kJ\,mol^{-1}$ and $S^{\ominus} = 301\,J\,K^{-1}\,mol^{-1}$; trans-but-2-ene has $\Delta H_F^{\ominus} = -10.1\,kJ\,mol^{-1}$ and $S^{\ominus} = 296\,J\,K^{-1}\,mol^{-1}$. Calculate

a) $\Delta G^{\ominus}$ for the transition cis-but-2-ene $\longrightarrow$ trans-but-2-ene and

b) for the transition trans-but-2-ene $\longrightarrow$ cis-but-2-ene

Which is the more stable isomer?

## EXERCISE 55    Questions from A-level Papers

1. a) Describe, with the aid of a diagram, the lattice structure of crystalline sodium chloride and show how it accounts for the characteristic physical properties of the substance.

b) Define: i) *enthalpy change of formation*, ii) *lattice energy*. What factors determine the magnitude of a *lattice energy*?

Draw a Born–Haber cycle for the formation of caesium chloride and use it to calculate a value for the lattice energy of this compound.

*Data*

Enthalpy change of atomisation of caesium:

$$Cs(s) \longrightarrow Cs(g); \qquad \Delta H = +79\,kJ\,mol^{-1}$$

Enthalpy change of atomisation of chlorine:

$$\tfrac{1}{2}Cl_2(g) \longrightarrow Cl(g); \qquad \Delta H = +121\,kJ\,mol^{-1}$$

First ionisation energy of caesium:

$$Cs(g) \longrightarrow Cs^+(g) + e^-; \qquad \Delta H = +376\,kJ\,mol^{-1}$$

Electron affinity of chlorine:

$$Cl(g) + e^- \longrightarrow Cl^-(g); \qquad \Delta H = -364\,kJ\,mol^{-1}$$

Enthalpy change of formation of caesium chloride:

$$Cs(s) + \tfrac{1}{2}Cl_2(g) \longrightarrow Cs^+Cl^-(s); \qquad \Delta H = -433\,kJ\,mol^{-1}.$$

(C80)

2. a) How and under what conditions do benzene and cyclohexene react with chlorine? Give equations and essential conditions and name the products for the reactions involved.

   b)  i) The heat of hydrogenation of cyclohexene is $-120\,kJ\,mol^{-1}$. Assuming the structural formula of naphthalene to be

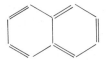

   what would you predict for its heat of hydrogenation? Write an equation for the reaction involved.

   ii) Use the bond energies below to calculate another value for the heat of hydrogenation of naphthalene, assuming that it has the structural formula shown above.

   $E(C—C)$ general $= 346\,kJ\,mol^{-1}$;
   $E(C\!=\!C)$ general $= 610\,kJ\,mol^{-1}$;
   $E(C—H)$ general $= 413\,kJ\,mol^{-1}$;
   $E(H—H) = 436\,kJ\,mol^{-1}$.

   iii) What is the difference between your answers to i) and ii)? How may this difference be explained?

   iv) The heat of hydrogenation of naphthalene, obtained by experiment, is much less than either of the theoretical values calculated in i) and ii). How are the differences between the experimental and theoretical values explained?        (SUJB80)

3. a) What do you understand by the term *entropy*?

   b) For each of the following reactions, indicate whether the entropy is likely to *increase*, *decrease* or *stay the same*, explaining your reasoning:

   i) $H_2(g) + C_2H_4(g) \longrightarrow C_2H_6(g)$
   ii) $N_2(g) + 3H_2(g) \longrightarrow 2NH_3(g)$
   iii) $2NaNO_3(s) \longrightarrow 2NaNO_2(s) + O_2(g)$

   c) What conditions must be satisfied for a change, either physical or chemical, to be spontaneous?

   d) The following is a list of standard entropies in $J\,K^{-1}\,mol^{-1}$:

   $H_2, 131;\quad C_2H_4, 220;\quad C_2H_6, 230;\quad N_2, 192;\quad NH_3, 193$

   Calculate the standard entropy changes of the reactions:

   i) $H_2(g) + C_2H_4(g) \longrightarrow C_2H_6(g)$
   ii) $N_2(g) + 3H_2(g) \longrightarrow 2NH_3(g)$

   e) For the reaction

   $$H_2(g) + Cl_2(g) \longrightarrow 2HCl(g)$$
   $$\Delta S = 20\,J\,K^{-1} \quad \text{and} \quad \Delta H = -185\,kJ$$

    i) Determine whether or not this reaction is thermodynamically feasible at 300 K and constant pressure.

    ii) What other factor is important in determining whether or not the reaction will actually take place?  (L80)

4. a) State Hess's law of constant heat summation (law of conservation of energy).

   b) Define precisely the terms:

     i) *enthalpy of formation (heat of formation)*;

     ii) *enthalpy of neutralisation (heat of neutralisation)*.

   c) Using the following data, collected at 25 °C and standard atmospheric pressure, in which the negative sign indicates heat evolved, calculate the enthalpy of formation of potassium chloride, KCl(s):

$$\Delta H \,(298\ \mathrm{K})/\mathrm{kJ\ mol^{-1}}$$

| | |
|---|---|
| i) $KOH(aq) + HCl(aq) = KCl(aq) + H_2O(l)$ | $-57.3$ |
| ii) $H_2(g) + \frac{1}{2}O_2(g) = H_2O(l)$ | $-285.9$ |
| iii) $\frac{1}{2}H_2(g) + \frac{1}{2}Cl_2(g) + aq = HCl(aq)$ | $-164.2$ |
| iv) $K(s) + \frac{1}{2}O_2(g) + \frac{1}{2}H_2(g) + aq = KOH(aq)$ | $-487.0$ |
| v) $KCl(s) + aq = KCl(aq)$ | $+18.4$ |

   d) The enthalpy of neutralisation of ethanoic acid (acetic acid) is $-55.8\ \mathrm{kJ\ mol^{-1}}$ while that of hydrochloric acid is $-57.3\ \mathrm{kJ\ mol^{-1}}$, both reactions being with potassium hydroxide solution. Explain the difference in these two values and make what deductions you can.

   e) Describe briefly an experiment by which the enthalpy change of a chemical reaction of *your own choice* may be determined, and outline how the various measurements would be used to calculate the final value.

   f) When solid potassium chloride is dissolved in water, heat is absorbed.

$$KCl(s) + aq = KCl(aq); \qquad \Delta H = +18.4\ \mathrm{kJ\ mol^{-1}}$$

Use the ideas formulated in the kinetic-molecular theory of matter to discuss what happens during the dissolution of a crystalline salt in water, and explain why heat is absorbed in the above reaction.  (SUJB80)

5. If the standard enthalpies of formation of carbon dioxide and water are respectively $-394\ \mathrm{kJ\ mol^{-1}}$ and $-286\ \mathrm{kJ\ mol^{-1}}$, and the standard enthalpy of combustion of propyne ($C_3H_4$) is $-1938\ \mathrm{kJ\ mol^{-1}}$, the standard enthalpy of formation of propyne is

a $-1258\ \mathrm{kJ\ mol^{-1}}$     b $-680\ \mathrm{kJ\ mol^{-1}}$     c $-184\ \mathrm{kJ\ mol^{-1}}$

d $+184\ \mathrm{kJ\ mol^{-1}}$     e $+1258\ \mathrm{kJ\ mol^{-1}}$  (NI82)

6. a) Draw a Born-Haber cycle for the formation of one mole of the hypothetical solid CaCl. Name the enthalpy change for each step in the cycle.

Use this cycle, and the following data, to calculate the standard enthalpy of formation of CaCl(s), $\Delta H_F^{\ominus}$ (CaCl(s)):

| Process | | $\Delta H$/kJ mol$^{-1}$ |
|---|---|---|
| Ca(s) $\longrightarrow$ Ca(g) | | 190 |
| Ca(g) $\longrightarrow$ Ca$^+$(g) | | 590 |
| $\frac{1}{2}$Cl$_2$(g) $\longrightarrow$ Cl(g) | | 120 |
| Cl(g) $\longrightarrow$ Cl$^-$(g) | | $-360$ |
| Ca$^+$(g) + Cl$^-$(g) $\longrightarrow$ CaCl(s) | | $-760$ |

Given that $\Delta H_F^{\ominus}$ (CaCl$_2$(s)) = $-780$ kJ mol$^{-1}$, comment on the two enthalpies of formation and upon whether it might be possible to prepare CaCl(s).

b) Draw a Born-Haber cycle for the formation of one mole of hydrated ions M$^+$(aq) from M(s). Calculate $\Delta H_F^{\ominus}$ (M$^+$(aq)) for M = Li and Cs from the following data:

| Process | | $\Delta H$/kJ mol$^{-1}$ for | |
|---|---|---|---|
| | | M = Li | M = Cs |
| M(s) $\longrightarrow$ M(g) | | 160 | 70 |
| M(g) $\longrightarrow$ M$^+$(g) | | 520 | 380 |
| M$^+$(g) $\longrightarrow$ M$^+$(aq) | | $-510$ | $-270$ |

Comment on the relative magnitudes of your answers and the way in which they reflect your knowledge of the chemistry of the alkali metals.

c) When 2.0 g of lithium chloride were dissolved in 100 g of water the temperature of the water changed from 20.0 °C to 24.2 °C. Estimate the enthalpy of solution of lithium chloride and predict, with reasons, the influence of temperature on the solubility of lithium chloride. (The heat capacity of water is 4.2 J K$^{-1}$g$^{-1}$.)     (JMB85)

7. The heat* (enthalpy) of solution of Na$_2$SO$_4$(s) and of Na$_2$SO$_4 \cdot 10H_2O$(s) are respectively $-2.4$ kJ mol$^{-1}$ and $+79.2$ kJ mol$^{-1}$ when the salts are separately dissolved in a large amount of water. (*A *negative sign indicates heat evolved.*)

Which of the following comments is/are correct?

a The heat of hydration of Na$_2$SO$_4$(s) to Na$_2$SO$_4 \cdot 10H_2O$(s) is $-81.6$ kJ mol$^{-1}$.

b The heat released from solvation of the ions in Na$_2$SO$_4$(s) is more than the energy needed to break down the crystal lattice into its ions.

c With increase in temperature the solubility of $Na_2SO_4(s)$ decreases while that of $Na_2SO_4 \cdot 10H_2O(s)$ (for the temperature range in which it exists) increases.

d Most salts absorb heat when dissolved in water but anhydrous sodium sulphate is a covalent macromolecular structure and this requires extra energy for its breakdown.                    (NI83)

8. The following is a Born-Haber cycle for the formation of magnesium chloride. The values of $\Delta H^{\ominus}$ for the changes $B$ to $G$ are given below.

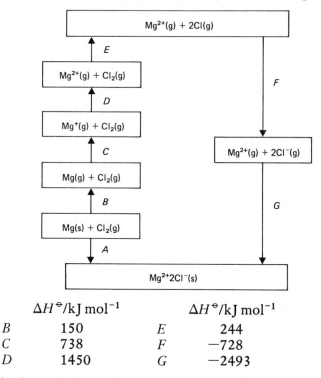

| | $\Delta H^{\ominus}/\text{kJ mol}^{-1}$ | | $\Delta H^{\ominus}/\text{kJ mol}^{-1}$ |
|---|---|---|---|
| $B$ | 150 | $E$ | 244 |
| $C$ | 738 | $F$ | −728 |
| $D$ | 1450 | $G$ | −2493 |

a) $A$ is the enthalpy of formation of magnesium chloride. Give the names of energy changes $B$, $E$, $F$ and $G$.

b) Calculate the enthalpy of formation of magnesium chloride.

c) i)   Using a similar Born-Haber cycle to the above, it is possible to calculate the enthalpies of formation of MgCl and $MgCl_3$. In order to do so, an *estimated* value for one energy change in the cycle must be used. Which energy change is this?

   ii)   The values obtained for the standard enthalpies of formation of MgCl and $MgCl_3$ are, approximately, $-110\,\text{kJ mol}^{-1}$ and $+4000\,\text{kJ mol}^{-1}$ respectively. Use your value for $\Delta H_F(MgCl_2)$ obtained above to deduce the relative energetic stabilities of the three possible magnesium chlorides with respect to the element.

   iii)   The enthalpy change for the reaction

$$2\text{MgCl(s)} \longrightarrow \text{MgCl}_2\text{(s)} + \text{Mg(s)}$$

is found by calculation to be about $-420\,kJ$. Use this information to deduce the relative stabilities of MgCl and $MgCl_2$ with respect to each other.

d) Given the following standard enthalpies of hydration

$$Mg^{2+}(g) \; + \; aq \; \longrightarrow \; Mg^{2+}(aq); \qquad \Delta H^\ominus = -1920\,kJ$$

$$Cl^-(g) \; + \; aq \; \longrightarrow \; Cl^-(aq); \qquad \Delta H^\ominus = -364\,kJ$$

and using the value of $\Delta H^\ominus$ for reaction $G$ already given, calculate the standard enthalpy of solution of magnesium chloride.    (L82)

9. a) State what is meant by the standard molar enthalpy change of neutralisation.

b) In an experimental determination of the molar enthalpy change of neutralisation between aqueous hydrochloric acid and aqueous sodium hydroxide, the following procedure was adopted. Separate $50\,cm^3$ portions of aqueous hydrochloric acid and sodium hydroxide each of concentration $1\,mol\,dm^{-3}$ were allowed to stand for 1 hour to attain laboratory temperature. The temperature of the acid was taken every 30 s for 2.5 min and at three minutes the two solutions were mixed in a polystyrene beaker of negligible thermal capacity. The temperature of the mixture was again recorded every 30 s. The results are given below.

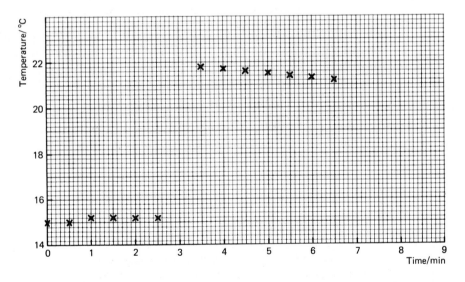

i) Using the results estimate the maximum temperature rise.

ii) Assuming that all the solutions have a density of $1.0\,g\,cm^{-3}$ and a specific heat capacity of $4200\,J\,kg^{-1}\,K^{-1}$, calculate the molar enthalpy change of neutralisation.

c) What temperature rise would be expected if aqueous sulphuric acid of the same concentration had been used? Give reasons for your answer.    (AEB85)

10. The enthalpy changes of precipitation of the carbonates of Group II metals were investigated by mixing metal nitrate solutions with sodium carbonate solution.

In a first experiment $10.0 \, cm^3$ of $1.0 \, M$ magnesium nitrate solution at $20.0 \, °C$ was placed in a plastic beaker of negligible heat capacity and an equal volume of $1.0 \, M$ sodium carbonate solution at $20.0 \, °C$ was added. The temperature rose to $22.1 \, °C$. The experiment was performed a *second* time, using $50.0 \, cm^3$ of each $1.0 \, M$ solution, and then performed a *third* time, using $50.0 \, cm^3$ of $0.20 \, M$ solutions.

a) What range and what sensitivity should you select for the thermometer to be used in this experiment?

b) i)   Calculate the temperature rise you would expect in the second experiment, and the third experiment.

   ii)  Which of the three experiments is likely to lead to the *most accurate* determination of a temperature rise? Give your reason(s).

   iii) Which of the three experiments is likely to lead to the *least accurate* determination of a temperature rise? Give your reason(s).

c) i)   Calculate the number of joules produced in the first experiment. (Specific heat capacity of solutions $= 4.2 \, J \, g^{-1} K^{-1}$.)

   ii)  Hence calculate $\Delta H_{reaction}$ for the change

   $$Mg^{2+}(aq) \; + \; CO_3{}^{2-}(aq) \longrightarrow MgCO_3(s).$$

d) The other Group II metals have the following values for $\Delta H_{reaction}$ for the precipitation of their carbonates:

| | |
|---|---|
| calcium carbonate, | $+9.24 \, kJ \, mol^{-1}$ |
| strontium carbonate, | $+2.9 \, kJ \, mol^{-1}$ |
| barium carbonate, | $-5.0 \, kJ \, mol^{-1}$ |

How might changes in the properties of these metal cations account for the trend in enthalpy changes of precipitation?          (L(N)82)

11. a) The ionisation energy of hydrogen atoms is $1310 \, kJ \, mol^{-1}$ and the electron affinity of chlorine atoms is $-347 \, kJ \, mol^{-1}$.

   i) Write the equations for the two processes.

   ii) Given the following additional information

   $$H(g) \; + \; Cl(g) \rightleftharpoons HCl(g) \qquad \Delta H^{\ominus} = -432 \, kJ \, mol^{-1}$$

   $$HCl(g) \rightleftharpoons H^+(aq) \; + \; Cl^-(aq)$$

   $$\Delta H^{\ominus} = -75 \, kJ \, mol^{-1}$$

   calculate the standard enthalpy change for the process

   $$H^+(g) \; + \; Cl^-(g) \rightleftharpoons H^+(aq) \; + \; Cl^-(aq).$$

b) The table below gives the enthalpies of hydration of some ions.

| Ion | $Cl^-$ | $Br^-$ | $I^-$ | $Li^+$ | $Na^+$ | $K^+$ |
|---|---|---|---|---|---|---|
| Enthalpy of hydration/kJ mol$^{-1}$ | $-380$ | $-350$ | $-310$ | $-520$ | $-400$ | $-320$ |

    i) Using the result from a), obtain the enthalpy of hydration of the proton.

    ii) Compare this value with those of the ions in the table and explain briefly why the proton has the value it does.   (JMB84)

12. a) Define standard enthalpy of combustion of ethane ($C_2H_6$) and standard enthalpy of formation of propane ($C_3H_8$).

  b) i) Give a *brief* account of the determination of the standard enthalpy of combustion of a solid or liquid of known relative molecular mass.

    ii) 5.6 dm$^3$ (measured at s.t.p.) of a mixture of propane and butane on complete combustion evolved 654 kJ of heat. Calculate the percentage composition of the mixture by volume.

  (The enthalpies of combustion of propane and butane are 2220 and 2877 kJ mol$^{-1}$ respectively and the molar volume is 22.4 dm$^3$ at s.t.p.)

  c) Using the following data which give the energy liberated when 1 g of each substance is burned in excess oxygen, calculate the standard enthalpy of formation of ethene:

| Substance | Energy liberated kJ |
|---|---|
| Graphite | 32 |
| Hydrogen | 143 |
| Ethene | 50 |

(H $= 1$; C $= 12$.)

  d) Explain *briefly either* why calcium chloride has a high melting point *or* why it is very soluble in water.   (SUJB85)

13. a) For an ionic crystal explain what is meant by the *crystal lattice* and define the term *lattice energy*.

  b) Explain *briefly* what is involved in the calculation of the lattice energy of an ionic crystal based on the forces between the ions in the lattice.

  c) The calculated lattice energies for the ionic compound $MgCl_2$ and the hypothetical ionic compound $MgCl$ are $-2502$ and $-753$ kJ mol$^{-1}$ respectively. Use the data given below in an energy cycle to calculate the standard enthalpy of formation for each of the compounds.

| Enthalpy of atomisation of magnesium | $+146\,kJ\,mol^{-1}$ |
|---|---|
| First ionisation energy of magnesium | $+738\,kJ\,mol^{-1}$ |
| Second ionisation energy of magnesium | $+1450\,kJ\,mol^{-1}$ |
| Enthalpy of atomisation of chlorine | $+121\,kJ\,mol^{-1}$ |
| Electron affinity of chlorine | $-364\,kJ\,mol^{-1}$ |

d) Use your answers from c) to calculate the standard enthalpy change for the decomposition of the hypothetical solid magnesium(I) chloride into magnesium(II) chloride and magnesium metal.

(O86,S)

14. a) Explain the meaning of the term *electronegativity*. Suggest reason(s) for fluorine being more electronegative than iodine.

b) Define the term *electron affinity*.

For the process

$$\tfrac{1}{2}X_2 + e = X^-(aq),$$

use the data given below to compare the enthalpy change ($\Delta H$) values when X = Cl and I. Use your values,

i) to account for what is observed when chlorine is passed into aqueous potassium iodide, and

ii) to compare the relative values of $E^{\ominus}_{X_2|X^-}$ for Cl and I.

| | $\Delta H^{\ominus}_F(X(g))$ /kJ mol$^{-1}$ | Electron affinity of X(g) $(-\Delta H_{EA})$/kJ mol$^{-1}$ | Enthalpy of hydration of X$^-$(g)/kJ mol$^{-1}$ |
|---|---|---|---|
| X = Cl | 121 | 362 | $-380$ |
| X = I | 107 | 315 | $-310$ |

c) If $\Delta H^{\ominus}_F(Na^+(g))$ is $602\,kJ\,mol^{-1}$ and $\Delta H^{\ominus}_F(NaCl(s))$ is $-411\,kJ\,mol^{-1}$ and $\Delta H^{\ominus}_F(NaI(s))$ is $-288\,kJ\,mol^{-1}$, compare the lattice energies of NaCl and NaI suggesting reasons for the difference.

$(\Delta H^{\ominus}_F(Na^+(g)) = \Delta H^{\ominus}_{At} + \text{first ionisation energy of sodium.})$

(WJEC84)

15. a) What is meant by *the average bond enthalpy* for the nitrogen-hydrogen bond in ammonia?

b) The standard molar enthalpies of formation of four hydrocarbons are given below.

| Hydrocarbon | $\Delta H^{\ominus}_F$/kJ mol$^{-1}$ |
|---|---|
| Ethane, $C_2H_6(g)$ | $-84.7$ |
| Ethene, $C_2H_4(g)$ | $+52.3$ |
| Ethyne, $C_2H_2(g)$ | $+227$ |
| Benzene, $C_6H_6(g)$ | $+82.9$ |

Given that the H—H and C—H bond enthalpies are +436 and +412 kJ mol$^{-1}$ respectively and that the standard molar enthalpy change of atomisation of graphite is +715 kJ mol$^{-1}$, calculate values for the carbon–carbon bond enthalpies in the four hydrocarbons and comment upon their values. (AEB85,S)

16. The heat of combustion of naphthalene ($C_{10}H_8$) was measured by burning it in an excess of oxygen within a sealed container known as a bomb calorimeter. The heat capacity of the calorimeter assembly and contents was 3.87 kJ K$^{-1}$. In a particular experiment, the temperature of the calorimeter assembly increased from 25.00 °C to 29.48 °C due to the combustion of 3.36 × 10$^{-3}$ mol of naphthalene.

   a) Write the equation for the reaction which takes place inside the calorimeter.

   b) Which of the two quantities $\Delta H$ or $\Delta U$ is derived directly from the use of the bomb calorimeter? Explain your answer.

   c) For the combustion of naphthalene, calculate the heat change you have selected in b).

   d) For the combustion of naphthalene, which of the two quantities $\Delta H$ or $\Delta U$, measured at 25 °C, is numerically smaller? Explain your answer.

   e) Give the equation, including state symbols, of a combustion for which the values of $\Delta H$ and $\Delta U$ are equal when measured at 25 °C; assume ideal gas behaviour. (JMB83)

17. a) By referring to ammonia, $NH_3$, define the term *average N-H bond enthalpy*.

   b) Use the following bond enthalpies, at 298 K, to estimate the enthalpy change for the reaction:

$$2NH_3(g) \longrightarrow N_2(g) + 3H_2(g)$$

| Bond | Bond enthalpy/kJ mol$^{-1}$ |
|------|------------------------------|
| H—H | 436 |
| N—H | 388 |
| N≡N | 944 |

   c) Using the answer you obtained in b), predict how the proportion of ammonia in an equilibrium mixture of ammonia, nitrogen and hydrogen, would change when the temperature is increased.

   d) For the reaction

$$2NH_3(g) \longrightarrow N_2(g) + 3H_2(g),$$

   the value of $\Delta S$ in the equation

$$\Delta G = \Delta H - T\Delta S,$$

   is +199 J K$^{-1}$ mol$^{-1}$.

Assuming that both $\Delta H$ and $\Delta S$ are independent of temperature, calculate the minimum temperature at which the thermal decomposition of ammonia becomes spontaneous.          (AEB82)

18. a) Explain what is meant by the term *lattice energy*.

Discuss the factors that determine the magnitude of a lattice energy.

Draw a Born-Haber cycle for the formation of calcium chloride and use it to calculate a value for the lattice energy of this compound.

*Data*

| Enthalpy change of atomisation of calcium | $= +193 \, kJ \, mol^{-1}$ |
| Enthalpy change of atomisation of chlorine | $= +121 \, kJ \, mol^{-1}$ (i.e. per mole of Cl atoms) |
| First ionisation energy of calcium | $= +590 \, kJ \, mol^{-1}$. |
| Electron affinity of chlorine | $= -364 \, kJ \, mol^{-1}$ |
| Second ionisation energy of calcium | $= +1150 \, kJ \, mol^{-1}$ |
| Enthalpy change of formation of calcium chloride | $= -795 \, kJ \, mol^{-1}$ |

How would you expect the magnitude of the lattice energy of
i) potassium chloride,
ii) calcium fluoride,
to compare with that of calcium chloride?

b) How do the solubilities of the alkaline earth metal sulphates vary down the group? Account for this variation in terms of the appropriate enthalpy changes.          (C85,S)

19. a) Describe a simple experiment to measure the enthalpy of combustion of ethanol. Indicate any significant sources of experimental error and state how these errors may affect the result.

b) Carbon disulphide burns in air according to the equation

$$CS_2(l) \ + \ 3O_2(g) \ \longrightarrow \ CO_2(g) \ + \ 2SO_2(g)$$
$$\Delta H^{\ominus} = -1075 \, kJ \, mol^{-1}$$

i) Give the general expression which relates enthalpy change ($\Delta H$) to internal energy change ($\Delta U$). Explain briefly how the expression is derived. State, with reasons, the value of $\Delta U^{\ominus}$ for the combustion of carbon disulphide.

ii) Calculate the standard enthalpy of formation of carbon disulphide using the enthalpies of combustion provided. Comment on the possibility of making carbon disulphide directly from its elements.

(Standard enthalpies of combustion: $C(s)$, $-394\,kJ\,mol^{-1}$; $S(s)$, $-297\,kJ\,mol^{-1}$.)                                                                 (JMB82)

20. a) State Hess' Law and define
      i) *standard state* of a substance and
      ii) *standard enthalpy change* for a reaction.

b) Describe briefly how you would determine experimentally the enthalpy of neutralisation of a dilute solution of hydrochloric acid by a dilute solution of sodium hydroxide.

c) The standard enthalpy of combustion of propane ($C_3H_8$) at 298 K is $-2220.0\,kJ\,mol^{-1}$. Calculate the standard enthalpy of formation ($\Delta H_F^{\ominus}$) of propane at 298 K given that at this temperature $\Delta H_F^{\ominus} = -285.9\,kJ\,mol^{-1}$ for $H_2O(l)$ and $\Delta H_F^{\ominus} = -393.5\,kJ\,mol^{-1}$ for $CO_2(g)$.

d) Discuss briefly the following enthalpy of hydrogenation data for ethene ($C_2H_4$) and benzene ($C_6H_6$) in relation to the bonding in the benzene molecule.

$C_2H_4(g) + H_2(g) = C_2H_6(g)$ $\qquad \Delta H^{\ominus}(298\,K) = -132\,kJ\,mol^{-1}$

$C_6H_6(g) + 3H_2(g) = C_6H_{12}(g)$ $\qquad \Delta H^{\ominus}(298\,K) = -246\,kJ\,mol^{-1}$

# 10 Reaction Kinetics

Reaction Kinetics is the study of the factors which affect the rates of chemical reactions.

## REACTION RATE

The rate of a chemical reaction is the rate of change of concentration. Consider a reaction of the type A ⟶ B, where one molecule of the reactant forms one molecule of the product. Fig. 10.1 shows how the concentration of product, $x$, increases as the time, $t$, which has passed since the start of the reaction increases. The initial concentration of reactant (the concentration at the start of the reaction) is $a$, and at any time after the start of the reaction, the concentration of reactant is $(a - x)$.

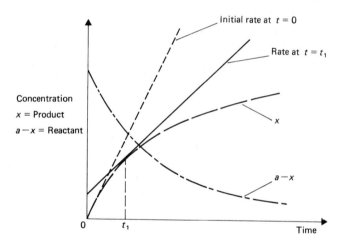

Fig. 10.1   Variation of concentrations of reactant and product with time

You can see that the rate of reaction is decreasing as the reaction proceeds and the reactant is being used up. One can only state the rate of reaction between certain times.

One can calculate the average rate of reaction over a certain interval of time in this way.

To $1 \, dm^3$ of solution containing $0.300 \, mol$ methyl ethanoate is added a small amount of mineral acid. This catalyses the hydrolysis reaction

$$CH_3CO_2CH_3(aq) + H_2O(l) \rightleftharpoons CH_3CO_2H(aq) + CH_3OH(aq)$$

After 100 seconds, the concentration has decreased to $0.292 \, \text{mol} \, \text{dm}^{-3}$. This means that $0.008 \, \text{mol} \, \text{dm}^{-3}$ of methyl ethanoate has reacted, and $0.008 \, \text{mol} \, \text{dm}^{-3}$ of methanol and ethanoic acid have been formed.

$$\left(\begin{array}{l}\text{Average rate of reaction}\\ \text{over this time interval}\end{array}\right) = \frac{\text{Change in concentration}}{\text{Time}}$$

$$= \frac{0.008}{100} = 8 \times 10^{-5} \, \text{mol} \, \text{dm}^{-3} \, \text{s}^{-1}$$

Since the rate of reaction varies with time, it is usual to quote the initial rate of the reaction. This is the rate at the start of the reaction when an infinitesimally small amount of the reactant has been used up. In Fig. 10.1, the gradient of the tangent to the curve at $t = 0$ gives the initial rate of the reaction.

## THE EFFECT OF CONCENTRATION ON RATE OF REACTION

Consider a reaction between A and B to form C:

$$A + B \longrightarrow C$$

The rate of formation of C depends on the concentrations of A and B, but one cannot simply say that the rate of formation of C is proportional to the concentration of A and proportional to the concentration of B. The relationship is

$$\text{Rate of formation of C} \propto [A]^m [B]^n$$

where $m$ and $n$ are usually integers, often 0, 1 or 2, and are characteristic of the reaction. One says that the reaction is of order $m$ with respect to A and of order $n$ with respect to B. The order of reaction is $(m + n)$. One cannot tell the order simply by looking at the chemical equation for the reaction. For example, the reaction between bromate(V) ions and bromide ions and acid to give bromine

$$BrO_3^-(aq) + 5Br^-(aq) + 6H^+(aq) \longrightarrow 3Br_2(aq) + 3H_2O(l)$$

has a rate of disappearance of $BrO_3^-$

$$\frac{-d[BrO_3^-]}{dt} \propto [BrO_3^-][Br^-][H^+]^2$$

It is first order with respect to bromate(V), first order with respect to bromide, second order with respect to hydrogen ion and fourth order overall. The negative sign means that $[BrO_3^-]$ decreases with time.

If        Reaction rate $\propto [A]^m [B]^n$        it follows that

**Reaction rate** $= k[A]^m [B]^n$

The proportionality constant $k$ is called the *rate constant* for the reaction or the *rate coefficient* for the reaction.

## ORDER OF REACTION

As a reaction proceeds, the concentrations of the reactants decrease, and the rate of reaction decreases, as shown in Fig. 10.1. The shape of the curve depends on the order of the reaction. It is obtained by integrating the rate equation.

## FIRST-ORDER REACTIONS

If the reaction

$$A \longrightarrow Products$$

is a first-order reaction, the rate equation will be

$$\frac{-d[A]}{dt} = k[A]$$

If $a$ = initial concentration of $A$, and $x$ = decrease in concentration of $A$ in time $t$, so that $(a - x)$ = concentration of A at time $t$, then

$$\frac{-d(a - x)}{dt} = k(a - x)$$

Integrating this equation gives

$$kt = \ln[a/(a - x)] = 2.303 \lg[a/(a - x)]$$

The first-order rate constant can be obtained by plotting:

a) $\ln[a/(a - x)]$ against $t$ to give a straight line with gradient $k$,

b) $\lg[a/(a - x)]$ against $t$ to give a straight line with gradient $k/2.303$,

c) $\ln(a - x)$ against $t$ to give a straight line with gradient $-k$,

d) $\lg(a - x)$ against $t$ to give a straight line with gradient $-k/2.303$.

The units of $k$, the first-order rate constant, are $s^{-1}$.

## HALF-LIFE

Let $t_{1/2}$ be the time taken for half the amount of A to react. $t_{1/2}$ is called the *half-life* of the reaction.

After $t_{1/2}$ seconds,     $x = a/2$   and   $\dfrac{a}{a - x} = 2$

$\therefore$
$$kt_{1/2} = \ln 2 = 2.303 \lg 2$$

$$t_{1/2} = 0.693/k$$

The half-life of a first-order reaction is independent of the initial concentration of the reactant. Radioactive decay is an example of first-order kinetics.

## PSEUDO-FIRST-ORDER REACTIONS

The acid-catalysed hydrolysis of an ester, e.g. ethyl ethanoate,

$$CH_3CO_2C_2H_5(aq) + H_2O(l) \longrightarrow CH_3CO_2H(aq) + C_2H_5OH(aq)$$

is first order with respect to ester and first order with respect to water. If water is present in excess, so that the fraction of the water which is used up in the reaction is small, the concentration of water is practically constant, and, since the acid catalyst is not used up, the rate depends only on the concentration of ester:

$$\frac{-d[CH_3CO_2C_2H_5]}{dt} = k'[CH_3CO_2C_2H_5]$$

$k'$ is constant for a certain concentration of acid, and the reaction obeys a first-order rate equation.

## SECOND-ORDER REACTIONS, WITH REACTANTS OF EQUAL CONCENTRATIONS

In the simplest case, the reaction

$$A + B \longrightarrow Products$$

has a rate equation $\quad \dfrac{-d[A]}{dt} = \dfrac{-d[B]}{dt} = k[A][B]$

Let $\quad a = $ initial concentration of A and also of B

$x = $ decrease of concentration of A and of B in time $t$

$a - x = $ concentration of A and of B at time $t$.

The rate equation therefore becomes

$$\frac{-d(a-x)}{dt} = k(a-x)^2$$

Integration of this equation gives

$$kt = x/a(a-x)$$

A plot of $x/a(a-x)$ against $t$ gives a straight line of gradient $k$.

Since $\quad kt = \dfrac{x}{a(a-x)} = \dfrac{1}{a-x} - \dfrac{1}{a}$

a plot of $1/(a-x)$ against $t$ gives a straight line of gradient $k$.

## ZERO-ORDER REACTIONS

In a zero-order reaction, the rate is independent of the concentration of the reactant. In the reaction between propanone and iodine

$$CH_3COCH_3(aq) \ + \ I_2(aq) \ \longrightarrow \ CH_3COCH_2I(aq) \ + \ HI(aq)$$

the reaction rate does not change if the concentration of iodine is changed. The rate of reaction is independent of the iodine concentration, and the reaction is said to be zero order with respect to iodine.

## THE EFFECT OF TEMPERATURE ON REACTION RATES

An increase in temperature increases the rate of a reaction by increasing the rate constant. A plot of the logarithm of the rate constant, $k$, against $1/T$ is a straight line, with a negative gradient (see Fig. 10.2).

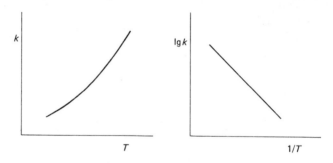

Fig. 10.2   Dependence of rate constant on temperature

The variation of rate constant with temperature obeys the Arrhenius equation

$$k \ = \ Ae^{-E/RT}$$

$A$ and $E$ are constant for a given reaction; $R$ is the gas constant. In order to react, two molecules must collide with a minimum amount of energy $E$, which is called the *activation energy*. The fraction of molecules possessing energy $E$ is given by $e^{-E/RT}$. The constant $A$, called the *pre-exponential factor*, represents the maximum rate which the reaction can reach when all the molecules have energy equal to or greater than $E$.

The Arrhenius equation can be written as

$$\lg k \ = \ \lg A - \frac{E}{2.303RT}$$

A plot of $\ln k$ against $1/T$ is a straight line of gradient $-E/R$.

A plot of $\lg k$ against $1/T$ is linear with a gradient of $-E/2.303R$ and intercept $\lg A$. The values of $A$ and $E$ can thus be found from a plot of $\lg k$ against $1/T$. If only two values of $k$ have been found, then

$$\lg \frac{k_2}{k_1} = -\frac{E}{2.303R}\left(\frac{1}{T_2} - \frac{1}{T_1}\right)$$

$$= \frac{E(T_2 - T_1)}{2.303RT_1T_2}$$

The activation energy can be obtained from the ratio of the two rates.

**EXAMPLE 1** The rate constant of a first-order reaction is $2.0 \times 10^{-6}\,\text{s}^{-1}$. The initial concentration of the reactant is $0.10\,\text{mol dm}^{-3}$. What is the value of the initial rate in $\text{mol dm}^{-3}\text{s}^{-1}$?

**METHOD** The rate equation has the form

$$dx/dt = k[A]$$

Putting the values of $[A]$ and $k$ into this equation gives

**ANSWER** $$dx/dt = 2.0 \times 10^{-6} \times 0.10 = 2.0 \times 10^{-7}\,\text{mol dm}^{-3}\text{s}^{-1}.$$

**EXAMPLE 2** A first-order reaction is 40% complete at the end of 20 min. a) What is the value of the rate constant? b) In how many minutes will the reaction be 80% complete?

**METHOD** a) For a first-order reaction,

$$k = \frac{2.303}{t}\lg\frac{a}{a-x}$$

$a$ = initial concentration = 100%

$x$ = decrease in concentration = 40% at time $t$ = 20 min = 1200 s

$a-x$ = remaining concentration = 60% at time $t$ = 1200 s

$$k = \frac{2.303}{1200}\lg\frac{100\%}{60\%}$$

**ANSWER** $$k = 4.3 \times 10^{-4}\,\text{s}^{-1}$$

b) For 80% reaction, $(a-x) = 20\%$, and

$$4.25 \times 10^{-4} = \frac{2.303}{t}\lg\frac{100\%}{20\%}$$

**ANSWER** $$t = 3790\,\text{s} = 63\,\text{minutes}$$

**EXAMPLE 3** The half-life for the radioactive decay of thorium-234 is 24 days. a) Calculate the rate constant for the decay. b) What time will elapse before 3/4 of the thorium has decayed?

**METHOD** a) Radioactive decay follows the first-order law:

$$kt_{1/2} = 2.303 \lg 2$$

**ANSWER**
$$k = \frac{2.303 \lg 2}{24 \times 24 \times 60 \times 60} = 3.34 \times 10^{-7} s^{-1}$$

b) When 3/4 of the thorium has decayed, the initial amount of thorium, $a$, has become $a/4$.

$$kt = 2.303 \lg \frac{a}{a-x}$$

$$3.34 \times 10^{-7} \times t = 2.303 \lg \frac{a}{a/4}$$

**ANSWER**
$$t = 4.15 \times 10^6 s = 48 \, \text{days}$$

**EXAMPLE 4** Carbon-14 is radioactive. The half-life is 5600 years. Calculate the age of a piece of wood which gives 10 counts per minute per gram of carbon, compared with 15 c.p.m. per gram of carbon from a sample of new wood.

**METHOD** a) First, find the first-order rate constant for the decay.
b) Use the rate constant to find the age of the sample.

a)
$$2.303 \lg \frac{a}{a-x} = kt$$

∴
$$2.303 \lg \frac{N_0}{N} = kt$$

where        $N_0$ = Count rate for new wood and

$N$ = Count rate after time $t$

At the half-life,        $2.303 \lg 2 = kt_{1/2}$

Since        $t_{1/2} = 5600 \, \text{years}$,        $k = 1.24 \times 10^{-4} \text{year}^{-1}$

b) Since $N_0/N = 15/10$,
$$2.303 \lg 1.5 = 1.24 \times 10^{-4} t$$

**ANSWER** Age of wood,        $t = 3270 \, \text{years}$.

**EXAMPLE 5** The initial rate of a second-order reaction is $8.0 \times 10^{-3} \, \text{mol dm}^{-3} \, s^{-1}$. The initial concentrations of the two reactants, A and B, are 0.20 mol $\text{dm}^{-3}$. What is the rate constant in $\text{dm}^3 \, \text{mol}^{-1} \, s^{-1}$?

**METHOD** Since
$$-\frac{d[A]}{dt} = k[A][B]$$

putting

$$\frac{d[A]}{dt} = 8.0 \times 10^{-3} \, \text{mol dm}^{-3} \, \text{s}^{-1} \text{ gives } [A] = [B] = 0.20 \, \text{mol dm}^{-3}$$

$$8.0 \times 10^{-3} = k \times 0.20 \times 0.20$$

**ANSWER**  Rate constant,  $k = 0.20 \, \text{dm}^3 \, \text{mol}^{-1} \, \text{s}^{-1}$

**EXAMPLE 6** In the alkaline hydrolysis of an ester, both ester and alkali had the initial concentration of $0.0500 \, \text{mol dm}^{-3}$. The following results were obtained.

| Time/s | 100 | 200 | 300 | 400 | 600 |
|---|---|---|---|---|---|
| % of ester hydrolysed | 29.5 | 44.2 | 55.5 | 62.2 | 70.3 |

Find the order of reaction, and calculate the rate constant.

**METHOD** If the reaction is first order, a plot of $\lg(a-x)$ against $t$ will be linear with a gradient of $-k/2.303$. If the reaction is second order, a plot of $1/(a-x)$ against $t$ will be linear with a gradient of $k$. The table below shows values of $\lg(a-x)$ and $1/(a-x)$ calculated from the percentage hydrolysis at given times. If the original concentration of ester $= a$, then $x =$ % hydrolysis $\times a$.

| Time/s | 0 | 100 | 200 | 300 | 400 | 600 |
|---|---|---|---|---|---|---|
| Hydrolysis/% | 0 | 29.5 | 44.2 | 55.5 | 62.2 | 70.3 |
| $x$ | 0 | 0.0148 | 0.0221 | 0.0276 | 0.0311 | 0.0351 |
| $a-x$ | 0.0500 | 0.0352 | 0.0279 | 0.0224 | 0.0189 | 0.0149 |
| $\lg(a-x)$ | $-1.30$ | $-1.45$ | $-1.55$ | $-1.65$ | $-1.72$ | $-1.83$ |
| $1/(a-x)$ | 20.0 | 28.4 | 35.8 | 44.6 | 52.9 | 67.1 |

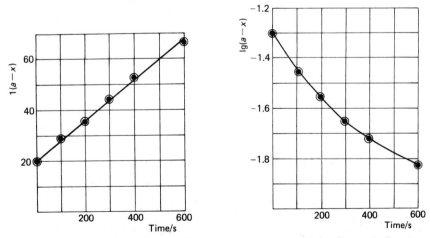

Fig. 10.3   Plots of $\lg(a-x)$ and $1/(a-x)$ against $t$ for Example 6

The plots in Fig. 10.3 show that the second-order plot is linear.

The gradient is $40.0/500 = 8.00 \times 10^{-2}\,dm^3\,mol^{-1}\,s^{-1}$.

This is the value of $k$.

The reaction is second order, with a rate constant of $8.00 \times 10^{-2}\,dm^3$ $mol^{-1}\,s^{-1}$.

**EXAMPLE 7**   The reaction

$$2N_2O_5(g) \longrightarrow 2N_2O_4(g) + O_2(g)$$

was studied at a number of temperatures, and the following values for the rate constant were obtained:

| Temperature/K | 293 | 308 | 318 | 338 |
|---|---|---|---|---|
| Rate constant/$s^{-1}$ | $1.76 \times 10^{-5}$ | $1.35 \times 10^{-4}$ | $4.98 \times 10^{-4}$ | $4.87 \times 10^{-3}$ |

Calculate the activation energy and the pre-exponential factor for the reaction.

**METHOD**   Since

$$\lg k = \lg A - \frac{E}{2.303RT}$$

a plot of $\lg k$ against $1/T$ gives a straight line with a gradient of $-E/2.303R$ and an intercept on the $\lg k$ axis of $\lg A$.

The table below gives the values which must be plotted:

| T/K | 293 | 308 | 318 | 338 |
|---|---|---|---|---|
| $k/s^{-1}$ | $1.76 \times 10^{-5}$ | $1.35 \times 10^{-4}$ | $4.98 \times 10^{-4}$ | $4.87 \times 10^{-3}$ |
| $1/T$ / 1/K | $3.41 \times 10^{-3}$ | $3.25 \times 10^{-3}$ | $3.14 \times 10^{-3}$ | $2.96 \times 10^{-3}$ |
| $\lg k$ | $-4.755$ | $-3.870$ | $-3.303$ | $-2.312$ |

Fig. 10.4 shows the plot of $\lg k$ against $1/T$. The gradient is $-5\,550\,K$.

$$\therefore \qquad -5\,500 = -\frac{E}{2.303 \times 8.31}$$

$$E = 106\,000\,J\,mol^{-1} = 106\,kJ\,mol^{-1}$$

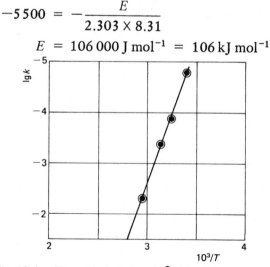

Fig. 10.4   Plot of $\lg k$ against $10^3/T$ for Example 6

$A$ can be found from the intercept or by substituting values of $\lg k$ and $1/T$ in the equation. At $1/T = 3.28 \times 10^{-3}$, $\lg k = -4.00$.

$$\therefore \quad \lg A = -4.00 + \frac{106 \times 10^3 \times 3.28 \times 10^{-3}}{2.303 \times 8.314} = 14.16$$

$$A = 1.45 \times 10^{14}\,\text{s}^{-1}$$

**ANSWER**    The activation energy is $106\,\text{kJ mol}^{-1}$; the pre-exponential factor is $1.45 \times 10^{14}\,\text{s}^{-1}$.

**EXAMPLE 8**    Calculate the activation energy for the second-order reaction

$$2N_2O(g) \longrightarrow 2N_2(g) + O_2(g)$$

given the rate constants $1.10 \times 10^{-3}\,\text{dm}^3\,\text{mol}^{-1}\,\text{s}^{-1}$ at 828 K and $1.67\,\text{dm}^3\,\text{mol}^{-1}\,\text{s}^{-1}$ at 1053 K.

**METHOD**    The equation $\qquad \lg\dfrac{k_2}{k_1} = \dfrac{E(T_2 - T_1)}{2.303RT_1T_2}$

gives $\qquad \lg\dfrac{1.67}{1.10 \times 10^{-3}} = \dfrac{E \times 215}{2.303 \times 8.314 \times 1053 \times 838}$

**ANSWER**    Activation energy, $\quad E = 250\,\text{kJ mol}^{-1}$.

## SUMMARY

### The effect of concentration on rate

The type of linear plot tells the order of the reaction.

The slope of the linear plot gives the rate constant for the reaction.

| Order | Rate equation | Linear plots | Gradient | Units of $k$ |
|---|---|---|---|---|
| 0 | $k = x/t$ | $x$ against $t$ | $k$ | $\text{mol dm}^{-3}\,\text{s}^{-1}$ |
| 1 | $k = \dfrac{2.303}{t}\lg\dfrac{a}{a-x}$ | $\lg(a-x)$ against $t$ | $-k/2.303$ | $\text{s}^{-1}$ |
|  |  | $\ln(a-x)$ against $t$ | $-k$ | $\text{s}^{-1}$ |
| 2 | $k = \dfrac{x}{ta(a-x)}$ | $1/(a-x)$ against $t$ | $k$ | $\text{dm}^3\,\text{mol}^{-1}\,\text{s}^{-1}$ |

### The effect of temperature on rate constant

| Equation | Linear plot | Gradient | Intercept |
|---|---|---|---|
| $k = A\,e^{-E/RT}$ |  |  |  |
| $\lg k = \lg A - \dfrac{E}{2.303RT}$ | $\lg k$ against $1/T$ | $-\dfrac{E}{2.303R}$ | $\lg A$ |
|  | $\ln k$ against $1/T$ | $-\dfrac{E}{R}$ | $\ln A$ |

**EXERCISE 56**    Problems on Finding the Order of Reaction

1. X and Y react together. For a three-fold increase in the concentration of X, there is a nine-fold increase in the rate of reaction. What is the order of reaction with respect to X?

2. A and B react to form C. In one run, the concentration of A is doubled, while B is kept constant, and the initial rate is doubled. In a second run, the concentration of B is doubled while that of A is kept constant, and the initial rate is quadrupled. What can you deduce about the order of the reaction?

3. Fig. 10.5 shows that the rate of reaction is:

   a proportional to $[I_2]$
   b proportional to $[I_2]^2$
   c proportional to $1/[I_2]$
   d independent of $[I_2]$

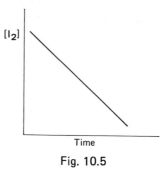

Time

Fig. 10.5

4. X decomposes to form Y + Z. The following results were obtained in a study of the reaction:

| Initial [X] /mol dm$^{-3}$ | $2.0 \times 10^{-3}$ | $4.0 \times 10^{-3}$ | $5.0 \times 10^{-3}$ |
|---|---|---|---|
| Initial rate/mol dm$^{-3}$ s$^{-1}$ | $1.3 \times 10^{-6}$ | $1.3 \times 10^{-6}$ | $1.3 \times 10^{-6}$ |

What is the rate expression? What is the order of the reaction?

5. The reaction A + B ⟶ C is first order with respect to A and to B. When the initial concentrations are $[A] = 1.5 \times 10^{-2}$ mol dm$^{-3}$ and $[B] = 2.5 \times 10^{-3}$ mol dm$^{-3}$, the initial rate of reaction is found to be $3.75 \times 10^{-4}$ mol dm$^{-3}$ s$^{-1}$. Calculate the rate constant for the reaction.

6. In the reaction

$$A + B \longrightarrow P + Q$$

the following results were obtained for the initial rates of reaction for different initial concentrations:

| [A] /mol dm$^{-3}$ | [B] /mol dm$^{-3}$ | Initial rate/mol dm$^{-3}$ s$^{-1}$ |
|---|---|---|
| 1.0 | 1.0 | $2.0 \times 10^{-3}$ |
| 2.0 | 1.0 | $4.0 \times 10^{-3}$ |
| 4.0 | 2.0 | $16 \times 10^{-3}$ |

Deduce the rate equation and calculate the rate constant.

7. The rate of a reaction depends on the concentrations of the reactants. In the reaction between X and Y, the following results were obtained for runs at the same temperature.

| Initial concentration of X/mol dm$^{-3}$ | Initial concentration of Y/mol dm$^{-3}$ | Initial rate/ mol dm$^{-3}$ h$^{-1}$ |
|---|---|---|
| $2 \times 10^{-3}$ | $3 \times 10^{-3}$ | $3.0 \times 10^{-3}$ |
| $2 \times 10^{-3}$ | $6 \times 10^{-3}$ | $1.2 \times 10^{-2}$ |
| $4 \times 10^{-3}$ | $6 \times 10^{-3}$ | $2.4 \times 10^{-2}$ |

Deduce the order of the reaction with respect to: a) X, b) Y. Calculate the rate constant for the reaction.

8.

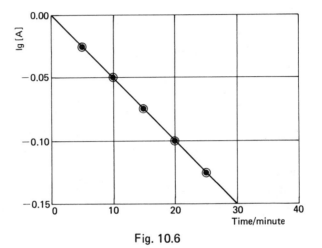

Fig. 10.6

Fig. 10.6 shows a plot of lg concentration of $A$ (in mol dm$^{-3}$) against time (in minutes) for the reaction A $\longrightarrow$ B. What is the order of reaction? Calculate the rate constant for the reaction.

9. The following results were obtained for the decomposition of nitrogen(V) oxide

$$2N_2O_5(g) \longrightarrow 4NO_2(g) + O_2(g)$$

| Concentration of N$_2$O$_5$/mol dm$^{-3}$ | Initial rate/mol dm$^{-3}$ s$^{-1}$ |
|---|---|
| $1.6 \times 10^{-3}$ | 0.12 |
| $2.4 \times 10^{-3}$ | 0.18 |
| $3.2 \times 10^{-3}$ | 0.24 |

What is the rate expression for the reaction? What is the order of reaction? What is the initial rate of reaction when the concentration of N$_2$O$_5$ is:

a) $2.0 \times 10^{-3}$ mol dm$^{-3}$    b) $2.4 \times 10^{-2}$ mol dm$^{-3}$?

**EXERCISE 57**    Problems on First-order Reactions

1. A isomerises to form B. The reaction is first order. If 75% of A is converted to B in 2.5 hours, what is the value of the rate constant for the isomerisation?

2. The reaction

$$2N_2O_5 \longrightarrow 4NO_2 + O_2$$

is first order. If the initial concentration of $N_2O_5$ is 1.25 mol dm$^{-3}$ and the initial rate is $1.38 \times 10^{-5}$ mol dm$^{-3}$ s$^{-1}$, what is the rate constant for the decomposition?

3.

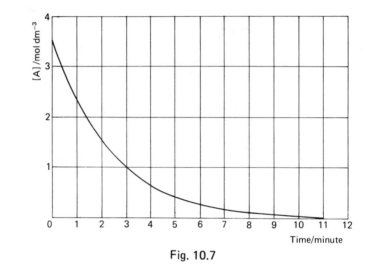

Fig. 10.7

The curve shown in Fig. 10.7 represents the decomposition of A at a certain temperature.

Calculate the gradients of the curve at: a) 1 minute, b) 3 minutes, c) 5 minutes, and d) 11 minutes. Plot a graph of the gradient against the concentration of A at 1, 3, 5 and 11 minutes. Calculate the rate constant for the reaction from the slope of the graph. What is the order of the reaction?

4. Fig. 10.8 shows the results of a study of the reaction

$$A + B \longrightarrow C$$

The experimental conditions were:

| | Initial concentrations/mol dm$^{-3}$ | |
| | [A] | [B] |
| --- | --- | --- |
| Curve 1 | 0.10 | 0.10 |
| Curve 2 | 0.10 | 0.20 |
| Curve 3 | 0.10 | 0.30 |

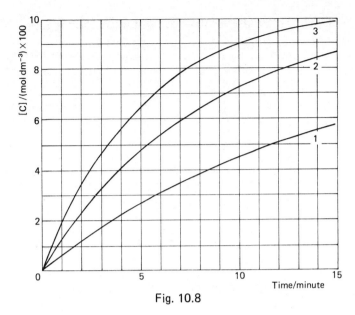

Fig. 10.8

a) Find the initial rates of curves 1, 2 and 3.
b) What is the order of reaction with respect to B?
c) Inspect curve 3. What is the time required for completion of $\frac{1}{2}$ reaction, and of $\frac{3}{4}$ reaction? What is the order with respect to A?
d) Write an overall rate equation for the reaction.
e) Find the rate constant for the reaction.

## EXERCISE 58    Problems on Second-order Reactions

1. The following results were obtained from a study of the reaction between P and Q.

| Concentrations/mol dm$^{-3}$ | | Initial rate/mol dm$^{-3}$ s$^{-1}$ |
|---|---|---|
| [P] | [Q] | |
| $2.00 \times 10^{-3}$ | $2.00 \times 10^{-3}$ | $2.00 \times 10^{-4}$ |
| $1.80 \times 10^{-3}$ | $1.80 \times 10^{-3}$ | $1.62 \times 10^{-4}$ |
| $1.40 \times 10^{-3}$ | $1.40 \times 10^{-3}$ | $9.80 \times 10^{-5}$ |
| $1.10 \times 10^{-3}$ | $1.10 \times 10^{-3}$ | $6.05 \times 10^{-5}$ |
| $0.80 \times 10^{-3}$ | $0.80 \times 10^{-3}$ | $3.20 \times 10^{-5}$ |

Prove that the reaction is second order. Calculate the rate constant.

2. A second-order reaction occurs between X and Y, which are present in solution at concentrations of $0.01 \, \text{mol dm}^{-3}$. The results in the table give the percentage reaction at various intervals of time after the start of the reaction. Calculate the second-order rate constant.

| Time/s | 40 | 80 | 120 | 160 | 200 |
|---|---|---|---|---|---|
| Percentage reaction | 37 | 53 | 63 | 69 | 74 |

3.  The following results were obtained for a reaction between A and B.

| Concentrations/mol dm$^{-3}$ | | Initial rate/mol dm$^{-3}$ s$^{-1}$ |
|---|---|---|
| [A] | [B] | |
| 0.5 | 1.0 | 2 |
| 0.5 | 2.0 | 8 |
| 0.5 | 3.0 | 18 |
| 1.0 | 3.0 | 36 |
| 2.0 | 3.0 | 72 |

What is the order of reaction with respect to A and with respect to B? What is the rate equation for the reaction? Calculate the rate constant. State the units in which it is expressed.

## EXERCISE 59    Problems on Radioactive Decay

1.  Plot a graph, using the following figures, to show the radioactive decay of krypton. From the graph, find the half-life.

| Time/minute | 0 | 20 | 40 | 60 | 80 | 100 | 120 |
|---|---|---|---|---|---|---|---|
| Activity/count per second | 100 | 92 | 85 | 78 | 72 | 66 | 61 |

2.  A sample of gold was irradiated in a nuclear reactor. It gave the following results when its radioactivity was measured at various intervals. Plot the results, and deduce the half-life of the radioactive isotope of gold formed.

| Time/hour | 0 | 1 | 5 | 10 | 25 | 50 | 75 | 100 |
|---|---|---|---|---|---|---|---|---|
| Radioactivity/count per minute | 300 | 296 | 285 | 270 | 228 | 175 | 133 | 103 |

3.  A sample of bromine was irradiated in a nuclear reactor. The following results were obtained when the radioactivity was measured after various time intervals. Plot the results, and deduce what you can about the decay of radioactive bromine.

| Time/hour | 0 | 0.1 | 0.2 | 0.5 | 1 | 2 | 5 | 10 | 25 | 50 | 75 | 100 |
|---|---|---|---|---|---|---|---|---|---|---|---|---|
| Radio-activity/count per minute | 500 | 442 | 399 | 320 | 268 | 242 | 225 | 204 | 154 | 95 | 55 | 35 |

4.  A radioactive source, after storing for 42 days, is found to have 1/8th of its original activity. What is the half-life of the radioactive isotope present in the source?

5.  Actinium B has a half-life of 36.0 min. What fraction of the original quantity of actinium remains after: a) 180.0 min, b) 1080 min?

6.  The half-life of carbon-14 is 5580 years. A 10 g sample of carbon prepared from newly cut timber gave a count rate of 2.04 s$^{-1}$. A 10 g sample of carbon from an ancient relic gave a count rate of 1.84 s$^{-1}$. Calculate the age of the relic.

7. A dose of $1.00 \times 10^{-4}$ g of astatine-211 is given to a patient for treatment of cancer of the thyroid gland. How much of this radioactive isotope $(t_{1/2} = 7.21 \text{ h})$ will remain in the body 24 hours later?

8. In 3 hours, the activity of a radioactive compound falls from 1112 c.p.m. to 678 c.p.m. What is the half-life of the isotope?

9. In 30 minutes, the activity of a radioactive source falls from 1173 c.p.m. to 724 c.p.m. What is the half-life of the radioisotope?

10. Tritium has a half-life of 12.3 years. When tritiated water is used in tracer experiments, what percentage of the original activity will remain after: a) 5 years, b) 50 years?

11. The half-life of carbon-14 is 5600 years. A piece of wood from an ancient ship gives a count of 10 counts per minute, while carbon obtained from new wood gives 15 counts per minute. What is the age of the ship?

# EXERCISE 60    Problems on Rates of Reaction

1. A first-order reaction is 50% complete at the end of 30 minutes. What is the value of the rate constant? In how many minutes is reaction 80% complete?

2. A second-order reaction is 60% complete in 1 hour. What is the value of the rate constant?

3. The half-life for the disintegration of bismuth-214 is 19.7 minutes. Calculate the rate constant for the decay in $s^{-1}$.

4. A second-order reaction occurs between reactants originally present at a concentration of $0.1 \text{ mol dm}^{-3}$. The reaction is 10% complete in 30 min. Calculate: a) the rate constant, b) the half-life, and c) the time for 10% completion of the reaction, if the original concentrations of the reactants are $0.01 \text{ mol dm}^{-3}$.

5. The half-life for the radioactive disintegration of bismuth-210 is 5.0 days. Calculate: a) the rate constant in $s^{-1}$, b) the time needed for 0.016 mg of bismuth-210 to decay to 0.001 mg.

6. Hydrogen and iodine combine to form hydrogen iodide. The reaction is first order with respect to hydrogen and first order with respect to iodine. The rate constant is $2.78 \times 10^{-4} \text{ mol dm}^{-3} \text{s}^{-1}$. If the concentrations are $[H_2] = 0.85 \times 10^{-2} \text{ mol dm}^{-3}$, and $[I_2] = 1.25 \times 10^{-2} \text{ mol dm}^{-3}$, what is the initial rate of reaction?

7. The reaction $2NO(g) + Cl_2(g) \longrightarrow 2NOCl(g)$ is third order. The rate constant is $1.7 \times 10^{-5} \text{ dm}^6 \text{mol}^{-2} \text{s}^{-1}$. If the concentrations of the reactants are each $0.20 \text{ mol dm}^{-3}$, what is the initial rate of reaction?

8. The decomposition of ethanal, $CH_3CHO \longrightarrow CH_4 + CO$, can be studied by measuring the increase in pressure of the mixture of gases at various times. Find: a) the order of the reaction, and b) the rate constant.

| Time/s | 0 | 50 | 100 | 150 | 250 | 350 | 500 |
|---|---|---|---|---|---|---|---|
| Pressure/$10^{-4}$ N m$^{-2}$ | 4.82 | 5.74 | 6.38 | 6.82 | 7.44 | 7.84 | 8.24 |

9. A and B react in the gas phase. In experiment 1, a glass vessel was used. In experiment 2, the glass was coated with another material. The results of the two experiments are shown below. Deduce the rate equations for the two experiments. Can you explain how they come to differ?

| | [A]/mol dm$^{-3}$ | [B]/mol dm$^{-3}$ | Initial rate/mol dm$^{-3}$ s$^{-1}$ |
|---|---|---|---|
| Experiment 1 | 0.20 | 0.12 | $2 \times 10^{-3}$ |
| | 0.40 | 0.12 | $8 \times 10^{-3}$ |
| | 0.20 | 0.24 | $4 \times 10^{-3}$ |
| Experiment 2 | 0.20 | 0.12 | $2 \times 10^{-3}$ |
| | 0.40 | 0.24 | $8 \times 10^{-3}$ |
| | 0.80 | 0.24 | $32 \times 10^{-3}$ |

10. Hydrogen peroxide decomposes in aqueous solution:

$$2H_2O_2(aq) \longrightarrow 2H_2O(l) + O_2(g)$$

The following results show how the rate of decomposition varies with the initial hydrogen peroxide concentration:

| Rate/mol dm$^{-3}$ s$^{-1}$ | [H$_2$O$_2$]/mol dm$^{-3}$ |
|---|---|
| $3.64 \times 10^{-5}$ | 0.05 |
| $7.41 \times 10^{-5}$ | 0.10 |
| $1.51 \times 10^{-4}$ | 0.20 |
| $2.21 \times 10^{-4}$ | 0.30 |

Plot the rate of decomposition against the concentration. Deduce: a) the order of the reaction, and b) the rate constant under the conditions of the experiment.

*11. A gaseous compound C dimerises to form D:

$$2C(g) \longrightarrow D(g)$$

There is a drop in pressure as two moles of C form one mole of D. From the values of pressure and time given below, find: a) the order of reaction, and b) the rate constant in terms of N m$^{-2}$ s$^{-1}$.

| Time/s | 0 | 5 | 10 | 15 | 20 | 25 | 30 |
|---|---|---|---|---|---|---|---|
| Pressure/kPa | 100 | 89 | 82 | 77 | 73 | 70.5 | 68 |

**\*EXERCISE 61**     Problems on Activation Energy

1. The reaction

$$2A(g) \; + \; B(g) \; \longrightarrow \; C(g)$$

was studied at a number of temperatures, and the following results were obtained:

| Temperature/°C | 12 | 60 | 112 | 203 | 292 |
|---|---|---|---|---|---|
| Rate constant/$dm^6 \, mol^{-2} s^{-1}$ | 2.34 | 13.2 | 52.5 | 316 | 1000 |

Calculate the activation energy.

2. The following results were obtained in the study of the effect of temperature on the rate of a chemical reaction.

| Temperature/K | 2000 | 1136 | 909 | 676 | 540 |
|---|---|---|---|---|---|
| Rate constant/$dm^3 \, mol^{-1} s^{-1}$ | 1.00 | 0.316 | 0.158 | 0.050 | 0.0158 |

From a plot of $\lg k$ against $1/T$, find the activation energy for the reaction.

3. The activation energy for a reaction is $60 \, kJ \, mol^{-1}$. How much faster will the reaction be at $30 \, °C$ than at $20 \, °C$?

4. If a reaction is 2.70 times faster at $40 \, °C$ than at $30 \, °C$, what is its activation energy?

5. Use the following results to calculate the energy of activation of the reaction:

| Temperature/K | 800 | 625 | 472 | 377 | 333 |
|---|---|---|---|---|---|
| Rate constant/$dm^3 \, mol^{-1} s^{-1}$ | 0.562 | 0.316 | 0.132 | 0.0562 | 0.0316 |


**EXERCISE 62**     Questions from A-level Papers

1. Write down a general equation for the rate of a reaction and define, or explain, the terms used.

An investigation of the decomposition of nitrogen pentoxide dissolved in tetrachloromethane (carbon tetrachloride) at $318 \, K$

$$N_2O_5 \; \longrightarrow \; N_2O_4 \; + \; \tfrac{1}{2}O_2$$
$$N_2O_4 \; \rightleftharpoons \; 2NO_2$$

gave the following information:

| Time/s | 0 | 250 | 500 | 750 | 1000 |
|---|---|---|---|---|---|
| $N_2O_5$ concentration/$mol \, dm^{-3}$ | 2.33 | 1.95 | 1.68 | 1.42 | 1.25 |

| Time/s | 1500 | 2000 | 2500 | 3000 |
|---|---|---|---|---|
| $N_2O_5$ concentration/$mol \, dm^{-3}$ | 0.95 | 0.70 | 0.50 | 0.35 |

(*Use graph paper.*)

a) Plot a graph of concentration of nitrogen pentoxide remaining against time.
   (*It is recommended that the time axis be extended to 4000s.*)

b) Measure the gradients of the curve at the following times, state the units in which these are measured and explain briefly what you have done:

   Time/s     500   1000   1500   2000   2500

   What information do you now possess?

c) Deduce the 1st order rate constant for the decomposition under these conditions by plotting a graph of the measured gradients against other appropriate data. State any assumption you are making. Give the units in which the rate constant is measured.

d) From the first graph, measure the time intervals for any concentration of nitrogen pentoxide to be successively halved (e.g. 2.0, 1.0, 0.5 mol dm$^{-3}$ . . .) and average your result. Comment on what you have found.
                                                                (SUJB80)

2. What do you understand by the term *order* when applied to a chemical reaction?

   Given data on the variation of concentration with time, give *two* methods which may be used to determine the order of that reaction.

   Propanone (acetone) reacts with iodine in the presence of acid according to the equation

   $$CH_3COCH_3 + I_2 \longrightarrow CH_3COCH_2I + H^+ + I^-$$

   In an experiment, 5 cm$^3$ of propanone, 10 cm$^3$ of sulphuric acid of concentration 1.0 mol dm$^{-3}$ and 10 cm$^3$ of a solution of iodine of concentration $1.0 \times 10^{-3}$ mol dm$^{-3}$ were mixed, made up to 100 cm$^3$ with distilled water and placed in a thermostat. Every five minutes samples were removed and titrated against a solution of sodium thiosulphate. The experiment was repeated but using 20 cm$^3$ of sulphuric acid in the reaction mixture. The following results were obtained.

| Time/minute | | 5 | 10 | 15 | 20 | 25 |
|---|---|---|---|---|---|---|
| Titre/cm$^3$ | for 10 cm$^3$ acid | 18.5 | 17.0 | 15.5 | 14.0 | 12.5 |
| | for 20 cm$^3$ acid | 17.0 | 14.0 | 11.0 | 8.0 | 5.0 |

   From plots of titre against time, determine the order of reaction with respect to iodine and with respect to [H$^+$].

   How would you determine the order with respect to propanone?
                                                                (O & C80)

3. Explain the meaning of *each* of the following terms: *law of mass action*; *order of reaction*; *rate constant*; *rate expression*.

   Briefly describe how you could follow the course of a chemical reaction of your choice.

For the hydrolysis of methyl ethanoate (acetate) in aqueous hydrochloric acid solutions, the following initial rates of reaction were measured at 30 °C.

| *Initial rate/* $mol\,dm^{-3}\,min^{-1}$ | *Initial concentrations/* $mol\,dm^{-3}$ | |
|---|---|---|
| | Methyl ethanoate | HCl |
| 0.001 | 1.0 | 0.5 |
| 0.002 | 1.0 | 1.0 |
| 0.001 | 0.5 | 1.0 |

Deduce the order of the reaction with respect to both methyl ethanoate and hydrogen ions. Explain the role of hydrogen ions in the reaction.

In a further experiment one mole amounts of both methanol and ethanoic (acetic) acid were mixed giving a total volume of 100 cm³. When the reaction was complete 0.35 moles of each remained. Make appropriate deductions from this observation.  (WJEC80)

4. a) The decomposition of dinitrogen oxide, $N_2O$, to nitrogen, $N_2$, and oxygen, $O_2$, is a first order reaction.

Use this example to answer the following:
   i) Write a rate equation for the decomposition and explain what is meant by the rate of the reaction, the order of the reaction and the rate constant.
   ii) Given that the half life of the reaction at 1000 K is 60 h, calculate the following.
      1. The rate constant, giving the units.
      2. The total gas pressure after 60 h at 1000 K, the initial pressure being $2 \times 10^5$ Pa, assuming the volume remains constant.

b) State, and explain, how each of the following affects the rate of a chemical reaction: activation energy, temperature, pressure and catalyst.

c) Given the following information about dinitrogen oxide:

| $\Delta H_F^{\ominus}/kJ\,mol^{-1}$ | $T\Delta S^{\ominus}/kJ\,mol^{-1}$ |
|---|---|
| +81.6 | −22.4 |

give an explanation of why the nitrogen and oxygen of the air would not be expected to combine under standard conditions to give dinitrogen oxide.  (AEB84)

5. a) The rates of chemical reactions may be determined by measuring changes in, for example, conductivity, pressure, colour or volume of the reaction mixture. Describe an experiment by which the rate of a chemical reaction of your choice may be investigated, showing how the measurements you make are related to the concentrations of the reactant.

b) The following data refer to the first-order decomposition of a compound $X$:

| Time/seconds | Concentration of $X$/ arbitrary units |
|---|---|
| 0 | 23.3 |
| 500 | 17.0 |
| 1200 | 11.2 |
| 2000 | 6.90 |
| 2800 | 4.30 |

By plotting a suitable graph, determine the half-life of the reaction.

c) Discuss the principal features of the 'collision theory' of reaction rates.                                                                (L83)

6. Explain the terms *half-life, rate-constant* and *order of reaction*.

The base-catalysed decomposition of a substance $A$ in aqueous solution was followed by measuring the volume increase of the solution at constant temperature. Some results were:

*Run 1.*    $[OH^-] = 0.500 \, mol \, dm^{-3}$

| time/min | 0 | 5 | 10 | 15 | 20 | ∞ |
|---|---|---|---|---|---|---|
| $\Delta V_t/cm^3$ | 0 | 0.5689 | 0.7782 | 0.8552 | 0.8835 | 0.9000 |

*Run 2.*    $[OH^-] = 0.100 \, mol \, dm^{-3}$

| time/min | 0 | | | | 20 | ∞ |
|---|---|---|---|---|---|---|
| $\Delta V_t/cm^3$ | 0 | | | | 0.4956 | 0.9000 |

Given that the concentration of unreacted $A$ is proportional to $\Delta V_\infty - \Delta V_t$, plot a suitable graph for Run 1 to show that the reaction is first-order with respect to $A$.

Calculate the first-order rate constants for both Runs 1 and 2.

Hence find the order of reaction with respect to $OH^-$.

What is the overall order of the reaction and the value of the overall rate constant for the reaction (give units)?                    (O&C83)

7. What is the effect of an increase in a) concentration, b) temperature on the rate of a reaction? Explain why these changes have such an effect and why a small increase in temperature has a much greater effect than a small increase in concentration.

Describe an experiment to measure the rate of a reaction of your choice and show how the results would be treated to enable the rate to be determined.

The data refer to the reaction

$$2A(g) \ + \ B(g) \longrightarrow C(g)$$

in a closed vessel at constant temperature.

| Experiment | Initial concentration/$mol\,l^{-1}$ | | Initial rate/$mol\,l^{-1}\,s^{-1}$ |
|---|---|---|---|
| | A | B | |
| 1 | 0.5 | 0.5 | $2 \times 10^{-4}$ |
| 2 | 1.0 | 0.5 | $8 \times 10^{-4}$ |
| 3 | 1.0 | 1.0 | $8 \times 10^{-4}$ |
| 4 | 1.5 | 1.5 | $18 \times 10^{-4}$ |

Deduce:

a) the rate expression for the reaction,

b) the value of the rate constant,

c) the initial rate if the initial concentration of A is $3.0\,mol\,l^{-1}$ and that of B is $2.5\,mol\,l^{-1}$,

d) the rate if experiment 2 were repeated with the same initial amounts of A and B in a vessel with twice the volume.          (JMB82)

8. a) Explain the terms *velocity constant* and *order of reaction*.

b) In an investigation of the recombination of $X$ atoms to give $X_2$ molecules in the gas phase in the presence of argon, that is the reaction

$$X + X + Ar = X_2 + Ar,$$

the following data were obtained.

With the concentration of argon fixed at $1.0 \times 10^{-3}\,mol\,dm^{-3}$:

| Initial concentration, $[X]$ in $mol\,dm^{-3}$ | Initial rate, $\dfrac{d[X_2]}{dt}$ in $mol\,dm^{-3}\,s^{-1}$ |
|---|---|
| $1.0 \times 10^{-5}$ | $8.70 \times 10^{-4}$ |
| $2.0 \times 10^{-5}$ | $3.48 \times 10^{-3}$ |
| $4.0 \times 10^{-5}$ | $1.39 \times 10^{-2}$ |

With the concentration of $X$ atoms fixed at $1.0 \times 10^{-5}\,mol\,dm^{-3}$:

| Initial concentration, $[Ar]$ in $mol\,dm^{-3}$ | Initial rate $\dfrac{d[X_2]}{dt}$ in $mol\,dm^{-3}\,s^{-1}$ |
|---|---|
| $1.0 \times 10^{-3}$ | $8.70 \times 10^{-4}$ |
| $5.0 \times 10^{-3}$ | $4.35 \times 10^{-3}$ |
| $1.0 \times 10^{-2}$ | $8.69 \times 10^{-3}$ |

Find the order of reaction with respect to i) $[X]$, ii) $[Ar]$ and hence determine the overall velocity constant for the formation of $X_2$ molecules.

c) Write a short account of the photochemical reaction between methane and chlorine.                                                    (O85)

9. a) Balance the following nuclear equations indicating clearly the type of radiation produced in each.

   i) $^{238}_{92}U \longrightarrow ^{234}_{90}Th +$

   ii) $^{40}_{19}K \longrightarrow ^{40}_{20}Ca +$

   iii) $^{1}_{0}n + ^{14}_{7}N \longrightarrow ^{14}_{6}C +$

   b) In a) ii), how many grams of potassium are required to produce 2.0 g of calcium?

   c) The following measurements relate to the decay of $^{40}_{19}K$ in a) ii) above.

   | Percentage of $^{40}_{19}K$ left | Time/years |
   |---|---|
   | 100 | 0 |
   | 80 | $0.45 \times 10^9$ |
   | 60 | $0.95 \times 10^9$ |
   | 40 | $1.75 \times 10^9$ |
   | 20 | $3.00 \times 10^9$ |
   | 13 | $4.00 \times 10^9$ |

   i) Plot these measurements on a copy of the grid shown below.

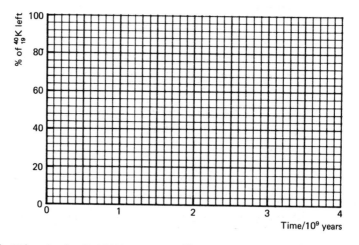

   ii) What is the half-life, $t_{1/2}$, of $^{40}_{19}K$?

   iii) What is the order of reaction of this decay?

   d) In a sample of the mineral mica whose age was to be determined, the calcium produced as a result of decay was estimated to be $128.5 \times 10^{15}$ atoms, while the $^{40}_{19}K$ content was $77.1 \times 10^{15}$ atoms ($n$).

i) What number of $^{40}_{19}K$ atoms was present originally ($n_0$)?

ii) The age of mica, $t$, can be calculated from the equation

$$\log_{10} n = \log_{10} n_0 - \frac{0.03t}{t_{1/2}}$$

where $n$ is the $^{40}_{19}K$ content at time $t$, $n_0$ is the original $^{40}_{19}K$ content, and $t_{1/2}$ is the half-life which you have found in c) iii). Calculate the age, $t$ of the mica sample.                    (O85)

**10.** The rate of the reaction between gaseous elements $X_2$ and $Y_2$

$$X_2(g) + Y_2(g) \longrightarrow 2XY(g)$$

has been found to obey the following equation:

$$\frac{-d[X_2]}{dt} = k[X_2][Y_2]$$

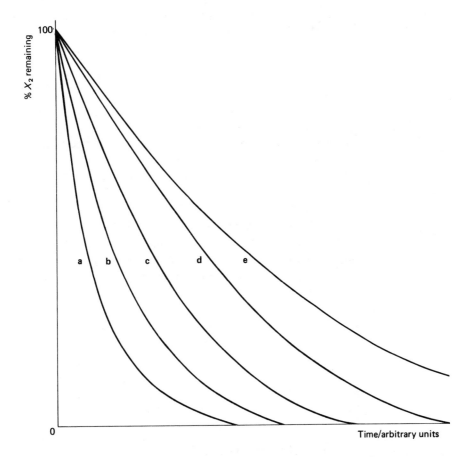

The curves show the percentage of $X_2$ remaining at different times under the following sets of initial experimental conditions, where partial pressures are measured in the non-SI unit, atmosphere.

| | $X_2$/atm | $Y_2$/atm | Temperature/ °C | Catalyst present |
|---|---|---|---|---|
| I | 0.5 | 0.5 | 25 | NO |
| II | 0.5 | 0.5 | 100 | NO |
| III | 0.1 | 0.1 | 25 | NO |
| IV | 0.5 | 0.5 | 25 | YES |
| V | 0.5 | 1.0 | 25 | NO |

Identify which of the curves a, b, c, d, e corresponds to the conditions described by I.

(NI83)

11. a) Mercury(II) chloride, $HgCl_2$, reacts with oxalate (ethanedioate) ions, $C_2O_4^{2-}$, in solution with the precipitaiton of $Hg_2Cl_2$.

$$2HgCl_2(aq) + C_2O_4^{2-}(aq) \longrightarrow 2Cl^-(aq) + 2CO_2(g)$$
$$+ Hg_2Cl_2(s)$$

Initial rates at 373 K for solutions of the concentration shown and expressed as $mol\,dm^{-3}\,min^{-1}$ of $HgCl_2$ reacted are given below.

| Experiment | $HgCl_2$/ $mol\,dm^{-3}$ | $K_2C_2O_4$/ $mol\,dm^{-3}$ | Rate $\times 10^4$/ $mol\,dm^{-3}\,min^{-1}$ |
|---|---|---|---|
| 1 | 0.0836 | 0.202 | 0.52 |
| 2 | 0.0836 | 0.404 | 2.08 |
| 3 | 0.0418 | 0.404 | 1.06 |

i) Write down the rate equation for the reaction, expressing the rate in differential form.

ii) Deduce the order of reaction with respect to each of the reactants.

iii) Calculate the value of the rate constant and state its units, if any.

iv) Compare the information expressed by the equation for the reaction with what has been found and comment.

b) Sketch curves of reactant concentration against time for the decomposition of a gas $A$ to show the relative rates for both first and second order decompositions. Label the graphs with explanatory notes.

c) What do you understand by the term *half-life* of a reaction? How would the half-life of the decomposition of the gas $A$ in b) depend on the initial concentration of $A$ according to the different orders?

d) The rate at which hydrogen peroxide decomposes may be followed by titration against potassium manganate(VII) solution (potassium permanganate) under acid conditions, equal volumes being withdrawn at specified times and titrated:

| Time/min | 0 | 10 | 20 | 30 |
|---|---|---|---|---|
| Volume of $KMnO_4$(aq)/$cm^3$ | 45.6 | 27.6 | 16.5 | 10.0 |

Determine the half-life of the decomposition, deduce the rate equation and explain how you might calculate the rate constant.

(SUJB84)

12. a) State *five* factors which may affect the rate of a chemical reaction.

 b) The reaction between iodide ions and persulphate (peroxodisulphate(VI)) ions in aqueous solution may be represented by the equation:

$$2I^-(aq) + S_2O_8{}^{2-}(aq) \longrightarrow I_2(aq) + 2SO_4{}^{2-}(aq)$$

Outline a method of investigating the rate of this reaction so that the order of the reaction with respect to persulphate ions may be determined.

 c) The hydrolysis of methyl ethanoate may be represented by the equation:

$$CH_3COOCH_3(l) + H_2O(l) \rightleftharpoons CH_3COOH(aq) + CH_3OH(aq)$$

When the hydrolysis was carried out in the presence of aqueous hydrochloric acid in a constant temperature bath, the following results were obtained.

| Time/s $\times 10^4$ | Concentration of ester/mol dm$^{-3}$ |
|---|---|
| 0 | 0.24 |
| 0.36 | 0.156 |
| 0.72 | 0.104 |
| 1.08 | 0.068 |
| 1.44 | 0.045 |

 i) State the reasons for using the hydrochloric acid and the constant temperature bath.
 ii) Plot a graph of $ln[ester]$ against *time*.
 iii) By reference to the graph, show that the reaction is first order with respect to the ester.
 iv) Given the equation $\dfrac{d[ester]}{dt} = -k[ester]$, use your graph to determine a value for $k$ and state its units. (AEB85)

13. a) Explain what is meant by the *activation energy* of a reaction, and show how the concept provides a qualitative explanation of the effect of heat on the velocity of a chemical reaction.

 b) The table below gives values for the velocity constant, $k$, of the reaction between potassium hydroxide and bromoethane in ethanol at a series of temperatures, $T$.

| $k$ in dm$^3$ mol$^{-1}$s$^{-1}$ | $T$ in K |
|---|---|
| 0.182 | 305.0 |
| 0.466 | 313.0 |
| 1.35 | 323.1 |
| 3.31 | 332.7 |
| 10.2 | 343.6 |
| 22.6 | 353.0 |

i) Use these data to obtain, by a graphical method, a value for the activation energy of the reaction. ($R = 8.31 \, J \, K^{-1} \, mol^{-1}$.)

ii) Calculate the initial rate of reaction in a solution at 318.0 K when the concentration of both potassium hydroxide and bromoethane are 0.100 mol dm$^{-3}$. (O82,S)

**\*14. a)** What is a photochemical reaction? Outline the essential steps in the photochemical gas-phase reaction of hydrogen with chlorine. Given that the energy content of radiation is directly proportional to its frequency, and that the constant of proportionality is the Planck constant, what can be deduced from the observation that this reaction proceeds smoothly in diffuse light of frequency 6.25 $\times$ 10$^{14}$ Hz, but that there is no reaction in light, however intense, of frequency 5.90 $\times$ 10$^{14}$ Hz?

**b)** Ethylamine decomposes at high temperature according to the following equation.

$$C_2H_5NH_2(g) \longrightarrow C_2H_4(g) + NH_3(g)$$

A quantity of ethylamine was introduced into a vessel of fixed volume at high temperature. The initial pressure, $p_0$, in the vessel at time $t = 0$ was found to be 7.33 kPa. The reaction was followed by observing the total pressure, $p_t$, during the decomposition, with the following results

| $t/s$ | 120 | 360 | 600 |
|-------|------|------|------|
| $p_t/kPa$ | 8.53 | 10.5 | 11.8 |

Write an expression for the pressure, $p_a$, of ethylamine after a fraction $x$ of the initial ethylamine has decomposed. Express $x$ in terms of $p_t$ and $p_0$ only.

It is thought that this reaction follows first-order kinetics, for which the rate law is $\ln p_a = -kt + constant$. Draw up a table giving values of $p_a$ at different times, and such other data as you will need to plot a graph from which you can

i) confirm that the reaction is first-order with respect to the pressure of ethylamine and

ii) obtain the value of the rate constant, $k$.

(Planck constant $h = 6.626 \times 10^{-34}$ J s, Avogadro constant $N_A = 6.022 \times 10^{23}$ mol$^{-1}$.) (JMB85,S)

**15.** Explain what is meant by

a) the velocity of a reaction,

b) the velocity constant for a reaction and

c) the order of a reaction.

In an investigation of the hydrolysis of ethyl ethanoate at 294 K the kinetic analysis was simplified by initially mixing equal volumes of

0.04 M aqueous solutions of ethyl ethanoate and sodium hydroxide. Integration of the appropriate rate expression leads to the equation

$$\frac{1}{c} = kt + \text{constant},$$

where $k$ is the velocity constant and $c$ is the concentration in mol dm$^{-3}$ at time $t$ (measured in seconds) of hydroxide ion.

If the hydroxide ion concentration 300 seconds after the start of the hydrolysis is 0.0128 mol dm$^{-3}$

    i) calculate the value of $k$, and

    ii) deduce the overall order for the reaction.

Explain briefly how measurements of $k$ at a series of temperatures would enable a value for the enthalpy of activation to be determined.

                                           (O81,S)

16. a) Radioactive isotopes may emit $\alpha$- or $\beta$-particles or $\gamma$-rays. What is the nature of these three types of emission and how can their differing properties be used to detect and identify them? (Technical details of equipment are not required.)

    b)   i) Explain the principles involved in the method of radio-carbon dating.

        ii) What assumptions have to be made in applying this method?

       iii) A sample of carbon prepared from a recently felled young tree showed 0.255 disintegrations s$^{-1}$g$^{-1}$. Another sample of carbon prepared from a wooden beam removed from an ancient Egyptian tomb showed 0.143 disintegrations s$^{-1}$g$^{-1}$. What is the age of the beam? (The half life of $^{14}$C is 5730 years.)

       iv) Why is this method inapplicable to the dating of a fossilised tree from a coal seam?

        v) Suggest a procedure for obtaining a sample of pure carbon from a piece of wood.               (O86,S)

# 11 Equilibria

## CHEMICAL EQUILIBRIUM

An example of a reversible reaction between gases is the reaction between hydrogen and iodine to form hydrogen iodide:

$$H_2(g) + I_2(g) \rightleftharpoons 2HI(g)$$

If the reaction takes place in a closed vessel, the combination of hydrogen and iodine gradually slows down as the concentrations of these gases decrease. At first, there is very little decomposition of hydrogen iodide into hydrogen and iodine, but, as the concentration of hydrogen iodide increases, the rate of decomposition of hydrogen iodide into hydrogen and iodine increases until the rates of the forward and reverse reactions are equal, and the concentration of each species is constant.

An example of a reversible reaction which takes place in solution is the reaction between ethanoic acid and ethanol to form ethyl ethanoate:

$$CH_3CO_2H(l) + C_2H_5OH(l) \rightleftharpoons CH_3CO_2C_2H_5(l) + H_2O(l)$$

As the concentrations of ester and water increase, the reverse reaction — hydrolysis of the ester to form the acid and alcohol — speeds up. At equilibrium, the rate of the forward reaction is equal to the rate of the reverse reaction. Esterification is catalysed by inorganic acids. The presence of a catalyst speeds up the rate at which equilibrium is established.

## THE EQUILIBRIUM LAW

If a reversible reaction is allowed to reach equilibrium, it is found that the product of the concentrations of the products divided by the product of the concentrations of the reactants has a constant value at a particular temperature. In the esterification reaction,

$$CH_3CO_2H(l) + C_2H_5OH(l) \rightleftharpoons CH_3CO_2C_2H_5(l) + H_2O(l)$$

it is found that

$$\frac{[CH_3CO_2C_2H_5]\,[H_2O]}{[CH_3CO_2H]\,[C_2H_5OH]} = K_c$$

where $K_c$ is the equilibrium constant for the reaction in terms of concentration. In the reaction between hydrogen and iodine,

$$H_2(g) + I_2(g) \rightleftharpoons 2HI(g)$$

and

$$\frac{[HI]^2}{[H_2]\,[I_2]} = K_c$$

Since this is a reaction between gases, the concentration of each gas can be expressed as a partial pressure. Then,

$$\frac{p_{HI}^2}{p_{H_2} \times p_{I_2}} = K_p$$

$K_p$ is the equilibrium constant in terms of partial pressures. In the reaction between iron and steam,

$$3Fe(s) + 4H_2O(g) \rightleftharpoons Fe_3O_4(s) + 4H_2(g)$$

The equilibrium constant is given by

$$\frac{p_{H_2}^4}{p_{H_2O}^4} = K_p$$

The solids do not appear in the expression. Their vapour pressures are constant (at a constant temperature) as long as there is some of each solid present. These constant vapour pressures are incorporated into the value of the constant $K_p$.

Another type of reaction which reaches an equilibrium position is thermal dissociation. For example, when phosphorus(V) chloride is heated, it dissociates partially to form phosphorus(III) chloride and chlorine:

$$PCl_5(g) \rightleftharpoons PCl_3(g) + Cl_2(g)$$

As explained in Chapter 6 (pp. 86-8) the dissociation increases the number of moles of substance present and causes an increase in volume, or, if the volume is kept constant, an increase in pressure. The result is that the experimental determinations of molar mass give an unexpectedly low value. The degree of dissociation, $\alpha$, can be obtained from the ratio,

$$\frac{\text{Molar mass calculated in the absence of dissociation}}{\text{Experimentally determined molar mass}} = 1 + \alpha$$

Inserting the value for $\alpha$ into the expression for the equilibrium constant, and putting $c$ = initial concentration of $PCl_5$, we get

$$K_c = \frac{[Cl_2][PCl_3]}{[PCl_5]}$$

$$K_c = \frac{\alpha c \times \alpha c}{(1-\alpha)c} = \frac{\alpha^2 c}{1-\alpha}$$

If the total pressure $= P$, the partial pressures of $PCl_3$ and $Cl_2$ are $P\alpha/(1+\alpha)$, the partial pressure of $PCl_5$ is $P(1-\alpha)/(1+\alpha)$, and

$$K_p = \frac{P^2\alpha^2/(1+\alpha)^2}{P(1-\alpha)/(1+\alpha)} = \frac{\alpha^2 P}{1-\alpha^2}$$

**EXAMPLE 1**  1.00 mole of ethanoic acid was allowed to react with: a) 0.50 mole, b) 1.00 mole, c) 2.00 mole, and d) 4.00 mole of ethanol. At equilibrium, the amount of acid remaining was a) 0.58 mole, b) 0.33 mole, c) 0.15 mole and d) 0.07 mole. Calculate the equilibrium constant for the esterification reaction.

**METHOD**  If the original amounts of acid and ethanol are $a$ mol and $b$ mol, then, at equilibrium, the amount of ester formed is $x$ mol, and the amounts of acid and ethanol remaining are $(a - x)$ and $(b - x)$ mol.

$$CH_3CO_2H(l) \ + \ C_2H_5OH(l) \ \rightleftharpoons \ CH_3CO_2C_2H_5(l) \ + \ H_2O(l)$$
$$\quad (a-x) \qquad\qquad (b-x) \qquad\qquad\qquad x \qquad\qquad\quad x$$

Since the equilibrium constant is given by

$$\frac{[CH_3CO_2C_2H_5] \ [H_2O]}{[CH_3CO_2H] \ [C_2H_5OH]} = K_c$$

then     $\dfrac{(x/V)(x/V)}{[(a-x)/V][(b-x)/V]} = K_c$    or    $\dfrac{x^2}{(a-x)(b-x)} = K_c$

In reaction a)

$$(a - x) = 0.58; \quad x = 0.42; \quad (b - x) = 0.08$$

$$\therefore \qquad \frac{(0.42)^2}{0.58 \times 0.08} = K_c = 3.8$$

Substituting the other values of $a$, $b$ and $x$ in the equation gives the following values of $K$:

| $a$ | $b$ | $a - x$ | $x$ | $b - x$ | $K_c$ |
|------|------|------|------|------|------|
| 1.00 | 0.50 | 0.58 | 0.42 | 0.08 | 3.8 |
| 1.00 | 1.00 | 0.33 | 0.67 | 0.33 | 4.1 |
| 1.00 | 2.00 | 0.15 | 0.85 | 1.15 | 4.4 |
| 1.00 | 4.00 | 0.07 | 0.93 | 3.07 | 4.0 |

**ANSWER**  The average value of the equilibrium constant is 4.1.


**EXAMPLE 2**  Calculate the amount of ethyl ethanoate formed when 1 mole of ethanoic acid and 3 moles of ethanol and 3 moles of water are allowed to come to equilibrium. The equilibrium constant for the reaction is 4.0.

**METHOD**  Let the amount of ethyl ethanoate $= x$ mol
Then     Equilibrium amount of acid $= (1 - x)$ mol
　　　　Equilibrium amount of ethanol $= (3 - x)$ mol
　　　　Equilibrium amount of water $= (3 + x)$ mol

Then, since    $\dfrac{[CH_3CO_2C_2H_5]\,[H_2O]}{[CH_3CO_2H]\,[C_2H_5OH]} = K_c = 4.0$

$$\frac{x(3+x)}{(1-x)(3-x)} = 4.0$$

$$3x^2 - 19x + 12 = 0$$

Solving this quadratic equation (see p. 8) gives

$$x = 5.6 \quad \text{or} \quad 0.72$$

The value $x = 5.6$ can be excluded because it is higher than the number of moles of ethanoic acid present initially. The solution $x = 0.72$ must be the practical one.

**ANSWER**    The amount of ethyl ethanoate formed is 0.72 mol.

**EXAMPLE 3**    A mixture of iron and steam is allowed to come to equilibrium at 600 °C. The equilibrium pressures of hydrogen and steam are 3.2 kPa and 2.4 kPa. Calculate the equilibrium constant $K_p$ for the reaction.

**METHOD**    The reaction is

$$3Fe(s) + 4H_2O(g) \rightleftharpoons 4H_2(g) + Fe_3O_4(s)$$

The equilibrium constant is given by

$$K_p = \frac{p_{H_2}{}^4}{p_{H_2O}{}^4}$$

Substituting in this equation gives

$$K_p = \left(\frac{3.2}{2.4}\right)^4$$

**ANSWER**    $K_p = 3.1$

**EXAMPLE 4**    A molar mass determination on dinitrogen tetraoxide, $N_2O_4$, gave a value of $60\,\text{g mol}^{-1}$ at 50 °C and $1.01 \times 10^5$ Pa. Find the equilibrium constant for the dissociation

$$N_2O_4(g) \rightleftharpoons 2NO_2(g)$$

**METHOD**    If the degree of dissociation is $\alpha$, then a total of $1 + \alpha$ moles of particles are formed from 1 mole of $N_2O_4$. $P$ is the total pressure.

$$\frac{\text{Molar mass}}{\text{Experimentally determined molar mass}} = \frac{92}{60} = 1 + \alpha$$

$$\alpha = 0.53$$

Since    $K_p = \dfrac{p_{NO_2}{}^2}{p_{N_2O_4}}$

and $\qquad p_{NO_2} = \dfrac{2\alpha}{1+\alpha} P$ and $p_{N_2O_4} = \dfrac{1-\alpha}{1+\alpha} P$

$$K_p = \dfrac{4\alpha^2}{1-\alpha^2} \times P = \dfrac{4(0.53)^2}{1-(0.53)^2} \times 1.01 \times 10^5 \, Pa$$

**ANSWER**       $K_p = 1.58 \times 10^5 \, Pa$

## EXERCISE 63    Problems on Equilibria

1. Write an expression for the equilibrium constant for the reaction,
   $A + B \rightleftharpoons C + D$:
   a) when A, B, C and D are gases
   b) when A, B, C and D are solutions
   c) when A and C are solids, B and D are gases
   d) when C is a solid, and A, B and D are solutions.

2. In the equilibrium
$$N_2O_4(g) \rightleftharpoons 2NO_2(g)$$
   3.20 g of dinitrogen tetraoxide occupy a volume 1.00 dm³ at $1.00 \times 10^5$ Pa and 25 °C. Calculate: a) the degree of dissociation, and b) the equilibrium constant.

3. Sulphur dichloride dioxide has an apparent molar mass of 75 g mol$^{-1}$ at 400 °C and a pressure of $10^5$ N m$^{-2}$. Calculate the equilibrium constant $K_p$ for the reaction
$$SO_2Cl_2(g) \rightleftharpoons SO_2(g) + Cl_2(g)$$
   at this temperature.

4. The equilibrium constant for the reaction
$$CO(g) + H_2O(g) \rightleftharpoons CO_2(g) + H_2(g)$$
   is 4.00. If 1 mole of CO and 1 mole of H$_2$O are allowed to come to equilibrium, what fraction of the carbon monoxide will remain?

5. 1 dm³ of dinitrogen tetraoxide, $N_2O_4$, weighs 2.50 g at 60 °C and $1.01 \times 10^5$ N m$^{-2}$ pressure. Find a) the degree of dissociation into $NO_2$ and b) the value of $K_p$.

6. Equimolar amounts of hydrogen and iodine are allowed to reach equilibrium:
$$H_2(g) + I_2(g) \rightleftharpoons 2HI(g)$$
   If 80% of the hydrogen can be converted to hydrogen iodide, what is the value of $K_p$ at this temperature?

7. A mixture of 3 moles of hydrogen and 1 mole of nitrogen is allowed to reach equilibrium at a pressure of $5 \times 10^6 \, N \, m^{-2}$. The composition of the gaseous mixture is then 8% $NH_3$, 23% $N_2$, 69% $H_2$ by volume. Calculate $K_p$.

8. a) Hydrogen and iodine react to form hydrogen iodide. When the initial concentrations of both reactants are $0.11 \, mol \, dm^{-3}$, the initial rate of reaction is $2.42 \times 10^{-5} \, mol \, dm^{-3} \, s^{-1}$. What is the rate constant for this second-order reaction?

   b) When 1 mol of hydrogen and 1 mol of iodine are allowed to reach equilibrium at 600 K, the equilibrium mixture contains 1.6 mol hydrogen iodide. Calculate the equilibrium constant for the reaction

$$H_2(g) \; + \; I_2(g) \; \rightleftharpoons \; 2HI(g)$$

9. At a certain temperature the reaction

$$C_2H_5CO_2H(l) \; + \; C_2H_5OH(l) \; \rightleftharpoons \; C_2H_5CO_2C_2H_5(l) \; + \; H_2O(l)$$

   has an equilibrium constant of 6.

   a) Calculate the amount of ester (in moles) that will be formed at equilibrium from 1 mol acid and 1 mol alcohol.

   b) What mass of propanoic acid must react with 46 g ethanol in order to give 51 g ethylpropanoate at equilibrium?

10. The equilibrium, $H_2(g) + Cl_2(g) \; \rightleftharpoons \; 2HCl(g)$ has $K_p = 2.5 \times 10^4$ at 1500 K. What percentage of hydrogen is converted to hydrogen chloride at this temperature in a 1:1 mixture of hydrogen and chlorine?

11. The equilibrium constant for the reaction $H_2(g) + I_2(g) \; \rightleftharpoons \; 2HI(g)$ is 60 at 450 °C. The number of moles of hydrogen iodide in equilibrium with 2 mol of hydrogen and 0.3 mol of iodine at 450 °C is:

   a 1/100     b 1/10     c 6     d 36     e 3

12. The esterification reaction

$$RCO_2H \; + \; R'OH \; \rightleftharpoons \; RCO_2R' \; + \; H_2O$$

   has an equilibrium constant of 10.0 at 25 °C. If 1 mole of ester is dissolved in 5 moles of water, what are the equilibrium amounts of:
   a) acid, b) alcohol, c) ester and d) water?

13. A mixture of 1 mol nitrogen and 3 mol hydrogen is allowed to reach equilibrium at $1.0 \times 10^7 \, Pa$ and 500 °C. The equilibrium mixture contains 20% of ammonia by volume. Calculate the value of $K_p$ for the reaction

$$N_2(g) \; + \; 3H_2(g) \; \rightleftharpoons \; 2NH_3(g)$$

   at 500 °C.

**14.** Sulphur dioxide and oxygen in the ratio 2 mol : 1 mol are allowed to reach equilibrium in the presence of a catalyst, at a pressure of 5 atm. At equilibrium, $\frac{1}{3}$ of the $SO_2$ was converted to $SO_3$. Calculate the equilibrium constant for the reaction

$$2SO_2(g) + O_2(g) \rightleftharpoons 2SO_3(g)$$

**15.** The equilibrium constant $K_p$ for the reaction

$$CO_2(g) + H_2(g) \rightleftharpoons CO(g) + H_2O(g)$$

is 0.72 at 1000 °C. Calculate the composition of the mixture which results when:

a) 0.5 mole $CO_2$ and 0.5 mole $H_2$ are mixed at a pressure of 1 atm and 1000 °C.

b) 5 moles $CO_2$ and 1 mole $H_2$ are mixed at a pressure of 1 atm and 1000 °C.

**16.** The oxidation of sulphur dioxide is a reversible process:

$$2SO_2(g) + O_2(g) \rightleftharpoons 2SO_3(g)$$

Calculate the value of the equilibrium constant, $K_p$, in terms of partial pressures from the following data, which were obtained at 1000 K:

| Partial pressures/N m$^{-2}$ | | |
|---|---|---|
| $p_{SO_2}$ | $p_{O_2}$ | $p_{SO_3}$ |
| 10 000 | 68 800 | 80 100 |

# EXERCISE 64     Questions from A-level Papers

**1.** a) The equilibrium

$$N_2 + 3H_2 \rightleftharpoons 2NH_3$$

was studied by passing a mixture of nitrogen and hydrogen in a fixed molar ratio over an appropriate catalyst and analysing the issuing gas mixture for ammonia.

The following data were obtained when the total pressure was 10.0 atm and the molar ratio of nitrogen to hydrogen was 1 : 3.

| Temperature in ° C | Partial pressure of $NH_3$ in atm |
|---|---|
| 350 | 0.735 |
| 450 | 0.204 |

i) Calculate for each temperature the partial pressures of hydrogen and nitrogen in atmospheres.

ii) Obtain an expression for $K_p$ in terms of partial pressures and hence calculate $K_p$ for the two temperatures.

iii) Deduce the sign of $\Delta H$ for the formation of ammonia.

iv) At $350\,^{\circ}C$ when the total pressure is $50.0$ atm the partial pressure of ammonia at equilibrium is $12.6$ atm. Calculate $K_p$ for these conditions and comment on the result.

b) State without explanation how a catalyst affects
   i) the rate,
   ii) the position of equilibrium,
   iii) the activation energy of a reaction.                 (O86)

2. a) State Le Chatelier's principle.

   b) For the reaction

$$H_2(g) \ + \ I_2(g) \ \rightleftharpoons \ 2HI(g)$$
$$\Delta H^{\ominus} \ = \ 52\,kJ\,mol^{-1}$$

give the expression for the equilibrium constant in terms of partial pressures $(K_p)$.

   c) What would be the effect, if any, on the equilibrium above of
      i) increasing the total pressure,
      ii) increasing the temperature?

   d) The following table gives the concentrations of the components in the above equilibrium at 600 K.

| Concentration of $H_2$ | Concentration of $I_2$ | Concentration of HI |
|---|---|---|
| $1.71 \times 10^{-3}\,mol\,l^{-1}$ | $2.91 \times 10^{-3}\,mol\,l^{-1}$ | $1.65 \times 10^{-2}\,mol\,l^{-1}$ |

Calculate the equilibrium constant in terms of concentrations $(K_c)$ at this temperature.                 (JMB84)

3. The two reversible reactions A and B, given below, are important in the production of substitute natural gas (SNG).

Reaction A    $C(s) \ + \ H_2O(g) \ \rightleftharpoons \ CO(g) \ + \ H_2(g)$

Reaction B    $CO(g) \ + \ 3H_2(g) \ \rightleftharpoons \ CH_4(g) \ + \ H_2O(g)$

a) Give the expression for $K_p$ for each reaction.

b) State Le Chatelier's principle.

c) State, giving explanations, whether the position of equilibrium will shift to favour products or reactants or remain unchanged, when the total pressure of each reaction (at constant volume) is increased for
   i) reaction A,
   ii) reaction B.

d) Consider reaction B when it is studied *at constant total pressure* in the presence and absence of an inert gas.
   i) State whether the partial pressure of each reactant is higher, lower or remains the same when the inert gas is present.

ii) State, giving an explanation, whether the presence of the inert gas produces a greater or lesser yield of product or has no effect on the equilibrium.

e) State the effect on the position of equilibrium of reaction A of the addition of more carbon (C(s)) and give an explanation of your answer.

f) Calculate the degree of conversion (dissociation), $\alpha$, of water vapour into $CO(g)$ and $H_2(g)$ when it is heated with excess carbon until equilibrium is attained at a constant total pressure of 1 atm and a temperature at which $K_p$ for reaction A is $10^{-4}$ atm.

g) Calculate the enthalpy change for reaction B. ($\Delta H_F$ values (kJ mol$^{-1}$) $CO(g)$, $-110.5$; $CH_4(g)$, $-74.8$; $H_2O(g)$, $-242.5$.)

h) The balance of products is determined not only by reaction B but also by the so called 'shift reaction', reaction C.

Reaction C     $CO(g) + H_2O(g) \rightleftharpoons CO_2(g) + H_2(g)$.

($\Delta H_F(CO_2(g))$) $= -393.5$ kJ mol$^{-1}$.)

State whether more of the products given below will be produced, assuming equilibrium conditions, at an operating temperature of 1000 °C (*Lurgi process*) or at 1500 °C (*Winkler process*).

i) $CH_4(g)$ (reaction B),

ii) $CO_2(g)$ (reaction C),

i) For reaction C, a nickel catalyst is often used. If the catalyst is changed (e.g. to a zeolite catalyst), state

i) *one* property of reaction C that will change,

ii) *one* property of reaction C that will not change.     (WJEC83)


4. Ammonium carbamate dissociates on heating above 30 °C, according to the following equation:

$$NH_2COONH_4(s) \rightleftharpoons 2NH_3(g) + CO_2(g)$$

The standard entropy and enthalpy changes, $\Delta S^\ominus$ and $\Delta H^\ominus$, respectively, and the equilibrium constant, $K_p$, for this dissociation are related in the following way:

$$2.303RT \lg K_p = T\Delta S^\ominus - \Delta H^\ominus,$$

where $T$ = temperature/K and $R$, the gas constant = 8.3 J K$^{-1}$ mol$^{-1}$.

a) Suggest a suitable experimental method of determining the dissociation pressure $P$ of ammonium carbamate at 100 °C.

b) Derive an expression for $K_p$ in terms of the dissociation pressure $P$.

c) Calculate $K_p$ at a pressure of 0.2 atm, stating the units.

d) At 330 K, $K_p$ and $\Delta S^{\ominus}$ have values of $7.25 \times 10^{-2}$ and $478\,\text{J}\,\text{K}^{-1}$ $\text{mol}^{-1}$, respectively. Calculate $\Delta H^{\ominus}$ and state the units.

e) Explain how, if at all,
   i) $P$, and
   ii) $(\Delta H^{\ominus} - T\Delta S^{\ominus})$ will change as the temperature is increased.
   (L82,S)

**5.** a) Describe briefly but with essential detail how you would determine in the laboratory the equilibrium constant for the reaction

$$CH_3COOH + C_2H_5OH \rightleftharpoons CH_3COOC_2H_5 + H_2O$$

at 25 °C.

b) For the decomposition of steam as represented by the equation

$$2H_2O(g) \rightleftharpoons 2H_2(g) + O_2(g)$$

obtain expressions for the partial pressures of steam, hydrogen and oxygen at equilibrium in terms of $\alpha$, the fraction of steam decomposed, and $P$ the total pressure.

Hence show that at a given temperature the equilibrium constant $K_p$ is given by

$$K_p = \frac{\alpha^3 P}{(2 + \alpha)(1 - \alpha)^2}.$$

At 2150 K and a pressure of $1.01 \times 10^5\,\text{N}\,\text{m}^{-2}$, $\alpha = 0.012$. Calculate $K_p$ at this temperature. (O85)

**6.** a) Describe, giving full practical details, an experiment to determine the dissociation constant $(K_a)$ of a weak monobasic acid and explain the basis of the method.

b) Discuss the way in which you would expect the magnitude of $K_a$ to vary with
   i) the concentration of the acid,
   ii) the temperature.

c) Ethanoic (*acetic*) acid reacts with ethanol as follows:

$$CH_3COOH + CH_3CH_2OH \rightleftharpoons CH_3COOCH_2CH_3 + H_2O$$

6.0 g of ethanoic acid and 4.6 g of ethanol were added to 4.4 g of ethyl ethanoate (*acetate*) and the mixture was allowed to reach equilibrium. It was found that 0.04 mol of the acid was present in the equilibrium mixture. Calculate the equilibrium constant for the reaction. (JMB83)

**7.** a) The following relate to a sample of gaseous dinitrogen tetraoxide in a sealed tube.

| Temperature/K | Pressure/atm | % of $N_2O_4$ dissociated to $NO_2$ |
|---|---|---|
| $T_1$ | $p_1$ | 0 |
| $T_2$ | $p_2$ | 50 |
| $T_3$ | $p_3$ | 100 |

   i) Derive an expression relating $p_1$ and $p_3$.
   ii) Derive an expression relating $K_p$ (at temperature $T_2$) to $p_2$.
   iii) Show by a sketch how the pressure in the tube varies with temperature.

   b) A liquid is to be exported in sealed drums which may burst if the internal pressure exceeds 2 atm. By using the data below consider whether it is safe to use these drums.
   i) The filling is carried out in the open, the drums being filled to 90% capacity. Filling may take place at temperatures as low as $0\,°C$, but during the journey the temperature may reach $30\,°C$.
   ii) The liquid expands by 0.1% per $°C$.
   iii) The vapour pressure of the liquid is 0.2 atm at $10\,°C$ and it doubles for every $10\,°C$ rise.                                    (C84,S)

**8.** What is meant by dynamic equilibrium? Using the Haber synthesis of ammonia from nitrogen and hydrogen as your example, discuss the factors which influence the position of such an equilibrium and the value of the equilibrium constant.

   a) Solid $NH_4HS$ dissociates on heating to give ammonia and hydrogen sulphide. The equilibrium was studied by adding various amounts of the two gases to the system, at constant pressure and volume, and measuring the equilibrium pressure of each gas. Show that the following results are consistent with the idea of equilibrium.

| $p_{NH_3}$/mm Hg | 189.0 | 211.8 | 250.4 | 263.5 |
|---|---|---|---|---|
| $p_{H_2S}$/mm Hg | 331.1 | 296.2 | 250.4 | 237.8 |

   b) 0.01 mole of phosphorus pentachloride was placed in a 1 litre vessel. At $210\,°C$ it was found to be 58.2% dissociated into phosphorus trichloride and chlorine. Calculate the equilibrium constant $K_c$ for the dissociation at this temperature.                                    (JMB82)

# 12 Organic Chemistry

All the techniques you need to enable you to tackle problems in organic chemistry have been covered in Chapter 2 in the sections on empirical formulae, calculations based on chemical equations and reacting volumes of gases, and in Chapter 3 on volumetric analysis.

Numerical problems in organic chemistry give you some quantitative data and ask you to use it in conjunction with your knowledge of the reactions of organic compounds. There is no set pattern for tackling such problems. They are solved by a combination of calculation, familiarity with the reactions of the compounds involved and logic. The following examples and problems will show you what to expect.

**EXAMPLE 1** When 0.2500 g of a hydrocarbon X burns in a stream of oxygen, it forms 0.7860 g of carbon dioxide and 0.3210 g of water. When 0.2500 g of X is vaporised, the volume which it occupies (corrected to s.t.p.) is 80.0 cm³. Deduce the molecular formula of X.

**METHOD** X burns to form carbon dioxide and water.

$$\text{Mass of C in 0.7860 g of } CO_2 = \frac{12.0}{44.0} \times 0.7860 = 0.2143 \text{ g}$$

$$\text{Mass of H in 0.3210 g of } H_2O = \frac{2.02}{18.0} \times 0.3210 = 0.0360 \text{ g}$$

Therefore 0.2500 g of X contains 0.2143 g of C and 0.0360 g of H

$$\text{These masses give the molar ratio for C:H of } \frac{0.2143}{12.0} \text{ to } \frac{0.0360}{1.01}$$

$$= 0.0178 \text{ to } 0.0360 = 1 \text{ to } 2$$

Thus, the empirical formula is $CH_2$.

Since 80.0 cm³ is the volume occupied by 0.2500 g of X,

$$22.4 \text{ dm}^3 \text{ is occupied by } \frac{22.4}{80.0 \times 10^{-3}} \times 0.2500 \text{ g of X} = 70.0 \text{ g of X}$$

The formula mass of $CH_2$ is 14. To give a molar mass of 70.0 g mol⁻¹, the empirical formula must be multiplied by 5. Therefore:

**ANSWER** The molecular formula is $C_5H_{10}$.

**EXAMPLE 2** An organic liquid, P, contains 52.2% carbon, 13.0% hydrogen and 34.8% oxygen by mass. Mild oxidation converts P to Q, and, on further oxidation, R is formed. P and Q react together in the presence of anhydrous calcium chloride to form S, which has a molecular formula of $C_6H_{14}O_2$. P and R react to give T, which has a molecular formula of $C_4H_8O_2$. Identify compounds P to T, and explain the reactions involved.

**METHOD**    First, calculate the empirical formula of P. This comes to $C_2H_6O$. This must be the molecular formula also as P and R combine to form T, which has 4C in the molecule. Since P contains one oxygen atom, it is an alcohol, an aldehyde, a ketone or possibly an ether. Other classes of compounds are ruled out by the absence of nitrogen and halogens. Oxidation proceeds in two stages, evidence that P is probably an alcohol, being oxidised first to an aldehyde and then to an acid.

According to the formulae, P would be $C_2H_5OH$, and Q would be $CH_3CHO$, and R would be $CH_3CO_2H$. The reaction between P and Q fits in with this theory as

**ANSWER**    $2C_2H_5OH(l) + CH_3CHO(l) \longrightarrow CH_3CH(OC_2H_5)_2(l) + H_2O(l)$

The reaction between P and R to form T is thus

**ANSWER**    $CH_3CO_2H(l) + C_2H_5OH(l) \longrightarrow CH_3CO_2C_2H_5(l) + H_2O(l)$

The molecular formula $C_6H_{14}O_2$ for S fits $CH_3CH(OC_2H_5)_2$, and the molecular formula $C_4H_8O_2$ for T fits $CH_3CO_2C_2H_5$.

**ANSWER**    P = $C_2H_5OH$, ethanol; Q = $CH_3CHO$, ethanal; R = $CH_3CO_2H$, ethanoic acid;  S = $CH_3CH(OC_2H_5)_2$, 1,1-diethoxyethane;  T = $CH_3CO_2C_2H_5$, ethyl ethanoate.

**EXAMPLE 3**    X is a liquid containing 31.5% by mass of C, 5.3% H and 63.2% O. An aqueous solution of X liberates carbon dioxide from sodium carbonate, with the formation of a solution from which a substance Y of formula $C_2H_3O_3Na$ can be obtained. X reacts with phosphorus(V) chloride to give hydrogen chloride and a compound, Z, of molecular formula $C_2H_2OCl_2$. Identify X, Y and Z. Write equations for the two reactions, X $\longrightarrow$ Y and X $\longrightarrow$ Z.

**METHOD**    The empirical formula of X is easily shown to be $C_2H_4O_3$. As X liberates carbon dioxide from a carbonate, it must be an acid. Taking $CO_2H$ from $C_2H_4O_3$ leaves $CH_3O$ as the formula for the rest of the molecule. This could be $CH_2OH$, making X $HOCH_2CO_2H$.

In the reaction with $PCl_5$, $C_2H_4O_3$ is converted into $C_2H_2OCl_2$. This would fit in with two hydroxyl groups being replaced by two chlorine atoms. This would be the case for the reaction

**ANSWER**    $HOCH_2CO_2H(l) + 2PCl_5(l) \longrightarrow ClCH_2COCl(l) + 2POCl_3(l) + 2HCl(g)$

If X is $HOCH_2CO_2H$, its reaction with sodium carbonate has an equation

**ANSWER**    $2HOCH_2CO_2H(aq) + Na_2CO_3(s) \longrightarrow 2HOCH_2CO_2Na(aq) + CO_2(g) + H_2O(l)$

and Y is $HOCH_2CO_2Na$.

**ANSWER**  All the information agrees with X = HOCH$_2$CO$_2$H, hydroxyethanoic acid, Y = HOCH$_2$CO$_2$Na, sodium hydroxyethanoate; Z = ClCH$_2$COCl, chloroethanoyl chloride.

**EXAMPLE 4**  1.220 g of a dicarboxylic aliphatic acid is dissolved in water and made up to 250 cm$^3$. A 25.0 cm$^3$ portion of the solution requires 21.0 cm$^3$ of 0.100 mol dm$^{-3}$ sodium hydroxide solution for neutralisation. Deduce the molecular formula of the acid, and write a structural formula for it.

**METHOD**  The equation for the neutralisation is

$$R(CO_2H)_2(aq) + 2NaOH(aq) \longrightarrow R(CO_2Na)_2(aq) + 2H_2O(l)$$

No. of moles of NaOH $= 21.0 \times 10^{-3} \times 0.100 = 2.10 \times 10^{-3}$ mol

From the equation, No. of moles of acid $= \frac{1}{2} \times$ No. moles of NaOH

$$= 1.05 \times 10^{-3} \text{ mol}$$

Mass of acid $= \frac{1}{10} \times 1.220$ g $= 0.1220$ g

$\therefore$ $\qquad\qquad 1.05 \times 10^{-3}$ mol $= 0.1220$ g

$$1 \text{ mol} = 116 \text{ g}$$

The molar mass is 116 g mol$^{-1}$. Subtracting 90 for (CO$_2$H)$_2$ leaves 26 g mol$^{-1}$ for the rest of the molecule. This is the mass of (CH$_2$)$_2$. The molecular formula is C$_4$H$_4$O$_4$, and the structural formula are

$$\begin{array}{ccc} \text{HCCO}_2\text{H} & \text{and} & \text{HO}_2\text{CCH} \\ \| & & \| \\ \text{HCCO}_2\text{H} & & \text{HCCO}_2\text{H} \end{array}$$

for cis- and trans-butenedioic acid.

## EXERCISE 65    Problems on Organic Chemistry

1. A compound X contains carbon and hydrogen only. 0.135 g of X, on combustion in a stream of oxygen, gave 0.410 g of carbon dioxide and 0.209 g of water. Calculate the empirical formula of X.

   X is a gas at room temperature, and 0.29 g of the gas occupy 120 cm$^3$ at room temperature and 1 atm. What is the molecular formula of X?

2. A monobasic organic acid C is dissolved in water and titrated with sodium hydroxide solution. 0.388 g of C require 46.5 cm$^3$ of 0.095 mol dm$^{-3}$ sodium hydroxide for neutralisation. Calculate the molar mass of C and deduce its formula.

3. An organic compound has a composition by mass of 83.5% C; 6.4% H; 10.1% O. The molar mass is 158 g mol$^{-1}$. Find the molecular formula and suggest a structural formula for the compound.

4. A compound A contains 5.20% by mass of nitrogen. The other elements present are carbon, hydrogen and oxygen. Combustion of 0.0850 g of A in a stream of oxygen gave 0.224 g of carbon dioxide and 0.0372 g of water. Calculate the empirical formula of A.

5. An alkene A contains one double bond per molecule. 0.560 g of bromine is required to react completely with 0.294 g of A. When A is treated with ozone and the product is hydrolysed and then oxidised, B, which is a monobasic carboxylic acid is formed. 0.740 g of this acid require $100 \text{ cm}^3$ of $0.100 \text{ mol dm}^{-3}$ sodium hydroxide for neutralisation. Deduce the formulae of A and B and the structural formula of A.

6. An organic compound, A, contains 70.6% carbon, 5.88% hydrogen and 25.5% oxygen, by mass. It has a molar mass of $136 \text{ g mol}^{-1}$. When A is refluxed with sodium hydroxide solution, and the resulting liquid is distilled, a liquid B distils over, and a solution of C remains. On addition of dilute hydrochloric acid to C, a white precipitate D forms. When this precipitate is mixed with soda lime and heated, the vapour burns with a smoky flame. B reacts with ethanoyl chloride to give a product with a fruity smell. B does not give a positive result in the iodoform test. Identify A, B and C.

7. A compound contains C, H, N and O. When 0.225 g of the compound was heated with sodium hydroxide solution, the ammonia evolved was passed into $25.0 \text{ cm}^3$ of sulphuric acid of concentration $0.100 \text{ mol dm}^{-3}$. The sulphuric acid that remained required $19.1 \text{ cm}^3$ of $0.100 \text{ mol dm}^{-3}$ sodium hydroxide for neutralisation.

A 0.195 g sample of the compound gave on complete oxidation 0.352 g of carbon dioxide and 0.168 g of water.

A solution of 9.12 g of the compound in $500 \text{ cm}^3$ of water froze at $-0.465\,^{\circ}\text{C}$. The cryoscopic constant for water is $1.86 \text{ K kg mol}^{-1}$.

a) Find the molecular formula of the compound, and b) suggest its identity.

8. Phenol reacts with bromine to form a crystalline product. $25.0 \text{ cm}^3$ of a solution of phenol of concentration $0.100 \text{ mol dm}^{-3}$ were added to $30.0 \text{ cm}^3$ of a solution of $0.100 \text{ mol dm}^{-3}$ potassium bromate(V). An excess of potassium bromide and hydrochloric acid were added to liberate bromine. The excess bromine was estimated by adding potassium iodide and titrating the iodine displaced with a solution of sodium thiosulphate. $30.0 \text{ cm}^3$ of a $0.100 \text{ mol dm}^{-3}$ solution of sodium thiosulphate were required.

a) Find the ratio of moles of $Br_2$ : moles of phenol, and b) deduce the equation for the reaction.

9. Measurements on an alkene showed that $100 \text{ cm}^3$ of the gas weighed 0.233 g at $25\,^{\circ}\text{C}$ and 1 atm. $25.0 \text{ cm}^3$ of the alkene reacted with $25.0 \text{ cm}^3$ of hydrogen.

a) Find the molar mass of the alkene, and b) give its molecular formula.

Give the names and structural formulae of two alkenes which have this molecular formula.

10. An organic acid has the percentage composition by mass: C, 41.4%; H, 3.4%; O, 55.2%. A solution containing 0.250 g of the acid, which is dibasic, required 26.6 cm$^3$ of 0.200 mol dm$^{-3}$ sodium hydroxide solution for neutralisation.

    Calculate: a) the empirical formula, and b) the molecular formula of the acid. c) Give its name and write its structural formula or formulae.

11. An organic liquid contains carbon, hydrogen and oxygen. On oxidation, 0.250 g of the liquid gave 0.595 g of carbon dioxide and 0.304 g of water. When vaporised, 0.250 g of the liquid occupied 131 cm$^3$ at 200 °C and 1 atm.

    Find: a) the empirical formula, and b) the molecular formula of the liquid. c) Write the structural formulae of compounds with this molecular formula.

12. A is an organic compound with the percentage composition by mass C, 71.1%; N, 10.4%; O, 11.8%; H, 6.7%, and a molar mass of 135 g mol$^{-1}$.

    On hydrolysis by aqueous sodium hydroxide, A gives an oily liquid, B. B has the percentage composition by mass: C, 77.1%; N, 15.1%; H, 7.5%, and a molar mass of 93 g mol$^{-1}$. B is basic and gives a precipitate with bromine water.

    Find the molecular formulae for A and B. From their reactions, deduce the identify of A and B.

**EXERCISE 66**    Questions from A-level Papers

1. An organic compound, W, on analysis, gave C = 40.0%, H = 8.5%, N = 23.7%, O = 27.1% by mass. Refluxing W with dilute hydrochloric acid produced a compound X which contained 40% carbon by mass and had the general formula $C_nH_{2n}O_2$.

   X was also prepared by the oxidation of a compound Y which had the general formula $C_nH_{2n}O$.

   a) Calculate the empirical formula of W.

   b) Determine the molecular formula of X and hence deduce the molecular formula of W.

   c) Write a balanced equation to represent the reaction which occurred when W was refluxed with dilute hydrochloric acid.

   d) Identify Y giving reasons for your answer.

   e) Give *two* chemical tests to distinguish between Y and butanone ($CH_3CH_2COCH_3$).                    (AEB80)

2. When the neutral compound A, $C_{10}H_{13}NO$, was refluxed with dilute acid it formed two products B, $C_2H_7N$ and C.

   On analysis, C was found to contain 70.59% carbon, 23.53% oxygen and 5.88% hydrogen by weight. The relative molecular mass of C was found to be 136.

   On reaction with alkaline potassium manganate(VII) (permanganate) solution, C was oxidized to D, $C_8H_6O_4$.

   D, which was acidific, was readily dehydrated to the neutral substance E, $C_8H_4O_3$.

   B reacted with gaseous hydrogen chloride to form the ionic solid F, $C_2H_8NCl$. When F was dissolved in dilute hydrochloric acid and sodium nitrite solution added, a yellow oil, G, was formed and no effervescence occurred.

   a) What is the empirical formula of C?
   b) What is the molecular formula of C?
   c) Write the structural formulae for substances A to G.
   d) What are the names of substances B, C and F?         (SUJB80)

3. a) An aliphatic alcohol, $X$, contains   C, 64.9%; H, 13.5%; O, 21.6% by mass. Calculate the empirical formula of $X$.

   b) 0.074 g of $X$, when vaporised at 177 °C and 740 mm Hg pressure, occupied a volume of 38.0 cm$^3$. Calculate the relative molecular mass of $X$. What is the molecular formula of $X$?

   c) Give the structures of all the isomers of $X$, which are alcohols.

   d) Describe the tests, (physical and/or chemical) you would carry out on $X$ to decide which of the possible structures it has.         (L85)

4. An organic compound contains 85.7% carbon and 14.3% hydrogen. The relative molecular mass is approximately 85 (relative atomic masses: H 1, C 12). The formula of the compound is

   a $C_6H_{14}$          b $C_6H_{12}$          c $C_6H_{10}$
   d $C_3H_6$          e $C_3H_8$          (O&C82)

5. a) State Raoult's law and explain briefly why this law is not obeyed by most binary solutions.

   Pure (anhydrous) nitric acid boils at 78.2 °C. An aqueous solution containing 68.2% of nitric acid boils at 121 °C, and fractionation of this mixture does not yield pure nitric acid. Explain this observation using a simple diagram.

   b) Compound $G$, a liquid at ordinary temperatures, is an alcohol and has the empirical formula $CH_3O$. At 196 °C, 6.2 g of gaseous $G$ exerts a pressure of 39 kPa when occupying a volume of 0.01 m$^3$.

   Compound $T$, a solid at room temperature, is a carboxylic acid which contains one or more carboxyl groups per molecule. The sodium salt of $T$ contains 45.72% of carbon, 1.905% of hydrogen and 30.48% of oxygen by mass. When a solution containing 2.80 g of this sodium salt is completely ionised in 1 kg of water it gives the same depression of freezing point as a solution containing 1.17 g of NaCl in 1 kg of water.

Compound $P$ (empirical formula $C_5H_4O_2$) is formed by the reaction of $G$ with $T$ under controlled conditions. A solution containing $2.30\,g\,l^{-1}$ of $P$ produces an osmotic pressure of 294 Pa at 22 °C.

Determine the molecular formulae and suggest structural formulae for $G$, $T$ and $P$ using, where appropriate, the data below. Comment on the nature of product $P$.

Gas constant $R = 8.314\,J\,mol^{-1}\,K^{-1}$ ($0.082\,05$ litre atm $mol^{-1}\,K^{-1}$) 1 atmosphere $= 101\,325\,Pa$.                                    (JMB84,S)

6. A sample of pure organic liquid $Q$ of mass 0.146 g was vaporised in a gas syringe at 127 °C and occupied a volume of 100 cm³ at a pressure of 101 kPa.

   a) Calculate the relative molecular mass of $Q$ from the experimental data.

   b) The accurate composition of $Q$ by mass is    C, 52.2%; H, 13.0%; O, 34.8%.

      i) Calculate the empirical formula of $Q$.

      ii) What are the likely molecular formula and relative molecular mass of $Q$?

   c) Suggest two possible reasons for the difference between the values of the relative molecular masses in a) and b) ii).

   d) $Q$ produced hydrogen when reacted with sodium. When $Q$ was warmed with aqueous sodium hydroxide and iodine, a yellow precipitate was formed.

      i) What is the likely identity of $Q$?

      ii) Write the equation for the reaction between $Q$ and sodium and name the organic product.

      iii) Name the yellow precipitate.                              (C83)

7. 4.706 g (0.05 mole) of a compound $A$ decolourised bromine water being quantitatively converted to 16.54 g of a white compound $B$ which contained C $= 21.78\%$, H $= 0.92\%$, Br $= 72.46\%$, O $= 4.84\%$ by mass. $A$ was sparingly soluble in water, dissolved in aqueous sodium hydroxide but did not react with aqueous sodium hydrogen carbonate.

   a) Deduce the empirical formula and molecular formula of $B$.

   b) Give structural formulae for $A$ and $B$.

   ($A_r(H) = 1.01$; $A_r(C) = 12.01$; $A_r(O) = 16.00$; $A_r(Br) = 79.90$.)
                                                          (WJEC84,p)

8. a) Compare the effectiveness of ROH and $RNH_2$ ($R = $ H or alkyl) as nucleophilic reagents in organic chemistry.

   b) Compare the properties of $H_2O$ and $NH_3$ as ligands both for the formation of metal complexes and in the properties conferred on the complexes.

c) $0.2000 \text{ cm}^3$ of a liquid of density $1655 \text{ kg m}^{-3}$ and boiling point $79\,^\circ C$ was introduced into a gas syringe which was then heated in boiling water. The gas volume increased by $85.0 \text{ cm}^3$. Calculate the molecular mass of the liquid compound.

d) $2.44 \text{ g}$ of an organic compound was reacted with $2.38 \text{ g}$ of the liquid compound in c) and complete conversion resulted in the loss of mass due to the evolution of gaseous products of $2.01 \text{ g}$. The high mass end of the mass spectrum of the product contained a peak at $m/e = 142$ and a peak at $m/e = 140$ approximately three times more intense. Give a full interpretation of these data.

(1 atm $= 101\,325 \text{ N m}^{-2}$; Gas constant, $R = 8.314 \text{ J K}^{-1} \text{mol}^{-1}$; $0\,^\circ C = 273.15 \text{ K}$; Appropriate $A_r$ values can be found elsewhere in this book.)                                                                 WJEC84)

9. A compound $A$ of molecular formula $C_{17}H_{19}O_3N$, known as 'piperine', may be extracted from black pepper. Aqueous hydrolysis of $A$ produces a basic compound $B$ and an acidic compound $C$.

Compound $B$ has a relative molecular mass of 85 and on combustion $0.425 \text{ g}$ gives $CO_2$, $1.10 \text{ g}$; $H_2O$, $0.50 \text{ g}$; and $N_2$, $60 \text{ cm}^3$ (under conditions at which the volume of one mole of nitrogen molecules is $24.0 \text{ dm}^3$). In compound $B$ the nitrogen reacts as in a secondary amine and is contained in a six-membered ring structure. From this information, calculate a molecular formula and draw a possible structural formula for $B$.

Compound $C$ has a relative molecular mass of 218 and on combustion $0.271 \text{ g}$ of $C$ gives $CO_2$, $0.656 \text{ g}$; and $H_2O$, $0.112 \text{ g}$. From this information, calculate a molecular formula for $C$.

Use the following information to deduce a possible structural formula for $C$.

The oxidation of $C$ by potassium manganate(VII) destroys a side chain and produces a monobasic acid $D$ of molecular formula $C_8H_6O_4$. $D$ is also produced by the oxidation of the aldehyde piperonal,

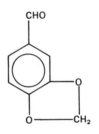

In one mole of $C$ the original side chain would neutralise one mole of sodium hydroxide, and would add four moles of bromine atoms on treatment with bromine dissolved in tetrachloromethane.

Use the two structural formulae you have deduced for $B$ and $C$ to draw a possible structural formula for the original compound piperine, $A$.                                                                 (L(N)81)

10. Compound $C$ contains neither nitrogen nor oxygen. Combustion analysis shows it to contain 52.2% of carbon and 3.7% of hydrogen. The peaks of highest mass/charge ratio in the mass spectrum of $C$ are at $m/z$ 160, 162 and 164, with relative intensities $9:6:1$. When $C$ is boiled with aqueous sodium hydroxide it gives a compound $D$. Oxidation of $D$ with alkaline potassium manganate(VII) followed by acidification gives compound $E$. Reaction of $E$ with phosphorus pentachloride gives as the organic product a compound $F$ containing 40.6% of chlorine. In the presence of aluminium chloride $F$ reacts with benzene to give a compound $G$. When treated with an acidic solution of 2,4-dinitrophenylhydrazine $G$ gives an orange derivative.

Give possible structures for compounds $C$ to $G$, interpret the analytical and spectroscopic data given in the question, and elucidate the reactions which take place. (O86,S)

11. $X$ contains C, H, and O only. When it is refluxed with aqueous acid for several hours, two components may be separated from the mixture.

The lower boiling fraction ($A$) is a liquid at room temperature, 1.0 g of which gave 2.2 g $CO_2$ and 1.2 g water on complete combustion. 0.1 g $A$ occupies 51.0 cm³ at 100 °C and 1 atm., and $A$ gives an oxidation product which yields an orange precipitate with 2:4 dinitrophenylhydrazine but does not react with Fehling's solution.

The higher boiling fraction ($B$) from the original reflux, when neutralised with ammonia and the product heated with phosphorus(V) oxide, yields a product of molecular mass 97, whose composition by mass is 74.2%C, 11.3%H, and 14.4%N. Identify the substances involved and write the reaction scheme. Explain the reaction of $X$ with aqueous acid. (Molar volume $= 22.4$ dm³ at s.t.p. H $= 1$, C $= 12$, N $= 14$, O $= 16$.)

Write an account of the isomerism of the alcohols of formula $C_5H_{11}OH$. (SUJB84)

12. Compound $A$ is a monoacid base and has the composition, 78.5% carbon, 8.4% hydrogen, and 13.08% nitrogen. 2.14 g of $A$ react with 20.0 cm³ of 1.0 mol dm⁻³ hydrochloric acid. With nitric(III) acid (nitrous acid), $A$ gives $B$, $C_7H_8O$. $B$, on oxidation, forms $C$, $C_7H_6O_2$. (H $= 1$; C $= 12$; N $= 14$.)

a) Deduce the molecular and structural formula of $A$, explaining the reaction with hydrochloric acid.

b) Write the structural formulae of $B$ and $C$, explaining the reactions involved.

c) By giving reagents and reaction conditions, show the steps by which $A$ can be obtained from phenylethanoic acid.

d) Write the structural formula of $D$ which is an isomer of $A$. Mention *two* reactions in which $D$ will behave differently from $A$.

e) How would you distinguish chemically between $A$ and $D$? Describe the observations which will enable you to decide. (SUJB84)

# Answers to Exercises

## CHAPTER 1

### Exercise 1

1. a) $2.3678 \times 10^4$    b) $4.376 \times 10^2$    c) $1.69 \times 10^{-2}$    d) $3.45 \times 10^{-4}$
   e) $6.72891 \times 10^5$
2. a) $5.85 \times 10^4$    b) $2.66 \times 10^6$    c) $6.35 \times 10^7$    d) $1.21 \times 10$
   e) $1.34 \times 10^2$
3. a) $3.32 \times 10^6$    b) $2.72 \times 10^5$    c) $1.86 \times 10^{-4}$    d) $6.44 \times 10^{-5}$
   e) $6.11 \times 10^{-3}$
4. a) $2.001 \times 10^4$    b) $5.648 \times 10^3$    c) $1.29 \times 10^5$    d) $-1.12 \times 10^{-3}$
   e) $6.252 \times 10^4$
5. a) $4 \times 10^{10}$    b) $2 \times 10^2$    c) $5 \times 10^8$    d) $1 \times 10^3$
   e) $2 \times 10^{10}$
6. a) $3.6753$    b) $3.7052$    c) $-2.8771$    d) $1.0033$
   e) $-5.6356$
7. a) $2.862 \times 10^3$    b) $1.135$    c) $6.969 \times 10^7$    d) $3.3791 \times 10^{-7}$
   e) $8.7680 \times 10^{-3}$
8. a) $4.264 \times 10^{-3}$    b) $2.867 \times 10^{-7}$    c) $4.037 \times 10^2$    d) $2.055 \times 10^{-10}$
   e) $3.781 \times 10^4$
9. a) $x = 7$ or $-13$    b) $x = 4$ or $-0.4$    c) $y = 5$    d) $z = 14$ or $2$
   e) $x = -\frac{1}{2}$ or $-4\frac{1}{2}$

## CHAPTER 2

### Exercise 2

1. a) $14.3\%$    b) $43.5\%$    c) $29.2\%$    d) $47.4\%$
   e) $5.70\%$
2. a) $MgO$    b) $CaCl_2$    c) $FeCl_3$    d) $CuS$
   e) $LiH$
3. a) $FeO$    b) $Fe_2O_3$    c) $Fe_3O_4$    d) $K_2CrO_4$
   e) $K_2Cr_2O_7$    f) $CH$    g) $C_3H_8$
4. a) $a = 5$    b) $b = 6$    c) $c = 2$    d) $d = 3$
   e) $e = 6$    f) $f = 12$
5. a) $C_5H_{10}O$    b) $C_5H_{10}O$    6. a) $C_2H_4O$    b) $C_2H_4O$
7. a) $C_5H_{10}O_2$    b) $C_5H_{10}O_2$    8. $C_9H_{10}O_2$

### Exercise 3

1. $9.78\,kg$    2. $522.7$ tonnes    3. $0.1435\,g$    4. $15.63\,kg$
5. $2.808\,g$    6. a) $7.00$ tonnes    b) $7.24$ tonnes    7. $63.1\,g$
8. $36.46\,g$    9. $4.481\,g$
10. $50.00\,g\ Ca_3(PO_4)_2$    $29.03\,g\ SiO_2$    $4.839\,g\ C$
11. $40\%\ CaCO_3$    $60\%\ MgCO_3$    12. $3.5000\,g\ NaHCO_3$    $6.5000\,g\ Na_2CO_3$

### Exercise 4

1. a) $2\,dm^3$    b) $750\,cm^3$    c) $625\,cm^3$    d) $937.5\,cm^3$    e) $2\,dm^3$
2. $500\,cm^3\ SO_2$    3. $50\%$    4. $C_3H_8$    5. $C_4H_6$
6. $a = 30\,cm^3$    $b = 40\,cm^3$    7. d

**Exercise 5**

1. 267 dm³     2. 3.50 dm³    3. 1.107 g    4. 2.388 g
5. 3.646 g    6. 11.5 dm³    7. 2460 dm³

**Exercise 6**

1. 92.9%    2. 90.5%    3. 89.0%    4. 91.0%
5. 99.2%

**Exercise 7**

1. 436 tonnes    2. 46 kg    3. 2.7 kg    4. 304 kg
5. 93.5%

**Exercise 8**

1. $S_2O_8^{2-}(aq) + 2I^-(aq) \longrightarrow I_2(aq) + 2SO_4^{2-}(aq)$
2. $H_2NSO_3^-(aq) + OH^-(aq) \longrightarrow NH_3(g) + SO_4^{2-}(aq)$
3. $Na_2S_2O_3(aq) + AgCl(s) \longrightarrow NaCl(aq) + NaAgS_2O_3(aq)$
4. $C_6H_8 + 2Br_2 \longrightarrow C_6H_8Br_4$
5. $C_6H_5NH_2 + 3Br_2 \longrightarrow C_6H_2Br_3NH_2 + 3HBr$

**Exercise 9**

1. a) $m = 2, n = 6$    b) $Mg(OH)Cl$
2. B
3. a) 353    b) $Ga_2Cl_6$
4. b    5. b) $C_4H_8O_2$    6. a) $SnI_4$    7. c
8. d    9. d    10. e) $C_5H_{10}$
11. b) $C_4H_8$    12. $x = 2$    13. $A_r(M) = 88$

# CHAPTER 3

**Exercise 10**

1. a) 5.90 g    b) 5.30 g    c) 9.45 g    d) 42.0 g
   e) 19.1 g
2. a) 0.500 dm³    b) 0.0750 dm³ or 75.0 cm³    c) 11.4 cm³
   d) 192 cm³    e) 192 cm³
3. a) i) 0.325 g    ii) 0.560 g    iii) 0.618 g    iv) 1.43 g
   b) i) 50.0 cm³    ii) 100 cm³    iii) 10.0 cm³    iv) 40.0 cm³
4. 34.0 cm³    5. $4.76 \times 10^{-2}$ mol dm⁻³    6. 133 cm³
7. 400 cm³    8. 95.0%    9. 87.6%    10. 90.6%
11. $[Na_2CO_3] = [NaHCO_3] = 0.120$ mol dm⁻³    12. 18.0%    13. 46.1%
14. 3    15. 68.2%    16. 10

**Exercise 11**

1. a) +2    b) +2    c) 0    d) +1
   e) +3    f) +1    g) +3    h) +5
   i) +5    j) +4    k) 0    l) +7
   m) +2    n) −1    o) +3    p) +6
   q) +5    r) −1    s) +6    t) +4
   u) −1    v) +6    w) +4    x) +1
   y) +4    z) +5

**2. a)** Sn is oxidised from $+2$ to $+4$; Pb is reduced from $+4$ to $+2$.
   **b)** Mn is oxidised from $+2$ to $+7$; Bi is reduced from $+5$ to $+3$.
   **c)** As is oxidised from $+3$ to $+7$; Mn is reduced from $+7$ to $+2$.

**3. a)** F is reduced from 0 to $-1$.
   **b)** Cl disproportionates from 0 to $+5$ and $-1$.
   **c)** N disproportionates from $-3$ and $+5$ to $-1$.
   **d)** Cr is reduced from $+6$ to $+3$.
   **e)** C is oxidised from $+3$ to $+4$.

**4. a)** $-2$                    **b)** $+2$

**5. a)** $IO_4^- + 7I^- + 8H^+ \longrightarrow 4I_2 + 4H_2O$
   **b)** $BrO_3^- + 6I^- + 6H^+ \longrightarrow Br^- + 3I_2 + 3H_2O$
   **c)** $2V^{3+} + H_2O_2 \longrightarrow 2VO^{2+} + 2H^+$
   **d)** $SO_2 + 2H_2O + Br_2 \longrightarrow 4H^+ + SO_4^{2-} + 2Br^-$
   **e)** $4NH_3 + 3O_2 \longrightarrow 2N_2 + 6H_2O$
   **f)** $2NH_3 + 2O_2 \longrightarrow N_2O + 3H_2O$
   **g)** $4NH_3 + 5O_2 \longrightarrow 4NO + 6H_2O$
   **h)** $Fe^{2+}C_2O_4^{2-} + 3Ce^{3+} \longrightarrow 2CO_2 + 3Ce^{2+} + Fe^{3+}$

**6.** $Cr_2O_7^{2-} + 6I^- + 14H^+ \longrightarrow 3I_2 + 2Cr^{3+} + 7H_2O$

# Exercise 12

**1. a)** $NO_2^- + H_2O \longrightarrow NO_3^- + 2H^+ + 2e^-$
   **b)** $AsO_3^{3-} + H_2O \longrightarrow AsO_4^{3-} + 2H^+ + 2e^-$
   **c)** $Hg_2^{2+} \longrightarrow 2Hg^{2+} + 2e^-$
   **d)** $H_2O_2 \longrightarrow 2H^+ + 2e^- + O_2$
   **e)** $V^{3+} + H_2O \longrightarrow VO^{2+} + 2H^+ + e^-$

**2. a)** $NO_3^- + 2H^+ + e^- \longrightarrow NO_2 + H_2O$
   **b)** $NO_3^- + 4H^+ + 3e^- \longrightarrow NO + 2H_2O$
   **c)** $NO_3^- + 10H^+ + 8e^- \longrightarrow NH_4^+ + 3H_2O$
   **d)** $2BrO_3^- + 12H^+ + 10e^- \longrightarrow Br_2 + 6H_2O$
   **e)** $PbO_2 + 4H^+ + 2e^- \longrightarrow Pb^{2+} + 2H_2O$

**3. a)** $2MnO_4^-(aq) + 5H_2O_2(aq) + 6H^+(aq) \longrightarrow 5O_2(g) + 2Mn^{2+}(aq) + 8H_2O(l)$
   **b)** $MnO_2(s) + 4H^+(aq) + 2Cl^-(aq) \longrightarrow Mn^{2+}(aq) + Cl_2(g) + 2H_2O(l)$
   **c)** $2MnO_4^-(aq) + 5C_2O_4^{2-}(aq) + 16H^+(aq) \longrightarrow 2Mn^{2+}(aq) + 10CO_2(g) + 8H_2O(l)$
   **d)** $Cr_2O_7^{2-}(aq) + 3C_2O_4^{2-}(aq) + 14H^+(aq) \longrightarrow 2Cr^{3+}(aq) + 6CO_2(g) + 7H_2O(l)$
   **e)** $Cr_2O_7^{2-}(aq) + 6I^-(aq) + 14H^+(aq) \longrightarrow 2Cr^{3+}(aq) + 3I_2(aq) + 7H_2O(l)$
   **f)** $H_2O_2(aq) + NO_2^-(aq) \longrightarrow NO_3^-(aq) + H_2O(l)$

**4. a)** $1.5 \times 10^{-2}$ mol    **b)** $7.5 \times 10^{-3}$ mol    **c)** $7.5 \times 10^{-3}$ mol    **d)** $7.5 \times 10^{-3}$ mol
   **e)** $1.5 \times 10^{-2}$ mol

**5. a)** $6.0 \times 10^{-4}$ mol    **b)** $3.0 \times 10^{-4}$ mol    **c)** $6.0 \times 10^{-4}$ mol    **d)** $3.0 \times 10^{-4}$ mol
   **e)** $3.0 \times 10^{-4}$ mol

**6. a)** $4.0 \times 10^{-3}$ mol    **b)** $2.0 \times 10^{-3}$ mol    **c)** $2.0 \times 10^{-3}$ mol    **d)** $4.0 \times 10^{-3}$ mol
   **e)** $6.7 \times 10^{-4}$ mol

**7. a)** $62.5$ cm$^3$    **b)** $250$ cm$^3$    **c)** $5.00$ cm$^3$    **d)** $12.5$ cm$^3$
   **e)** $8.3$ cm$^3$

**8. a)** $45.0$ cm$^3$    **b)** $12.0$ cm$^3$    **c)** $7.2$ cm$^3$    **d)** $4.50$ cm$^3$
   **e)** $9.0$ cm$^3$

**9.** $0.090$ mol dm$^{-3}$    **10.** $0.0195$ mol dm$^{-3}$    **11.** $0.0447$ mol dm$^{-3}$    **12.** $99.5\%$

**13.** $90.6\%$            **14.** $1.64 \times 10^{-2}$ mol dm$^{-3}$            **15.** $0.103$ mol dm$^{-3}$

**16. a)** $[Fe^{2+}] = 0.0600$ mol dm$^{-3}$            **b)** $[Fe^{3+}] = 0.0160$ mol dm$^{-3}$

**17. a)** $20.0$ cm$^3$        **b)** $22.4$ cm$^3$        **18.** $7.63 \times 10^{-2}$ mol dm$^{-3}$

**19.** $+4$            **20.** $2.7 \times 10^{-2}$ mol dm$^{-3}$

**21. a)** $1NH_2OH : 2Fe^{3+}$        **b)** $-1$            **c)** $-1$            **d)** $+1$
   **e)** $N_2O$            **f)** $2NH_2OH + 4Fe^{3+} \longrightarrow 4Fe^{2+} + N_2O + H_2O + 4H^+$

**22.** $82.3\%$            **23. b)** $0.74$ mol dm$^{-3}$

# Exercise 13

**1.** $9.37 \times 10^{-3}$ mol dm$^{-3}$            **2.** $78$ ppm
**3.** $60.0\%$ CaO    $40.0\%$ MgO            **4.** $18$

**Exercise 14**

1. $1.48 \times 10^{-2}\,\text{mol}\,\text{dm}^{-3}$  
2. 49.7%  
3. 34.6% NaCl  65.4% NaBr  
4. $2.77 \times 10^{-2}\,\text{mol}\,\text{dm}^{-3}$  
5. a) $1.58 \times 10^{-2}\,\text{mol}\,\text{dm}^{-3}$  
b) $4.97 \times 10^{-2}\,\text{mol}\,\text{dm}^{-3}$  
6. 55.3%

**Exercise 15**

1. a) $2\text{Fe}^{3+}:1\text{NH}_3\text{OH}^+$  b) $4\text{Fe}^{3+} + 2\text{NH}_3\text{OH}^+ \longrightarrow 4\text{Fe}^{2+} + \text{N}_2\text{O} + \text{H}_2\text{O} + 6\text{H}^+$  
2. $A$, $2.10 \times 10^{-2}\,\text{mol}\,\text{dm}^{-3}$, $B$, $1.76 \times 10^{-2}\,\text{mol}\,\text{dm}^{-3}$  
$6\text{Fe}^{2+} + \text{ClO}_3^- + 6\text{H}^+ \longrightarrow 6\text{Fe}^{3+} + \text{Cl}^- + 3\text{H}_2\text{O}$  
3. c  4. e  
5. a) $0.40\,\text{mol}$; CuI  b) $1.60\,\text{mol}$  c) $4:1$  d) $\text{Cu(NH}_3)_4\text{SO}_4$  
6. a) i) $1.00 \times 10^{-2}\,\text{mol}$  ii) $0.180\,\text{g}$  
b) i) $1.00 \times 10^{-2}\,\text{mol}$  ii) $0.960\,\text{g}$  
c) i) $5.00 \times 10^{-3}\,\text{mol}$  ii) $0.318\,\text{g}$  
d) i) $0.543\,\text{g}$  ii) $0.0301\,\text{mol}$  e) $\text{Cu(NH}_4)_2(\text{SO}_4)_2 \cdot 6\text{H}_2\text{O}$  
7. ii) 96%  
8. a) $0.50\,\text{mol}\,\text{O}_2$  b) $2\text{Mn}^{3+} + 2\text{I}^- \longrightarrow \text{I}_2 + 2\text{Mn}^{2+}$  
c) $3.0 \times 10^{-4}\,\text{mol}\,\text{dm}^{-3}$  
9. i) $1.67 \times 10^{-2}\,\text{mol}\,\text{dm}^{-3}$  ii) $2\text{IO}_3^- + 3\text{N}_2\text{H}_4 \longrightarrow \text{I}_2 + 2\text{NO}_3^- + 4\text{NH}_3$  
iii) $\text{I}:+5 \longrightarrow 0$; $\text{N}:-2 \longrightarrow +5$ and $-3$  
10. e  
11. $\text{S}_2\text{O}_3^{2-} + 4\text{Cl}_2 + 5\text{H}_2\text{O} \longrightarrow 2\text{SO}_4^{2-} + 8\text{Cl}^- + 10\text{H}^+$  
$\text{Cl}:0 \longrightarrow -1; \text{S}:+2 \longrightarrow +6; \text{I}:0 \longrightarrow -1; \text{S}:+2 \longrightarrow +2.5$  
12. h) $1.27\,\text{g}$  
13. $3\text{Cu}:1\text{Zn}$  
14. $[\text{HIO}_3] = 0.107\,\text{mol}\,\text{dm}^{-3}$, $[\text{H}_2\text{SO}_4] = 0.268\,\text{mol}\,\text{dm}^{-3}$, $[\text{H}_2\text{SO}_4] = 2.5\,[\text{HIO}_3]$  
$2\text{HIO}_3 + 5\text{SO}_2 + 4\text{H}_2\text{O} \longrightarrow \text{I}_2 + 5\text{H}_2\text{SO}_4$  
15. $\text{H}_3\text{NSO}_3 + \text{HNO}_2 \longrightarrow \text{N}_2 + \text{SO}_4^{2-}$  
one $\text{H}^+$ per molecule; therefore

$$\begin{array}{c} \text{H--N} \quad\quad\; \text{O} \\ \diagdown \quad \diagup\!\!\diagup \\ \text{S} \\ \diagup \quad \diagdown\!\!\diagdown \\ \text{HO} \quad\quad\; \text{O} \end{array}$$

16. Ethanedioic acid, $2.50 \times 10^{-2}\,\text{mol}\,\text{dm}^{-3}$;  
Sulphuric acid, $5.00 \times 10^{-2}\,\text{mol}\,\text{dm}^{-3}$

# CHAPTER 4

**Exercise 16**

1. $85.6\,m_u$  2. $69.8\,m_u$  3. 24.3  
4. $3\,^{35}\text{Cl}:1\,^{37}\text{Cl}$; $35.5\,m_u$  5. 6.93  
6. 1, $^1\text{H}$  2, $^2\text{H}$  3, $^1\text{H}^2\text{H}$  4, $^2\text{H}_2$  17 $^{16}\text{O}^1\text{H}$  18, $^{16}\text{O}^2\text{H}$ and $^1\text{H}_2{}^{16}\text{O}$  19, $^1\text{H}^2\text{H}^{16}\text{O}$  
20, $^2\text{H}_2{}^{16}\text{O}$  
7. $39.1\,m_u$  
8. $^{63}\text{Cu}$  $^{65}\text{Cu}$  $^{63}\text{CuO}$  $^{65}\text{CuO}$  $^{63}\text{CuNO}_3$  $^{65}\text{CuNO}_3$  $^{63}\text{Cu(NO}_3)_2$  $^{65}\text{Cu(NO}_3)_2$

**Exercise 17**

1. $A_r(\text{Ne}) = 20.2$. The peaks at 32 and multiples of 32 are due to S, $\text{S}_2$, $\text{S}_3$, $\text{S}_4$, $\text{S}_5$, $\text{S}_6$, $\text{S}_7$  
and $\text{S}_8$. There is no evidence here for any isotope other than sulphur-32.  
2. c

3. i) $^{79}Br$ and $^{81}Br$ in the ratio $1:1$
  ii) $124 = {}^{79}BrCH_2CH_2OH$
      $126 = {}^{81}BrCH_2CH_2OH$
      $142 = {}^{79}BrCH_2CH_2{}^{35}Cl$
      $144 = {}^{81}BrCH_2CH_2{}^{35}Cl$ and ${}^{79}BrCH_2CH_2{}^{37}Cl$
      $146 = {}^{81}BrCH_2CH_2{}^{37}Cl$
4. b) $72.5\%\ {}^{63}Cu + 27.5\%\ {}^{65}Cu$
  c) Peaks at $220, 222, 224, 226$ in the ratio $3:4:4:1$
5. ii) $M = 92$
  iii) $C_6H_5CH_3 \longrightarrow C_6H_5CH_2 + H$
      $CH_3COCH_2CH_3 \longrightarrow CH_3CO + CH_3CH_2$

## Exercise 18

1. a) $a = 1,\ b = 0$    b) $a = 17,\ b = 8$    c) $a = 4,\ b = 2$    d) $a = 210,\ b = 83$
  e) $a = 4,\ b = 2$    f) $a = 4,\ b = 2$    g) $a = 14,\ b = 6$    h) $a = 16,\ b = 7$
  i) $a = 207,\ b = 83$    j) $a = 4,\ b = 11$    k) $q = 0,\ p = 1,\ r = 35$
  l) $c = 3,\ d = 1$

## Exercise 19

1. a) $A_r(X) = 18.8$    b) $A(X) = 19$      c) Group $= 7$      d) 10 neutrons
2. a) i) $+7\,{}_{-1}^{0}e + 2\,{}_{0}^{1}n$
3. ${}_{85}^{212}At$
4. a) i) $X = {}_{1}^{2}H,\ Y = {}_{1}^{1}H,\ Z = {}_{2}^{3}He$      ii) ${}_{2}^{4}He$
  b) i) $3.97 \times 10^{-15}$ J    ii) 2.39 TJ

# CHAPTER 5

## Exercise 20

1. a) $187\ cm^3$      b) $387\ cm^3$      c) $6.23\ dm^3$      d) $132\ cm^3$
  e) $2.43\ dm^3$
2. $1.75 \times 10^5\ N\,m^{-2}$    3. $0.943\ dm^3$    4. $484$ K    5. $586\ cm^3$
6. a) $185\ cm^3$      b) $36.7\ cm^3$      c) $6.46\ dm^3$      d) $3.83\ dm^3$
  e) $436\ cm^3$

## Exercise 21

1. a) $46$      b) $NO_2$      2. $59$ cm      3. $160$
4. $4$      5. $A = H_2$   $B = O_3$      6. $48$ s
7. $65.3$    $59\%$ of the $NO_2$ molecules are dimerised      8. $100.5$ s
9. $16.3\ cm^3\ min^{-1}$    10. $24.9\ cm^3$      11. $44\ g\,mol^{-1}$
12. $CO, 25\%$   $CO_2, 75\%$      13. $NO_2, 43.7\%$    $N_2O_4, 56.3\%$

## Exercise 22

1. $44.1\ g\,mol^{-1}$    2. $39.9\ g\,mol^{-1}$    3. $6.01\ dm^3$    4. $83.8\ g\,mol^{-1}$
5. $0.583$ mol    6. $44.0\ g\,mol^{-1}$    7. $6.18\ dm^3$    8. $7.54 \times 10^{-2}$ mol

## Exercise 23

1. a) $176.5\ cm^3$      b) $207\ cm^3$      c) $26.3\ dm^3$
2. $2.73 \times 10^5\ N\,m^{-2}$    3. $2.53 \times 10^4\ N\,m^{-2}$    4. $4.50 \times 10^5\ N\,m^{-2}$
5. a) $p(N_2) = 3.00 \times 10^4\ N\,m^{-2}$   $p(O_2) = 2.63 \times 10^4\ N\,m^{-2}$,   $p(CO_2) = 1.88 \times 10^4\ N\,m^{-2}$
  b) $p(N_2) = 3.00 \times 10^4\ N\,m^{-2}$   $p(O_2) = 2.63 \times 10^4\ N\,m^{-2}$
6. a) $p(NH_3) = 6.00 \times 10^4\ N\,m^{-2}$   $p(H_2) = 3.75 \times 10^4\ N\,m^{-2}$   $p(N_2) = 5.25 \times 10^4\ N\,m^{-2}$
  b) $p(H_2) = 3.75 \times 10^4\ N\,m^{-2}$   $p(N_2) = 5.25 \times 10^4\ N\,m^{-2}$

**Exercise 24**

1. a) 67.25 cm³  b) 16.7% $N_2$  8.9% $O_2$  74.3% $CO_2$
2. a) 48.12 cm³  b) 64.3% $N_2$  34.0% $O_2$  1.63% $CO_2$
3. 3.7% CO   96.3% $CO_2$   4. 95% $Cl_2$   5% $O_2$

**Exercise 25**

1. 84.4 g mol⁻¹   2. 1840 m s⁻¹   3. 3410 J mol⁻¹   4. 413 m s⁻¹
5. 44.1 g mol⁻¹   6. 2.02   7. 242 m s⁻¹
8. a) 4.47   b) 5460 K

**Exercise 26**

1. b) i) Y      ii) $Cl_2$      c)

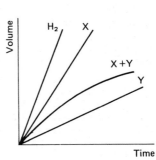

2. d
3. d
4. c) i) $1.99 \times 10^{-3}$ mol   ii) 190      iii) $XCl_4$, transition metal
5. c) i) 450 R J mol⁻¹   ii) 119R$^{1/2}$ m s⁻¹   iii) 2      iv) 40 s
6. a
7. b) 71      c) 1.29 atm      d) 132 m s⁻¹
8. b) i) $p(CH_4) = 13.5$ kPa, $p(C_2H_4) = 27.0$ kPa, $p(C_4H_{10}) = 60.8$ kPa
   iii) 745 cm³   iv) 82.5 kPa
9. a) M = 190      b) 75%, $A_r(X) = 48$   c) $A = MCl_4$, $B = MCl_3$

**CHAPTER 6**

**Exercise 27**

1. 343 g mol⁻¹   2. $PF_5$   3. 90 g mol⁻¹   4. 64.6 g mol⁻¹
5. 134 g mol⁻¹
6. a) 46 g mol⁻¹      b) 58 g mol⁻¹      c) 74 g mol⁻¹

**Exercise 28**

1. 0.400   2. 0.500   3. 73.9%   4. 0.300
5. 0.68

**Exercise 29**

1. 3.46 g   2. 1.98 g   3. $1.62 \times 10^{17}$ cm⁻³   4. 3.17 kPa

**Exercise 30**

1. 20 690 N m⁻²   2. a) 40 000 N m⁻²   b) 0.300
3. a) $1.60 \times 10^3$ Pa   b) $6.38 \times 10^2$ Pa   4. a) 38.0 kPa   b) 33.6 kPa

**Exercise 31**

1. 156 g mol⁻¹   2. 26.9%   3. 182 g mol⁻¹   4. 27.8%
5. 32.3 g   6. 123 g mol⁻¹

## Exercise 32

1. 13.6 g     2. 4.76 g     3. 20.0     4. 0.0514 mol
5. a) 4.74 g     b) 4.95 g     6. 25.0     7. $1.18 \times 10^{-2}$
8. a) 3.0 g     b) 3.5 g     9. a) 4.75 g     b) 4.95 g
10. $Cu(NH_3)_4^{2+}$

## Exercise 33

1. a) $1.67 \times 10^{25}$     b) $4.31 \times 10^{20}$
2. c) i) 4.80 g     ii) 5.33 g
3. 103 mg
4. b) i) 3.33 mol     ii) 3.75 mol     c) $67\%N_2$, $33\%O_2$
5. $K_c = 726 \, dm^3 \, mol^{-1}$
6. b) i) $p$(methanol) $= 22.5$ mm Hg, $p$(ethanol) $= 47.5$ mm Hg
    ii) 70.0 mm Hg
    iii) $x$(ethanol) $= 0.68$, $x$(methanol) $= 0.32$
7. b) i) $C_3H_6O$     ii) $C_3H_6O$
8. e
9. a) iii) $3.36 \times 10^{-4} \, mol \, dm^{-3}$
    b) ii) 1, $K_c = 712.5$, 2, $K_c = 708.3$, 3, $K_c = 714.3 \, mol^{-1} \, dm^3$
    d) $Cu^{2+} + 4NH_3 \longrightarrow Cu(NH_3)_4^{2+}$
10. i) 120     ii) $SOCl_2$, therefore accurate $M = 119$
11. e) 0.259     g) $93 \, g \, mol^{-1}$
12. a) i) $A$ 80 °C, $B$ 110 °C     ii) $A$ 136 kN m$^{-2}$, $B$ 54 kN m$^{-2}$
    b) vi) 0.481     c) ii) 60 kN m$^{-2}$
13. c) i) 0.20     ii) 0.86     iii) 0.40     e) ii) 0.4

# CHAPTER 7

## Exercise 34

1. a) $2.33 \times 10^3 \, N \, m^{-2}$     b) $2.30 \times 10^3 \, N \, m^{-2}$     c) $2.29 \times 10^3 \, N \, m^{-2}$
    d) $1.80 \times 10^3 \, N \, m^{-2}$
2. e     3. $63.7 \, g \, mol^{-1}$     4. $3.09 \times 10^3 \, Pa$

## Exercise 35

1. a) 100.21 °C     b) 100.31 °C     c) 100.59 °C     d) 100.12 °C
    e) 100.35 °C     f) 100.53
2. a) $60.0 \, g \, mol^{-1}$     b) $342 \, g \, mol^{-1}$     c) $73 \, g \, mol^{-1}$     d) $90 \, g \, mol^{-1}$
    e) $180 \, g \, mol^{-1}$

## Exercise 36

1. a) $52.5 \, g \, mol^{-1}$     b) $80.0 \, g \, mol^{-1}$     c) $59.9 \, g \, mol^{-1}$     d) $180 \, g \, mol^{-1}$
    e) $342 \, g \, mol^{-1}$
2. $180 \, g \, mol^{-1}$

## Exercise 37

1. a) 0.36 °C     b) 82.6 °C     c) $1.23 \times 10^4 \, N \, m^{-2}$
2. a) 100.13 °C     b) $-0.47$ °C     c) $2.33 \times 10^3 \, N \, m^{-3}$
3. $146 \, g \, mol^{-1}$
4. a) $-1.86$ °C   100.52 °C     b) $-0.37$ °C   100.10 °C
    c) $-0.37$ °C   100.10 °C     d) $-2.07$ °C   100.58 °C
    e) $-4.04$ °C   101.13 °C     5. 4.00 kg

## Exercise 38

1. a) $5.19 \times 10^5 \, N \, m^{-2}$   b) $1.98 \times 10^6 \, N \, m^{-2}$   c) $7.87 \times 10^5 \, N \, m^{-2}$
   d) $1.41 \times 10^6 \, N \, m^{-2}$   e) $2.44 \times 10^6 \, N \, m^{-2}$
2. a) $68.6 \, g \, mol^{-1}$   b) $166 \, g \, mol^{-1}$   c) $259 \, g \, mol^{-1}$
3. $-0.58 \, ^\circ C$   4. 50   5. 2 400
6. $1.66 \times 10^4 \, g \, mol^{-1}$   7. a) $2.01 \times 10^5 \, Pa$   b) $5.00 \, g \, dm^{-3}$

## Exercise 39

1. $-0.818 \, ^\circ C$   $100.229 \, ^\circ C$   2. $0.578 \, mol \, dm^{-3}$   3. $0.439 \, mol \, dm^{-3}$
4. $3.70 \, g$   5. $101.793 \, ^\circ C$   6. $0.0215$   7. $0.785$
8. $0.779$   9. $-0.0772 \, ^\circ C$   10. $C_6H_4Br_2$   11. $0.763$
12. $-0.389 \, ^\circ C$   13. $C_7H_6O_2$ in water   $C_{14}H_{12}O_4$ in benzene
14. $0.333 \, kg$ per kg water   15. $3.16 \times 10^3 \, Pa$
16. $M_r = 74$ in water, 146 in benzene   The acid is almost completely dimerised in benzene
    Degree of dimerisation $= 0.99$
17. a) $80 \, g \, mol^{-1}$   b) $5.66 \times 10^5 \, Pa$   18. $1.95 \times 10^5 \, N \, m^{-2}$

## Exercise 40

1. $1.24 \times 10^3 \, N \, m^{-2}$   2. a) $1.39 \times 10^4$
3. a) $107, CH_3C_6H_4NH_2$   b) f.p. $= -0.0544 \, ^\circ C$, $\pi = 7.24 \times 10^4 \, N \, m^{-2}$
4. c) $22.9 \, mm \, Hg$   d) $0.274 \, mol \, dm^{-3} : 16.0 \, g \, dm^{-3}$
5. a) i) $0.0269 \, mol$   ii) $180 \, g \, mol^{-1}$
   b) ii) $3\alpha$   iii) $1 - \alpha$   iv) $1 + 2\alpha$   v) $0.186 \, ^\circ C$
   vi) $0.81$
6. b) $p = 1.013 \times 10^5 \, Pa$, $M_r(X)$ in water $= 210$, in benzene $= 420 : X$ dimerises in benzene
   c) i) $\pi$(water)$/\pi$(benzene) $= 2$   ii) same f.p.
7. d) i) $-0.272 \, ^\circ C$,   ii) $-0.374 \, ^\circ C$, Solution i)
8. c) NaCl   d) $6.13 \times 10^3$
9. c) i) $0.75$   ii) $0.85$

# CHAPTER 8

## Exercise 41

1. $0.265 \, g$
2. $0.403 \, g$   a) doubled   b) doubled   c) unchanged
3. $1.24 \, g \, Ca$   $2.21 \, g \, Cl_2$   4. $0.0560 \, g$
5. a) $0.0672 \, A$   b) $23.1 \, cm^3$   c) $0.195 \, g$   6. 268 minutes
7. $0.454 \, A$   8. $1.77 \, mol \, dm^{-3}$   9. 2482 hours
10. $2.14 \, dm^3 \, O_2$   $4.28 \, dm^3 \, H_2$   11. $1.84 \times 10^4 \, C$   12. 1050 s
13. $0.292 \, dm^3$

## Exercise 42

1. a) $0.100 \, \Omega \, m$ or $10.0 \, \Omega \, cm$   b) $10.0 \, \Omega^{-1} m^{-1}$   $0.100 \, \Omega^{-1} cm^{-1}$
2. a) $0.060 \, \Omega \, m$ or $6.00 \, \Omega \, cm$   b) $16.0 \, \Omega^{-1} m^{-1}$   $0.160 \, \Omega^{-1} cm^{-1}$
3. $1.77 \, m^{-1}$ or $1.77 \times 10^{-2} \, cm^{-1}$
4. $1.63 \times 10^{-2} \, \Omega^{-1} m^2 \, mol^{-1}$ or $163 \, \Omega^{-1} cm^2 \, mol^{-1}$
5. $0.210 \, \Omega^{-1} m^2 \, mol^{-1}$

## Exercise 43

1. $1.75 \times 10^{-5} \, mol \, dm^{-3}$
2. a) $0.0271$   b) $6.78 \times 10^{-4} \, mol \, dm^{-3}$   c) $1.89 \times 10^{-5} \, mol \, dm^{-3}$
3. a) $0.230$   b) $1.37 \times 10^{-3} \, mol \, dm^{-3}$
4. $1.75 \times 10^{-4} \, mol \, dm^{-3}$
5. a) $0.0256$   b) $2.02 \times 10^{-5} \, mol \, dm^{-3}$

## Exercise 44

**1.**

| | pH | pOH | | pH | pOH | | pH | pOH | | pH | pOH |
|---|---|---|---|---|---|---|---|---|---|---|---|
| a) | 8 | 6 | b) | 4 | 10 | c) | 7 | 7 | d) | 2.2 | 11.8 |
| e) | 4.5 | 9.5 | f) | 1.5 | 12.5 | g) | 0.60 | 13.4 | h) | 8.3 | 5.7 |
| i) | 6.2 | 7.8 | j) | 1.0 | 13.0 | | | | | | |

**2.** a) 12         b) 11         c) 6.0         d) 12.7
    e) 11         f) 12.9         g) 12         h) 9.7
    i) 6.8         j) 4.6

**3.** In mol dm$^{-3}$, the values are:
    a) 1.00         b) $5.01 \times 10^{-5}$         c) $4.47 \times 10^{-3}$         d) 0.0132
    e) $7.08 \times 10^{-5}$         f) $1.45 \times 10^{-8}$         g) $6.17 \times 10^{-10}$         h) $2.00 \times 10^{-14}$
    i) $3.16 \times 10^{-1}$         j) $2.34 \times 10^{-3}$

**4.** a) 0.784         b) 1.05         c) 13.3         d) 12.7
    e) 13.4

**5.** 2.52                 **6.** a) $1.00 \times 10^{-6}$ mol dm$^{-3}$         b) 6.00

**7.** 9.92

**8.** a) $2.04 \times 10^{-4}$         b) $1.77 \times 10^{-5}$         c) $3.96 \times 10^{-10}$         d) $1.84 \times 10^{-4}$
    e) $3.37 \times 10^{-2}$         (all in mol dm$^{-3}$)

**9.** a) $1.81 \times 10^{-5}$         b) $3.97 \times 10^{-10}$         c) $1.38 \times 10^{-3}$         d) $1.43 \times 10^{-5}$
    e) $2.00 \times 10^{-9}$         (all in mol dm$^{-3}$)

**10.** a) $2.28 \times 10^{-11}$         b) $5.62 \times 10^{-10}$         c) $1.86 \times 10^{-11}$         d) $1.44 \times 10^{-11}$
    e) $4.24 \times 10^{-10}$         (all in mol dm$^{-3}$)

**11.** a) $4.46 \times 10^{-4}$         b) $1.89 \times 10^{-2}$         c) $2.41 \times 10^{-2}$         d) $1.89 \times 10^{-2}$
    e) $7.93 \times 10^{-2}$         (all in mol dm$^{-3}$)

**12.** a) 2.00         b) 12.0         c) 2.30         **13.** 0.0110%

**14.** 11.1         **15.** a) $10^{-6}$         b) 6

**16.** pH = 3.0         a) 9.0 cm$^3$         b) 0.90 cm$^3$

## Exercise 45

**1.** a         **2.** 1.00 mole         **3.** a) 3.34         b) 3.94
**4.** a) 4.73         b) 0.117 mol         **5.** 2.86
**6.** a) A; 4.75 $\longrightarrow$ 4.66         B: 4.45 $\longrightarrow$ 4.31
        C: 4.05 $\longrightarrow$ 3.71         D: 4.75 $\longrightarrow$ 4.71
    b) Buffering capacity decreases as [Salt]/[Acid] decreases, and increases as concentration increases
**7.** a) 2.40         b) 3.49         c) 3.75
**8.** a) 2.72         b) 4.44         c) 4.38

## Exercise 46

**1.** a) 5.13         b) 6.32         c) 11.2         d) 8.38
    e) 8.38
**2.** a) 8.72         b) 4.93         c) 11.35         d) 8.32
    e) 5.08         f) 0.569 mol dm$^{-3}$
**3.** a) 8.88         b) 8.03         c) 10.81         d) 5.91
    e) 3.72         f) 2.96

## Exercise 47

**1.** 3.0         **2.** $4.1 \times 10^{-17}$ mol dm$^{-3}$
**3.** $1.1 \times 10^{-3}$ mol dm$^{-3}$         **4.** $2.8 \times 10^{-7}$ mol$^4$ dm$^{-12}$
**5.** $4.7 \times 10^{-13}$ mol$^4$ dm$^{-12}$

## Exercise 48

**1.** a) $1.69 \times 10^{-28}$ mol$^2$ dm$^{-6}$         b) $3.99 \times 10^{-19}$ mol$^2$ dm$^{-6}$
    c) $5.93 \times 10^{-51}$ mol$^3$ dm$^{-9}$         d) $1.08 \times 10^{-93}$ mol$^5$ dm$^{-15}$
    e) $5.00 \times 10^{-16}$ mol$^3$ dm$^{-9}$

**2.** a) $2.52 \times 10^{-27} \, mol^2 \, dm^{-6}$     b) $8.29 \times 10^{-17} \, mol^2 \, dm^{-6}$
  c) $1.07 \times 10^{-10} \, mol^2 \, dm^{-6}$     d) $2.68 \times 10^{-11} \, mol^3 \, dm^{-6}$
  e) $1.25 \times 10^{-16} \, mol^2 \, dm^{-6}$

**3.** a) $6.3 \times 10^{-34} \, mol \, dm^{-3}$     b) $6.3 \times 10^{-43} \, mol \, dm^{-3}$
  c) $3.2 \times 10^{-14} \, mol \, dm^{-3}$     d) $2.2 \times 10^{-30} \, mol \, dm^{-3}$
  e) $1.6 \times 10^{-46} \, mol \, dm^{-3}$     f) $6.3 \times 10^{-12} \, mol \, dm^{-3}$

**4.** a) Yes      b) No      c) Yes      d) No
  e) Yes      f) No

**5.** CdS and NiS

**6.** $K_{sp}(AgCl) = 1.96 \times 10^{-10} \, mol^2 \, dm^{-6}$
  $K_{sp}(Ag_2CrO_4) = 3.61 \times 10^{-12} \, mol^3 \, dm^{-9}$
  a) $[Ag^+] = 1.96 \times 10^{-9} \, mol \, dm^{-3}$     b) $[Ag^+] = 2.69 \times 10^{-5} \, mol \, dm^{-3}$

**7.** $CaCO_3$    9.99 g         **8.** 0.3 ion $dm^{-3}$       **9.** $PbI_2$     45.6 g

**10.** a) $8.12 \times 10^{-3} \, g \, dm^{-3}$   b) $6.38 \times 10^{-8} \, g \, dm^{-3}$   c) $9.63 \times 10^{-4} \, g \, dm^{-3}$

**11.** a) $6.3 \times 10^{-4} \, mol \, dm^{-3}$     b) $4.0 \times 10^{-6} \, mol \, dm^{-3}$

**12.** a) $1.2 \times 10^{-3}$     b) $1.8 \times 10^{-7} \, mol \, dm^{-3}$

**13.** a) $2.15 \times 10^{-4} \, mol \, dm^{-3}$     b) $4.00 \times 10^{-5} \, mol \, dm^{-3}$
  c) $7.12 \times 10^{-8} \, mol \, dm^{-3}$

# Exercise 49

**1.** Ag, $I^-$        **2.** $I_2$, $Fe^{3+}$

**3.** a) $+0.40 \, V$      b) $+0.26 \, V$      c) $-0.27 \, V$

**4.** a) $+0.94 \, V$      b) $+0.44 \, V$      c) $-0.67 \, V$      d) $+0.78 \, V$
  e) $+0.63 \, V$

**5.** $Fe(s) + Fe^{3+}(aq) \longrightarrow 2Fe^{2+}(aq)$

**6.** $Cr_2O_7^{2-}(aq) + 14H^+(aq) + 6Fe^{3+}(aq) \longrightarrow 2Cr^{3+}(aq) + 6Fe^{3+}(aq) + 14H_2O$

**7.** a) $2Fe^{3+}(aq) + 2I^-(aq) \longrightarrow 2Fe^{2+}(aq) + I_2(aq)$
  b) $2Ag^+(aq) + Cu(s) \longrightarrow 2Ag(s) + Cu^{2+}(aq)$
  c) No reaction      d) No reaction
  e) $Br_2(aq) + 2Fe^{2+}(aq) \longrightarrow 2Br^-(aq) + 2Fe^{3+}(aq)$
    i) $Cl_2$ and $Br_2$        ii) $Cl_2$, $Br_2$ and $I_2$

# Exercise 50

**1.** a) $5.45 \times 10^{-4} \, mol \, dm^{-3}$, 11.7       **2.** b) $4.71 \times 10^{-6} \, mol \, dm^{-3}$

**3.** b) i) $1.59 \times 10^{-2} \, mol \, l^{-1}$      ii) $4 \times 10^{-3} \, mol$
  d) iii) $1.23 \times 10^{-15} \, mol \, l^{-1}$

**4.** c) i) $0.180 \, g$, $96 \, 300 \, C \, mol^{-1}$     ii) $2+$     iii) $6.011 \times 10^{23} \, mol^{-1}$     iv) $Y$

**5.** b) 4.74            c) 1.34%       d) pH = 2.87, 3.05, 3.35, 3.83, 12.0

**6.** b) ii) $1.74 \times 10^{-5} \, mol \, dm^{-3}$      iii) 2.68

**7.** c) i) $4.31 \times 10^{-2} \, mol \, dm^{-3}$      ii) $8.62 \times 10^{-2} \, mol \, dm^{-3}$
  iii) $3.20 \times 10^{-4} \, mol^3 \, dm^{-9}$     iv) $2.43 \, g \, dm^{-3}$       v) $83.5 \, cm^3$

**8.** a) Solubilities: $AgCl = 1.27 \times 10^{-5} \, mol \, dm^{-3}$, $Ag_2CrO_4 = 1.00 \times 10^{-4} \, mol \, dm^{-3}$
  b) $[Ag^+] = 4.0 \times 10^{-3} \, mol \, dm^{-3}$, mass $= 0.680 \, g$     c) $30.0 \, g$

**9.** b) i) $5.56 \times 10^{-10} \, mol \, l^{-1}$      ii) 3.02

**10.** b) $\alpha = 0.478$, $K_a = 2.86 \times 10^{-2} \, mol \, dm^{-3}$
  c) i) $140.3 \, ohm^{-1} \, cm^2 \, mol^{-1}$     ii) $8.84 \times 10^{-7} \, mol \, dm^{-3}$     iii) $7.81 \times 10^{-13} \, mol^2 \, dm^{-6}$

**11.** b) i) $MnO_4^-(aq) + H^+(aq)$ and $Ce^{4+}(aq)$
  ii) $MnO_4^-(aq) + 5VO^{2+}(aq) + H_2O(l) \longrightarrow Mn^{2+}(aq) + 5VO_2^+(aq) + 2H^+(aq)$

**12.** a) i) $2Fe^{3+} + 2I^- \longrightarrow 2Fe^{2+} + I_2$     ii) no reaction
  iii) $2[Fe(CN)_6]^{4-} + I_2 \longrightarrow 2[Fe(CN)_6]^{3-} + 2I^-$
  b) i) $+7$      ii) $Cl_2O_7$

**13.** a) $2.74 \, g \, Sr$, $4.98 \, g \, Br_2$        b) i) 3.37      ii) 5.73
  c) i) 5.61 kJ      ii) 5.61 kJ, ratio = 1.41

**14.** a) iv) $IO_3^- + 6H^+ + 5I^- \longrightarrow 3I_2 + 3H_2O$
  v) $2Cr^{2+} + 2H^+ \longrightarrow 2Cr^{3+} + H_2(g)$

**15.** c) i) $+1.14 \, V$      ii) no

# CHAPTER 9

## Exercise 51

1. $-52.9\,kJ\,mol^{-1}$    2. $-58.4\,kJ\,mol^{-1}$    3. $-49.3\,kJ\,mol^{-1}$    4. $-56.4\,kJ\,mol^{-1}$
5. $-59.1\,kJ\,mol^{-1}$

## Exercise 52

1. a) $-3.23\,MJ\,mol^{-1}$    b) $-3.24\,MJ\,mol^{-1}$  2. a) $-9.91\,kJ\,mol^{-1}$    b) $-3520\,kJ\,mol^{-1}$
3. a) $1.24\,kJ\,mol^{-1}$    b) $\Delta U^{\ominus} - \Delta H^{\ominus} = 1.24\,kJ\,mol^{-1}$
4. a) $-4210\,kJ\,mol^{-1}$    b) $-2220\,kJ\,mol^{-1}$

## Exercise 53

1. a) $-484\,kJ\,mol^{-1}$    b) $108\,kJ\,mol^{-1}$    c) $-75$  and  $+33\,kJ\,mol^{-1}$
   d) $-118\,kJ\,mol^{-1}$    e) $-11\,kJ\,mol^{-1}$    f) $-106$  and  $+504\,kJ\,mol^{-1}$
   g) $-251, -246, -471$ and $-152\,kJ\,mol^{-1}$
2. $-78.2\,kJ\,mol^{-1}$    3. a) $-297$    b) $-394$    c) $-262\,kJ\,mol^{-1}$
4. $-64\,kJ\,mol^{-1}$    5. $-847\,kJ\,mol^{-1}$  exothermic    6. $-0.30\,kJ\,mol^{-1}$
7. $184\,kJ\,mol^{-1}$    8. $-106\,kJ\,mol^{-1}$  Mg reduces $Al_2O_3$
9. a) $-890\,kJ\,mol^{-1}$    b) $+317\,kJ\,mol^{-1}$    c) $-55\,kJ\,mol^{-1}$
   d) i) $-1.43 \times 10^{5}\,kJ$    ii) $-2.97 \times 10^{4}\,kJ$    iii) $-4.84 \times 10^{4}\,kJ$
10. $416\,kJ\,mol^{-1}$
11. a) $C_2H_6, -76$; $C_2H_4, +48\,kJ\,mol^{-1}$    b) $-95\,kJ\,mol^{-1}$    c) $-200\,kJ\,mol^{-1}$
    d) $-343\,kJ\,mol^{-1}$
    e) $+264\,kJ\,mol^{-1}$    The difference is the 'bond delocalisation energy' of benzene
    f) $+136\,kJ\,mol^{-1}$    28 kJ mol$^{-1}$ higher than the value from combustion    Butadiene
    is more stable than it is calculated to be because it is stabilised by bond delocalisation
    g) i) $+74\,kJ\,mol^{-1}$    ii) $-20\,kJ\,mol^{-1}$  Reaction ii) will occur
12. $-362\,kJ\,mol^{-1}$    13. $-440\,kJ\,mol^{-1}$    14. $-372\,kJ\,mol^{-1}$    15. b) $-380\,kJ\,mol^{-1}$
16. $-775\,kJ\,mol^{-1}$
17. a) $-387\,kJ\,mol^{-1}$    b) $-220\,kJ\,mol^{-1}$
    $\Delta H_f^{\ominus}(CaCl_2)$ has a larger negative value than $\Delta H_f^{\ominus}(CaCl)$
18. a) Solubility (LiCl) > Solubility (NaCl) since $(\Delta H_{lattice}^{\ominus} + \Delta H_{hydration}^{\ominus})$ has a more
    negative value for LiCl than for NaCl.
    b) Solubility (NaCl) > Solubility (NaF) since $(\Delta H_{lattice}^{\ominus} + \Delta H_{hydration}^{\ominus})$ has a positive
    value for NaF and a negative value for NaCl.
19. $-367\,kJ\,mol^{-1}$
20. $\Delta H_c^{\ominus}$ is $< 3\Delta H_a^{\ominus}$ because bond delocalisation makes benzene more stable than
    calculated from bond energy terms, and less enthalpy than expected is released when
    benzene $\longrightarrow$ cyclohexane
    The hydrogenation of the first double bond destroys the bond delocalisation in
    benzene, and $\Delta H_b$ is therefore positive, showing that the enthalpy content of the system
    has increased

## Exercise 54

1. $87.8\,J\,K^{-1}\,mol^{-1}$    2. $8.5\,J\,K^{-1}\,mol^{-1}$
3. HF, $25.6\,J\,K^{-1}\,mol^{-1}$    HCl, $86.2\,J\,K^{-1}\,mol^{-1}$    HBr, $85.4\,J\,K^{-1}\,mol^{-1}$
   HI, $83.2\,J\,K^{-1}\,mol^{-1}$
4. $36.6\,J\,K^{-1}\,mol^{-1}$    5. $124\,J\,K^{-1}\,mol^{-1}$
6. In $J\,K^{-1}\,mol^{-1}$:    a) $+20.0$    b) $-199$    c) $-163.5$
                               d) $-121$    e) $+176$    f) $-90.1$
                               g) $+285$    h) $+681$
7. a) $+$    b) $+$    c) $-$    d) $-$
   e) $+$    f) $-$
8. a) $97.0\,J\,K^{-1}\,mol^{-1}$    b) $145\,J\,K^{-1}\,mol^{-1}$    c) $97.3\,J\,K^{-1}\,mol^{-1}$
9. a) $+0.202\,kJ\,mol^{-1}$ at $25\,°C$    $+0.127\,kJ\,mol^{-1}$ at $100°C$    above $227\,°C$
   b) $+2.56\,kJ\,mol^{-1}$ at $25\,°C$    above $491°C$
10. a) $-2.91\,kJ\,mol^{-1}$    b) $+2.91\,kJ\,mol^{-1}$    trans

**Exercise 55**

1. b) $-645\,\text{kJ mol}^{-1}$
2. b) i) $-600\,\text{kJ mol}^{-1}$     ii) $-630\,\text{kJ mol}^{-1}$
   iii) Bond energy is not the same in different compounds.
   iv) Bond delocalisation energy
3. b) i) decrease     ii) decrease     iii) increase
   c) $\Delta G$ negative
   d) i) $-121\,\text{J K}^{-1}\,\text{mol}^{-1}$     ii) $-199\,\text{J K}^{-1}\,\text{mol}^{-1}$
   e) i) $\Delta G^{\ominus} = -191\,\text{kJ mol}^{-1}$, yes     ii) in the presence of a catalyst
4. c) $-441\,\text{kJ mol}^{-1}$       5. d
6. a) $-220\,\text{kJ mol}^{-1}$
   The formation of $CaCl_2$ is more exothermic, therefore $CaCl_2$ will be formed.
   b) $\Delta H(\text{Li}) = 170\,\text{kJ mol}^{-1}$; $\Delta H(\text{Cs}) = 180\,\text{kJ mol}^{-1}$
   c) $\Delta H = -37.4\,\text{kJ mol}^{-1}$; solubility decreases with an increase in temperature
7. a, b and c
8. b) $-639\,\text{kJ mol}^{-1}$
   c) i) $G$     ii) $MgCl_2 > MgCl > Mg > MgCl_3$     iii) $MgCl_2 > MgCl$
   d) $-155\,\text{kJ mol}^{-1}$
9. b) i) $6.8\,^{\circ}\text{C}$     ii) $-57\,\text{kJ mol}^{-1}$       c) same
10. b) i) $2.1\,^{\circ}\text{C}$; $4.2\,^{\circ}\text{C}$     ii) 3rd       iii) 1st
    c) i) $176\,\text{J}$     ii) $17.6\,\text{kJ mol}^{-1}$
11. a) ii) $-1470\,\text{kJ mol}^{-1}$       b) i) $-1090\,\text{kJ mol}^{-1}$
12. b) ii) 40% propane, 60% butane     c) $60\,\text{kJ mol}^{-1}$
13. c) $MgCl_2 - 654\,\text{kJ mol}^{-1}$, $MgCl - 112\,\text{kJ mol}^{-1}$
    d) $-215\,\text{kJ (mol of MgCl)}^{-1}$
14. b) $\Delta H$ for $X = Cl$ is $-621\,\text{kJ mol}^{-1}$; $\Delta H$ for $X = I$ is $-518\,\text{kJ mol}^{-1}$
    c) $\Delta H_L(\text{NaCl}) = -772\,\text{kJ mol}^{-1}$; $\Delta H_L(\text{NaI}) = -682\,\text{kJ mol}^{-1}$
15. Ethane: $351\,\text{kJ mol}^{-1}$, ethene: $602\,\text{kJ mol}^{-1}$, ethyne: $815\,\text{kJ mol}^{-1}$,
    benzene: $507\,\text{kJ mol}^{-1}$
16. b) $\Delta U$       c) $-5161\,\text{kJ mol}^{-1}$     d) $\Delta H(\Delta V$ is negative)
17. b) $76\,\text{kJ mol}^{-1}$       c) decrease     d) $382\,\text{K}$
18. $\Delta H_L = -2242\,\text{kJ mol}^{-1}$
19. i) $-1075\,\text{kJ mol}^{-1}$     ii) $87\,\text{kJ mol}^{-1}$
20. c) $-104\,\text{kJ mol}^{-1}$

# CHAPTER 10

**Exercise 56**

1. 2       2. 1 w.r.t. A    2 w.r.t. B       3. d

4. $\dfrac{d[X]}{dt} = k[X]^0$     0       5. $10.0\,\text{mol}^{-1}\,\text{dm}^3\,\text{s}^{-1}$

6. $\dfrac{d[P]}{dt} = k[A][B]$     $2.0 \times 10^{-3}\,\text{dm}^3\,\text{mol}^{-1}\,\text{s}^{-1}$

7. a) 1    b) 2    $1.67 \times 10^5\,\text{mol}^{-2}\,\text{dm}^6\,\text{h}^{-1}$ or $46.4\,\text{mol}^{-2}\,\text{dm}^6\,\text{s}^{-1}$
8. 1     $1.92 \times 10^{-4}\,\text{s}^{-1}$

9. $\dfrac{d[N_2O_5]}{dt} = k[N_2O_5]$     a) $0.150\,\text{mol dm}^{-3}\,\text{s}^{-1}$    b) $1.80\,\text{mol dm}^{-3}\,\text{s}^{-1}$

**Exercise 57**

1. $1.54 \times 10^{-4}\,\text{s}^{-1}$       2. $1.10 \times 10^{-5}\,\text{s}^{-1}$
3. Gradients/$\text{mol dm}^{-3}\,\text{min}^{-1}$:       a) $-0.842$       b) $-0.386$
   c) $-0.200$     d) 0     $0.365\,\text{min}^{-1}$ or $6.10 \times 10^{-3}\,\text{s}^{-1}$     1
4. a) Initial rates/$\text{mol dm}^{-3}\,\text{min}^{-1}$:   1) $6.30 \times 10^{-3}$    2) $1.27 \times 10^{-2}$    3) $1.90 \times 10^{-2}$
   b) 1       c) $3.75\,\text{min}$    $7.50\,\text{min}$    1
   d) $\dfrac{d[C]}{dt} = k[A][B]$     e) $1.06 \times 10^{-2}\,\text{dm}^3\,\text{mol}^{-1}\,\text{s}^{-1}$

## Exercise 58

1. $50.0\,dm^3\,mol^{-1}\,s^{-1}$        2. $1.40 \times 10^{-4}\,dm^3\,mol^{-1}\,s^{-1}$

3. 1 w.r.t. A    2 w.r.t. B    Rate $= k\,[A]\,[B]^2$        $4.0\,dm^6\,mol^{-2}\,s^{-1}$

## Exercise 59

1. 170 min        2. 64.5 h

3. An isotope with $t_{1/2} = 0.85$ h decays to form an isotope with $t_{1/2} = 36$ h

4. 14 days        5. a) 0.031 25 $(= 1/2^5)$    b) $9.3 \times 10^{-10} (= 1/2^{30})$

6. 832 years        7. $1.00 \times 10^{-5}\,g$        8. 4.20 h

9. 43.1 min        10. a) 75.5%        b) 5.9%        11. 3276 years

## Exercise 60

1. $3.85 \times 10^{-4}\,s^{-1}$    70 min        2. $4.16 \times 10^{-6}\,s^{-1}$    3. $5.86 \times 10^{-4}\,s^{-1}$

4. a) $6.17 \times 10^{-4}\,mol^{-1}\,dm^3\,s^{-1}$        b) $1.62 \times 10^4\,s$ (4.5 h)

   c) $1.79 \times 10^4\,s$ (5.0 h)

5. a) $1.6 \times 10^{-6}\,s^{-1}$      b) 20 days        6. $2.95 \times 10^{-8}\,mol\,dm^{-3}\,s^{-1}$

7. $1.36 \times 10^{-7}\,mol\,dm^{-3}\,s^{-1}$

8. a) 2        b) $1.0 \times 10^{-7}\,m^2\,N^{-1}\,s^{-1}$

9. Experiment 1: $\dfrac{d[A]}{dt} = k\,[A]^2[B]$    $k = 0.417$

   Experiment 2: $\dfrac{d[A]}{dt} = k\,[A]^2$    $k = 0.05$    B is adsorbed on the surface

10. a) 1        b) $7.48 \times 10^{-4}\,s^{-1}$

11. a) 2        b) $5.96 \times 10^{-4}\,kN\,m^{-2}\,s^{-1}$

## Exercise 61

1. $29.0\,kJ\,mol^{-1}$        2. $25.5\,kJ\,mol^{-1}$        3. 2.25        4. $78.3\,kJ\,mol^{-1}$

5. $13.6\,kJ\,mol^{-1}$

## Exercise 62

1. c) $5.8 \times 10^{-4}\,s^{-1}$        2. 1 w.r.t. $[H^+]$, 0 w.r.t. $[I_2]$

3. 1 w.r.t. [ester], 1 w.r.t. $[H_3O^+]$, $K_c = 3.45$

4. ii) $1 : 3.21 \times 10^{-6}\,s^{-1}$, $2 : 2.5 \times 10^5\,Pa$        5. b) 1100 s

6. $1 : k = 0.200\,min^{-1}$, $2 : k = 0.0400\,min^{-1}$

   Order w.r.t. $[OH^-] = 1$; overall order $= 2$

   $k = 0.400\,min^{-1}\,dm^3\,mol^{-1}$

7. b) $k = 8 \times 10^{-4}\,s^{-1}\,l\,mol^{-1}$        c) Rate $= 7.2 \times 10^{-3}\,mol\,l^{-1}\,s^{-1}$

   d) Rate $= 2 \times 10^{-4}\,mol\,l^{-1}\,s^{-1}$

8. b) 2 w.r.t. [X], 1 w.r.t. [Ar]

   $k = 8.70 \times 10^9\,dm^6\,mol^{-2}\,s^{-1}$

9. a) i) $^4_2He$    ii) $^0_{-1}e$    iii) $^1_1H$        b) 2.0 g

   c) ii) $1.25 \times 10^9$ years    iii) 1

   d) i) $n_0 = 205.6 \times 10^{15}$        ii) $t = 1.78 \times 10^{10}$ years

10. d

11. a) ii) 1 w.r.t. $[HgCl_2]$ ; 2 w.r.t. $[C_2O_4^{2-}]$        iii) $k = 152\,dm^3\,mol^{-1}\,min^{-1}$

    c) 1st order: independent of $[A_0]$ ; 2nd order: $\propto 1/[A_0]$

    d) $t_{1/2} = 13.7$ min, $k = 5.05 \times 10^{-2}\,min^{-1}$

12. $k = 1.15 \times 10^{-4}\,s^{-1}$

13. $E = 91.2\,kJ\,mol^{-1}$, $k = 8.44 \times 10^{-3}\,mol\,dm^{-3}\,s^{-1}$

14. b) $p_a = p_0(1 - x)$

    $x = (p_t - p_0)/p_0$

    $k = 1.57 \times 10^{-3}\,s^{-1}$

15. i) $k = 0.177\,mol\,dm^{-1}\,s^{-1}$        ii) order $= 2$

16. b) iii) 4780 years

# CHAPTER 11

## Exercise 63

1. a) $K_p = \dfrac{p_C \times p_D}{p_A \times p_B}$    b) $K_c = \dfrac{[C][D]}{[A][B]}$    c) $K_{Het} = \dfrac{p_D}{p_B}$    d) $K_{Het} = \dfrac{[D]}{[A][B]}$

2. a) 0.16    b) $1.05 \times 10^4\,Pa$    3. $1.78 \times 10^5\,N\,m^{-2}$    4. $\frac{1}{3}$

5. a) 0.34    b) $8.1 \times 10^4\,N\,m^{-2}$

6. 64    7. $3.39 \times 10^{-15}\,N^{-2}\,m^4$

8. a) $2.00 \times 10^{-3}\,dm^3\,mol^{-1}\,s^{-1}$    b) 64.0

9. a) 0.71 mol    b) 43 g    10. 98.8%    11. c

12. a) 0.483 mol    b) 0.483 mol    c) 0.517 mol    d) 4.517 mol

13. $9.26 \times 10^{-15}\,Pa^{-2}$    14. $0.2\,atm^{-1}$

15. a) $CO_2, H_2$: 0.27 mol    $CO, H_2O$: 0.23 mol

    b) $CO_2$: 4.2 mol    $H_2$: 0.21 mol    $CO, H_2O$: 0.79 mol

16. $9.33 \times 10^{-4}\,N^{-1}\,m^2$

## Exercise 64

1. i) At $T = 350$, $p(N_2) = 2.32\,atm$, $p(H_2) = 6.95\,atm$

    At $T = 450$, $p(N_2) = 2.45\,atm$, $p(H_2) = 7.35\,atm$

  ii) At $T = 350$, $K_p = 6.95 \times 10^{-4}\,atm^{-2}$

    At $T = 450$, $K_p = 4.29 \times 10^{-5}\,atm^{-2}$

  iii) negative

  iv) $K_p = 7.70 \times 10^{-4}\,atm^{-2}$

    The yield increases with pressure.

2. d) 54.7

3. f) $\alpha = 10^{-2}$    g) $\Delta H = -206.8\,kJ\,mol^{-1}$

  h) i) less at $1500\,°C$ than at $1000\,°C$    ii) less at $1500\,°C$ than at $1000\,°C$

4. b) $K_p = 4P^3/27$    c) $K_p = 1.19 \times 10^{-3}\,atm^3$

  d) $\Delta H^{\ominus} = 165\,kJ\,mol^{-1}$

  e) i) $P$ increases with $T$    ii) $(\Delta H^{\ominus} - T\Delta S^{\ominus})$ decreases with $T$

5. $K_p = 8.89 \times 10^{-2}\,N\,m^{-2}$

6. $K_c = 8.10 \times 10^{-3}\,mol\,l^{-1}$    7. $K_c = 4.1$

8. a) i) $p_3 = 2p_1$    ii) $K_p = p_2/0.75$

  b) At $30\,°C$, $p\,(vapour) = 0.8\,atm$, $p\,(air) = 1.37\,atm$, and the sum exceeds 2 atm

# CHAPTER 12

## Exercise 65

1. $C_2H_5$    $C_4H_{10}$    2. $88\,g\,mol^{-1}$    $C_3H_7CO_2H$    3. $C_{11}H_{10}O$

4. $C_{16}H_{13}O_3N$    5. A is $C_6H_{12} = CH_3CH_2CH=CHCH_2CH_3$    B is $C_2H_5CO_2H$

6. A is $C_6H_5CO_2CH_3$,    B is $CH_3OH$    C is $C_6H_5CO_2Na$    D is $C_6H_5CO_2H$

7. a) $C_3H_7ON$    b) propanamide, $C_2H_5CONH_2$

8. a) 3 mol $Br_2$ : 1 mol $C_6H_5OH$

  b) $C_6H_5OH + 3Br_2 \longrightarrow C_6H_2Br_3OH + 3HBr$

9. a) $55.9\,g\,mol^{-1}$    b) $C_4H_8$

  but-1-ene, $CH_3CH_2CH=CH_2$ and but-2-ene, $CH_3CH=CHCH_3$

10. a) CHO    b) $C_4H_4O_4$

  c) butenedioic acid, cis $HCCO_2H$ and trans $HO_2CH$

$$\begin{array}{cc} \quad\;\; \| & \quad\;\; \| \\ HCCO_2H & HCCO_2H \end{array}$$

11. a) $C_4H_{10}O$    b) $C_4H_{10}O$    c) $C_2H_5OC_2H_5$    $CH_3OCH_2CH_2CH_3$

  $CH_3OCH(CH_3)_2$    $CH_3CH_2CH_2CH_2OH$    $CH_3CH_2C(OH)CH_3$

  $CH_3CH(CH_3)CH_2OH$    $CH_3CH_2CH(CH_3)OH$

12. A is $C_8H_9ON = C_6H_5NHCOCH_3$    B is $C_6H_7N = C_6H_5NH_2$

**Exercise 66**

1. a) $C_2H_5NO$        b) X is $C_2H_4O_2$   W is $CH_3CONH_2$
    c) $CH_3CONH_2(aq) + HCl(aq) + H_2O(l) \longrightarrow CH_3CO_2H(aq) + NH_4Cl(aq)$
    d) Y is $CH_3CHO$

2. a) $C_4H_4O$            b) $C_8H_8O_2$
    c) A is $CH_3C_6H_4CON(CH_3)_2$   B is $(CH_3)_2NH$   C is $CH_3C_6H_4CO_2H$

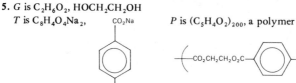

    D is $C_6H_5CO_2H$         E is $C_6H_4$        F is $(CH_3)_2NH_2^+Cl^-$
                                             G is $(CH_3)_2N—N=O$

3. a) $C_4H_{10}O$           b) $74; C_4H_{10}O$
4. b
5. G is $C_2H_6O_2$, $HOCH_2CH_2OH$
    T is $C_8H_4O_4Na_2$,        P is $(C_5H_4O_2)_{200}$, a polymer

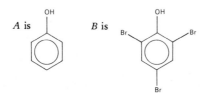

6. a) $M_r = 48$          b) i) $C_2H_6O$     ii) $C_2H_6O$, $M_r = 46$
7. $C_6H_3Br_3O$ is the empirical and the molecular formula

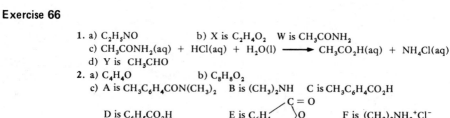

   A is             B is

8. c) 120
    d)

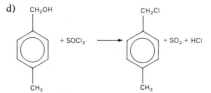

9. B is $C_5H_{11}N$,       C is $C_{12}H_{10}O_4$,       A is

 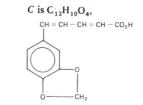 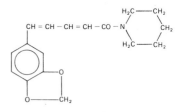

10. C is $1,4\text{-}ClC_6H_4CH_2Cl$
     D is $1,4\text{-}ClC_6H_4CH_2OH$
     E is $1,4\text{-}ClC_6H_4CO_2H$
     F is $1,4\text{-}ClC_6H_4COCl$
     G is $1,4\text{-}ClC_6H_4COC_6H_5$
11. X is $C_5H_{11}CO_2CH(CH_3)_2$
     A is $(CH_3)_2CHOH$
     B is $C_5H_{11}CO_2H$
12. A is $C_6H_5CH_2NH_2$
     B is $C_6H_5CH_2OH$
     C is $C_6H_5CO_2H$
     D is $CH_3C_6H_4NH_2$

# Table of Relative Atomic Masses

| Element | Symbol | Atomic number | Relative atomic mass | Element | Symbol | Atomic number | Relative atomic mass |
|---------|--------|---------------|----------------------|---------|--------|---------------|----------------------|
| Aluminium | Al | 13 | 27.0 | Lead | Pb | 82 | 207 |
| Antimony | Sb | 51 | 122 | Lithium | Li | 3 | 6.94 |
| Argon | Ar | 18 | 40.0 | Magnesium | Mg | 12 | 24.3 |
| Arsenic | As | 33 | 75.0 | Manganese | Mn | 25 | 54.9 |
| Barium | Ba | 56 | 137 | Mercury | Hg | 80 | 200 |
| Beryllium | Be | 4 | 9.0 | Neon | Ne | 10 | 20.2 |
| Bismuth | Bi | 83 | 209 | Nickel | Ni | 28 | 58.7 |
| Boron | B | 5 | 10.8 | Nitrogen | N | 7 | 14.0 |
| Bromine | Br | 35 | 80.0 | Oxygen | O | 8 | 16.0 |
| Cadmium | Cd | 48 | 112.5 | Phosphorus | P | 15 | 31.0 |
| Calcium | Ca | 20 | 40.1 | Platinum | Pt | 78 | 195 |
| Carbon | C | 6 | 12.0 | Potassium | K | 19 | 39.1 |
| Chlorine | Cl | 17 | 35.5 | Selenium | Se | 34 | 79.0 |
| Chromium | Cr | 24 | 52.0 | Silicon | Si | 14 | 28.1 |
| Cobalt | Co | 27 | 59.0 | Silver | Ag | 47 | 108 |
| Copper | Cu | 29 | 63.5 | Sodium | Na | 11 | 23.0 |
| Fluorine | F | 9 | 19.0 | Strontium | Sr | 38 | 87.6 |
| Germanium | Ge | 32 | 72.5 | Sulphur | S | 16 | 32.1 |
| Gold | Au | 79 | 197 | Tin | Sn | 50 | 119 |
| Helium | He | 2 | 4.00 | Titanium | Ti | 22 | 47.9 |
| Hydrogen | H | 1 | 1.01 | Vanadium | V | 23 | 50.9 |
| Iodine | I | 53 | 127 | Xenon | Xe | 54 | 131 |
| Iron | Fe | 26 | 55.8 | Zinc | Zn | 30 | 65.4 |
| Krypton | Kr | 36 | 83.8 | | | | |

# Periodic Table of the Elements

| 1 | 1.000 |
|---|---|
| **H** | |
| Hydrogen | |

| 3 | 6.90 | 4 | 9.00 |
|---|---|---|---|
| **Li** | | **Be** | |
| Lithium | | Beryllium | |

| 11 | 23.0 | 12 | 24.3 |
|---|---|---|---|
| **Na** | | **Mg** | |
| Sodium | | Magnesium | |

| 19 | 39.1 | 20 | 40.1 |
|---|---|---|---|
| **K** | | **Ca** | |
| Potassium | | Calcium | |

| 37 | 85.5 | 38 | 87.6 |
|---|---|---|---|
| **Rb** | | **Sr** | |
| Rubidium | | Strontium | |

| 55 | 133 | 56 | 137 |
|---|---|---|---|
| **Cs** | | **Ba** | |
| Caesium | | Barium | |

Key:

| Atomic number | 11 | 23.0 | Relative atomic mass |
|---|---|---|---|
| | **Na** | | |
| | Sodium | | |

| | | | | | | 2 | 4.00 |
|---|---|---|---|---|---|---|---|
| | | | | | | **He** | |
| | | | | | | Helium | |

| 5 | 10.8 | 6 | 12.0 | 7 | 14.0 | 8 | 16.0 | 9 | 19.0 | 10 | 20.2 |
|---|---|---|---|---|---|---|---|---|---|---|---|
| **B** | | **C** | | **N** | | **O** | | **F** | | **Ne** | |
| Boron | | Carbon | | Nitrogen | | Oxygen | | Fluorine | | Neon | |

| 13 | 27.0 | 14 | 28.1 | 15 | 31.0 | 16 | 32.1 | 17 | 35.5 | 18 | 39.9 |
|---|---|---|---|---|---|---|---|---|---|---|---|
| **Al** | | **Si** | | **P** | | **S** | | **Cl** | | **Ar** | |
| Aluminium | | Silicon | | Phosphorus | | Sulphur | | Chlorine | | Argon | |

| 31 | 69.7 | 32 | 72.6 | 33 | 74.9 | 34 | 79.0 | 35 | 79.9 | 36 | 83.8 |
|---|---|---|---|---|---|---|---|---|---|---|---|
| **Ga** | | **Ge** | | **As** | | **Se** | | **Br** | | **Kr** | |
| Gallium | | Germanium | | Arsenic | | Selenium | | Bromine | | Krypton | |

| 49 | 115 | 50 | 119 | 51 | 122 | 52 | 128 | 53 | 127 | 54 | 131 |
|---|---|---|---|---|---|---|---|---|---|---|---|
| **In** | | **Sn** | | **Sb** | | **Te** | | **I** | | **Xe** | |
| Indium | | Tin | | Antimony | | Tellurium | | Iodine | | Xenon | |

| 81 | 204 | 82 | 207 | 83 | 209 | 84 | 210 | 85 | 210 | 86 | 222 |
|---|---|---|---|---|---|---|---|---|---|---|---|
| **Tl** | | **Pb** | | **Bi** | | **Po** | | **At** | | **Rn** | |
| Thallium | | Lead | | Bismuth | | Polonium | | Astatine | | Radon | |

| | | | | | | | | | |
|---|---|---|---|---|---|---|---|---|---|
| 21  45.0 **Sc** Scandium | 22  47.9 **Ti** Titanium | 23  50.9 **V** Vanadium | 24  52.0 **Cr** Chromium | 25  54.9 **Mn** Manganese | 26  55.8 **Fe** Iron | 27  58.9 **Co** Cobalt | 28  58.7 **Ni** Nickel | 29  63.5 **Cu** Copper | 30  65.4 **Zn** Zinc |
| 39  88.9 **Y** Yttrium | 40  91.2 **Zr** Zirconium | 41  92.9 **Nb** Niobium | 42  95.9 **Mo** Molybdenum | 43  99 **Tc** Technetium | 44  101 **Ru** Ruthenium | 45  103 **Rh** Rhodium | 46  106 **Pd** Palladium | 47  108 **Ag** Silver | 48  112 **Cd** Cadmium |
| 57  139 **La** Lanthanum | 72  178 **Hf** Hafnium | 73  181 **Ta** Tantalum | 74  184 **W** Tungsten | 75  186 **Re** Rhenium | 76  190 **Os** Osmium | 77  192 **Ir** Iridium | 78  195 **Pt** Platinum | 79  197 **Au** Gold | 80  201 **Hg** Mercury |

# Index